NUCLEAR STRUCTURE IN CHINA 2014

Proceedings of the 15th National Conference on Nuclear Structure in China

NUCLEAR STRUCTURE IN CHINA 2014

15th National Conference on Nuclear Structure in China

Guilin, China, 25 – 28 October 2014

Editors

Jie Meng

Peking University, China

Ning Wang

Guangxi Normal University, China

Shan-Gui Zhou

Institute of Theoretical Physics, Chinese Academy of Sciences, China

World Scientific

NEW JERSEY • LONDON • SINGAPORE • BEIJING • SHANGHAI • HONG KONG • TAIPEI • CHENNAI • TOKYO

Published by

World Scientific Publishing Co. Pte. Ltd.

5 Toh Tuck Link, Singapore 596224

USA office: 27 Warren Street, Suite 401-402, Hackensack, NJ 07601

UK office: 57 Shelton Street, Covent Garden, London WC2H 9HE

Library of Congress Cataloging-in-Publication Data

Names: National Conference on Nuclear Structure in China (15th : 2014 :
Guilin, Guangxi Zhuangzu Zizhiqu, China), creator. | Meng, Jie, 1966 January 25– editor. |
Wang, Ning (Physics Professor), editor. | Zhou, Shan-Gui, 1971– editor.

Title: Nuclear structure in China 2014 : proceedings of the 15th National Conference on
Nuclear Structure in China, Guilin, China, 25–28 October 2014 / editors, Jie Meng
(Peking University, China), Ning Wang (Guangxi Normal University, China),
Shan-Gui Zhou (Chinese Academy of Sciences, China).

Description: Singapore ; Hackensack, NJ : World Scientific Publishing Co. Pte. Ltd., [2016] | 2016 |
Includes bibliographical references.

Identifiers: LCCN 2016002838| ISBN 9789813109629 (hardcover ; alk. paper) |
ISBN 9813109629 (hardcover ; alk. paper)

Subjects: LCSH: Nuclear structure--Congresses. | Nuclear physics--Congresses.

Classification: LCC QC793.3.S8 N356 2014 | DDC 539.7/4--dc23

LC record available at http://lccn.loc.gov/2016002838

British Library Cataloguing-in-Publication Data

A catalogue record for this book is available from the British Library.

In-house Editor: Ng Kah Fee

Typeset by Stallion Press
Email: enquiries@stallionpress.com

Printed in Singapore

PREFACE

The 15th National Conference on Nuclear Structure in China (NSC2014) was held on October 25–28, 2014 in Guilin, Guangxi Province and hosted by Guangxi Normal University. The series of National Conferences on Nuclear Structure in China is organized by the Nuclear Structure Sub-committee of Chinese Nuclear Physics Society and it is held biennially. This series of conferences is an important event in China for nuclear physicists working on the study of nuclear structure.

Apart from the Nuclear Structure Sub-committee of Chinese Nuclear Physics Society, NSC2014 was also organized by Institute of Theoretical Physics of Chinese Academy of Sciences, China Center of Advanced Science and Technology (CCAST), Institute of Modern Physics of Chinese Academy of Sciences, China Institute of Atomic Energy, Peking University, Beihang University, Shanghai Jiao Tong University, and Guangxi Normal Universit. NSC2014 was also partially supported by Chinese Nuclear Physics Society. There were about 250 participants from more than 60 universities or institutes, including some invited speakers from abroad.

Following the tradition of NSC conferences, NSC2014 presented a review of recent advances made in nuclear structure physics in China and provided a broad discussion forum on current and future research projects. NSC2014 received totally 165 abstracts. After experts recommendations and discussions within Nuclear Structure Sub-committee, 18 contributions were chosen as plenary talks. These presentations covered: nuclear spectroscopy of high-spin states, nuclear mass and half-life, nuclear astrophysics, superheavy nuclei, unstable nuclei, mean field theory, nuclear matter, neutron star and symmetry energy, nuclear shell model and other subjects related to nuclear structure.

In NSC2014, 35% of participants were students or post-doctoral researchers. Following the tradition starting from 11th National Conference on Nuclear Structure in China (2006), the excellent oral presentations in NSC2014 were selected from the students and junior researchers by an evaluation committee which was headed by Prof. Genming Jin (靳根明). The people who were awarded the excellent oral presentations were: Qibo Chen (陈启博) (Peking Univeristy), Pengcheng Chu (初鹏程) (Shanghai Jiao Tong University), Shipeng Hu (胡世鹏) (China Institute of Atomic Energy), Chen Liu (刘晨) (Shandong University-Weihai), Dongdong Ni (倪冬冬) (Nanjing University), Min Shi (仕敏) (Anhui Univeristy), Kaijia Sun (孙开佳) (Shanghai Jiao Tong University), Bing Wang (王兵) (Zhengzhou University/ Institute of Theoretical Physics of Chinese Academy of Sciences) and Huabin Yang (靳根明) (Institute of Modern Physics of Chinese Academy of Sciences).

We thank the members of the Nuclear Structure Sub-committee of Chinese Nuclear Physics Society and the members of Local Organizing Committee for their support and valuable advices on the organization of NSC2014. We would like to take this opportunity to extend our sincere gratitude to all participants for their attending, contributions and discussions. We also thank Min Liu (刘敏), Li Ou (欧立), Cai-Wan Shen (沈彩万) for their help in preparing this proceeding. We would especially like to thank Yong-Xu Yang (杨永栩) and Xiao-Jun Sun (孙小军) for their help in organizing the conference. Finally, the generous financial supports on the organizing institutions as well as the excellent and efficient work of World Scientific Publishing Co. are gratefully acknowledged.

Jie Meng (孟杰) — Peking University, Beijing
Ning Wang (王宁) — Guangxi Normal University, Guilin
Shan-Gui Zhou (周善贵) — Institute of Theoretical Physics, Chinese Academy of Sciences, Beijing

第十五届全国核结构大会合影
2014.10.25 广西师范大学
物理科学与技术学院
COLLEGE OF PHYSICS AND TECHNOLOGY

CONTENTS

EFFECTS OF THE DYNAMICAL DEFORMATION AND FISSION BARRIER IN THE SYNTHESIS OF SUPERHEAVY NUCLEI

Xiao Jun Bao[1], Jun Qing Li[1,2], Jie Ding Zhu[1] and Hong Fei Zhang[1,*]

[1] *School of Nuclear Science and Technology, Lanzhou University Lanzhou, 730000, China*
**Email: zhanghongfei@impcas.ac.cn*

[2] *Institute of Modern Physics, Chinese Academy of Sciences, Lanzhou 730000, China*

Within the dinuclear system concept, the time dependent dynamical deformation of each two nuclei are treated by F-P equation, and the resulted analytical solutions about the mean deformations are imported into a two variable M-E to describe the nucleon transfer and the fusion probability is evaluated. The agreement between the calculated results of the superheavy nuclear production cross sections and the available experimental data is improved. Meanwhile, by using different nuclear mass tables the production cross sections of superheavy nuclei are analyzed.

Keywords: Dynamical deformation; evaporation residue cross section; DNS.

1. Introduction

Producing superheavy nuclei (SHN) is one of the major aims of modern nuclear heavy-ion physics. Great success was achieved during the last two decades in experimental studies.[1] In recent years, many approaches are used to calculate the fusion probability.[2–4] The dinuclear system (DNS) model is one of them,[2,4,5] in which the nucleon transfer is coupled to the relative motion by solving a set of microscopically derived master equations(M-E) with distinguishing protons and neutrons. However, in the DNS concept it is assumed that each of the two touching nuclei always keeps its own identity with its ground state deformation; actually both nuclei in the DNS should be gradually deformed due to strong nuclear and Coulomb interactions between them. This deformation is not negligible, because it will alters the masses of nuclei, as well as the interactions between them, so that it will influence further evolution of the system. The DNS conception has to be improved. Due to the computation restriction for solving the multi-variable ME, and to further develop the DNS model, we combine the two variable ME (the variables of the proton and neutron number of the projectile-like fragment: Z_1,

N_1) with the analytical time dependent solutions of the FPE for mean fragment deformations[6] to treat the four variables in the process of the compound nucleus formation. Since the time dependent fragment deformations alter the interaction of the system, change the potential energy surface of the system, make it therefore also time dependent, affect the further nucleon transfer.

On the other hand, the basic nuclear data such as the nuclear mass, deformation, fission barrier and so on[7–9] play a key role for evaporation residue cross section (ERCS). The excited compound nucleus decays by the emission of light particles and γ rays against fission. The survival probability of the compound nucleus against fission in the deexcitation process is considered as one of the crucial factors for producing heavy and superheavy elements. Thus, precise knowledge of fission barriers and neutron separation energies is crucial for reasonable predictions of the synthesis cross sections.

2. Results and Discussion

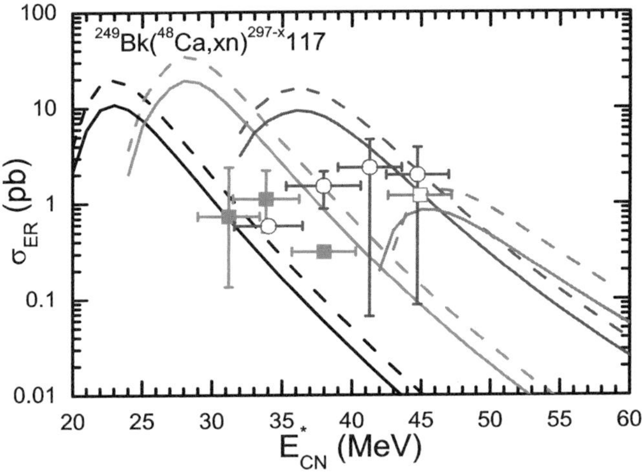

Fig. 1. Calculated excitation functions of ERCS by taking the dynamical deformation (solid line) and taking the ground deformation of the two nuclei (dashed line), compared with the available experimental data[10] for the reaction $^{48}Ca+^{249}Bk$.

It is found that the dynamical deformation increases the inner fusion barrier B_{fus}, therefore decreases te fusion probability. This decreasing is getting pronounced with the increasing excitation energy. Figure 1 shows the ERCSs of the reaction $^{48}Ca+^{249}Bk$ by considering (solid lines) and without considering (dashed lines) the dynamical deformation, and compared with the experimental data. The measured ERCS of the 3n, 4n, and 5n channels are denoted by red solid squares, blue open cycles, and dark cyan open squares, respectively. One may find that the solid lines are more close to the data. Actually, within the error bars, the experimental data are reproduced rather well, especially for the 4n channel. The dynamical deformation reduces the ERCS due to the fusion probability decreasing, and the reduction is more pronounced at higher excitation energy, implying larger deformation involved.

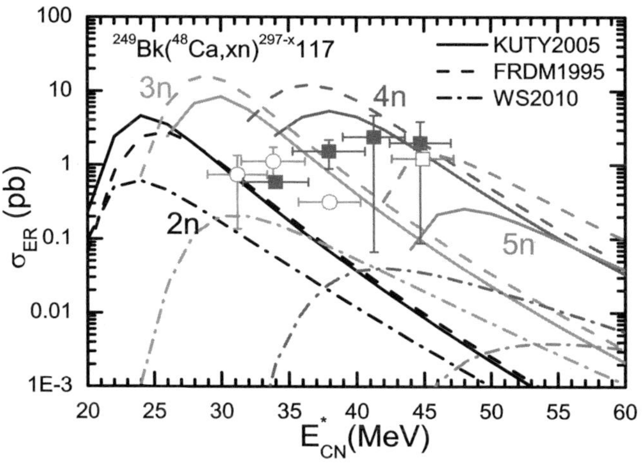

Fig. 2. Calculated evaporation residue excitation functions by using 2V-DNS based on different theoretical nuclear mass tables , compared with the available experimental data[10] in the reaction ^{48}Ca+^{249}Bk.

Theoretical results of the ERCSs for the ^{48}Ca+^{249}Bk reaction based on different nuclear mass tables are compared with experimental data in Fig. 2. The measured ERCS of the 3n, 4n, and 5n channels are denoted by red open circles, blue solid squares, and dark cyan open squares, respectively. The curves show theoretical results obtained in this work for the 2n (black lines), 3n (red lines), 4n (blue lines), and 5n (dark cyan lines) channels based on the mass tables of KTUY2005 (solid lines), FRDM1995 (dashed lines) and of WS2010 (dashed-dotted lines). One may find that the solid lines are more close to the data. Actually, within the error bars, the experimental data are reproduced rather well, especially for the 3n and 4n channels. Theoretical results of the ERCSs are obvious greater than experimental data that can be noticed when the fission barriers and the ground state masses are taken from FRDM1995. Meanwhile, the predicted positions of the maxima of the *xn* excitation functions are shifted by some 3-5 MeV toward lower energies as compared with the experimental data.

3. Summary

In summary, the analytical solution about the mean deformations are imported into a two variable ME to describe the nucleon transfer and the fusion probability is evaluated. In this way, the driving potential for nucleon transfer follows the time dependent nuclear deformation to become time dependent, therefore, actually evolutions of the four dynamical variables as well as their correlations are treated at the same time. It is found that the inclusion of the dynamical deformation improved the ER cross sections for the element $Z = 117$ as compared with the available data. We also have calculated the evaporation residue cross sections in the ^{48}Ca+^{249}Bk hot fusion reaction leading to formation of SHN $Z = 117$ using the different nuclear mass tables. One may find that the results from KTUY2005 are more close to the data.

4. Acknowledgments

The work is supported by the National Natural Science Foundation of China (Grants 11175074, 11475050, 11120101005 and 11105035), the Fundamental Research Funds for the Central Uni-versities (Grantlzujbky-2014-230).

References

1. Y. Oganessian *et al.*, *Radiochimica Acta International journal for chemical aspects of nuclear science and technology* **99**, 429 (2011).
2. Z.-Q. Feng, G.-M. Jin, F. Fu and J.-Q. Li, *Nuclear Physics A* **771**, 50 (2006).
3. C. Shen, Y. Abe, D. Boilley, G. Kosenko and E. Zhao, *International Journal of Modern Physics E* **17**, 66 (2008).
4. M. Huang, Z. Zhang, Z. Gan, X. Zhou, J. Li and W. Scheid, *Physical Review C* **84**, p. 064619 (2011).
5. G. Adamian, N. Antonenko, W. Scheid and V. Volkov, *Nuclear Physics A* **627**, 361 (1997).
6. L. Yu, Z. Gan, Z. Zhang, H. Zhang and J. Li, *Physics Letters B* **730**, 105 (2014).
7. P. Moller, J. Nix, W. Myers and W. Swiatecki, *Atomic Data and Nuclear Data Tables* **59**, 185 (1995).
8. H. Koura, T. Tachibana, M. Uno and M. Yamada, *Progress of theoretical physics* **113**, 305 (2005).
9. N. Wang, M. Liu and X. Wu, *Physical Review C* **81**, p. 044322 (2010).
10. Y. T. Oganessian, F. S. Abdullin, P. Bailey, D. Benker, M. Bennett, S. Dmitriev, J. G. Ezold, J. Hamilton, R. A. Henderson, M. Itkis *et al.*, *Physical review letters* **104**, p. 142502 (2010).

BEYOND MEAN FIELD STUDY OF PROPERTIES OF SINGLE-PARTICLE STATES

Li-Gang Cao[1,2,3,*], H. Sagawa[4], G. Colò[5,6], and P. F. Bortignon[5,6]

[1] *School of Mathematics and Physics, North China Electric Power University, Beijing 102206, China*

[2] *State Key Laboratory of Theoretical Physics, Institute of Theoretical Physics, Chinese Academy of Sciences, Beijing 100190, China*

[3] *Kavli Institute for Theoretical Physics China, CAS, Beijing 100190, China*

[4] *Center for Mathematics and Physics, University of Aizu, Aizu-Wakamatsu, Fukushima 965-8560, Japan*

[5] *Dipartimento di Fisica, Università degli Studi di Milano, via Celoria 16, Milano, Italy*

[6] *Istituto Nazionale di Fisica Nucleare, Sez. di Milano, via Celoria 16, Milano, Italy*

[*] *E-mail: caolg@ncepu.edu.cn*

The properties of single-particle states in magic nucleus ^{208}Pb, such as the energies, spectroscopic factors and the effective mass, have been studied with beyond mean field theory. All selected phonons are obtained by the random phase approximation (RPA) and the same Skyrme interaction is also used in the Particle-vibration coupling (PVC) vertex. We have paid more attention to the effect of the two-body spin-orbit and tensor interactions on the single-particle properties. We find that the contributions of those terms are important to improve the results. The calculated results are compared to available experimental data. The single-particle level density around the Fermi surface is significantly increased due to the effect of PVC.

Keywords: Beyond mean field theory; single-particle states; effective mass.

1. Introduction

In mean field theory, the nucleons move independently in the *static* average potential produced by the surrounding nucleons.[1] Of course, this is an ideal assumption. In practice, nucleons collide with other nucleons or couple to the collective vibrations of the whole system. This is related to the concept of the so-called 'dynamical effects' beyond the mean field approximation. To consider the fluctuations of the mean field potential, one must go beyond the mean field scheme, which means that the average potential is no more static or energy-independent, and is energy-dependent.

For finite nuclei, the fluctuations of the average potential are usually described by an effective theory denoted as particle-vibration coupling (PVC)[2] model. It has been shown that the pvc affects strongly the energies of single-particle states around the Fermi surface and increases the single-particle level density.[3] In the earlier times, the PVC calculations have been lacking any self-consistency since the Woods-Saxon potential was usually adopted to calculate the single-particle basis and the interactions at the PVC vertex have been chosen with a large degree of arbitrariness.[3] In few cases, the Skyrme interaction has been adopted both for the single-particle potential and PVC vertex, but using only the velocity-independent terms in the PVC vertex.[4] Recently, microscopic self-consistent PVC calculations have been performed within either the framework of the Skyrme energy density functionals[5,6] or the framework of relativistic (i.e., covariant) functionals.[7]

Recently, much attention has been devoted to the effect of tensor terms on the evolution of the single-particle levels in exotic nuclei.[8] Moreover, within HF plus self-consistent Random Phase Approximation (RPA), we have investigated the effect of the tensor force on the multipole response of finite nuclei.[9,10] The response function of uniform matter, and the occurence of possible instabilities, has been the subject of another recent study.[11] The non-central terms such as the two-body spin-orbit term or the tensor terms have been included recently in the PVC calculations.[12] In this study, we shall study the effect of the non-central terms of the Skyrme interactions at the PVC vertex on the single-particle properties of finite nucleus. We will discuss the sensitivity of the energy shifts associated with the single particle states, of the effective mass, and of the spectroscopic factors, when the tensor interaction and the spin-orbit interaction are included in the PVC vertex. The calculations are performed for the double magic nucleus ^{208}Pb. The Skyrme interaction T44[13] is adopted. For T44, the tensor parameters are fitted on the same footing as the other Skyrme parameters. The ground states and the various excited states of the nucleus ^{208}Pb are calculated on the basis of the fully self-consistent HF+RPA framework as in Ref.[14] The coupling of the particles to the vibrations is derived from the same Skyrme force in a consistent way.

2. Results and Discussion

In this part we will present and discuss the PVC results for nucleus ^{208}Pb. The PVC method has been discussed extensively in Ref. 5,12. The effective Skyrme interaction T44 is used in the calculations. The ground states and the various excited states of ^{208}Pb are calculated within a fully self-consistent HF plus RPA approach. We will stress, in our discussion, the effect of the non-central part of the Skyrme interaction (such as the spin-orbit and tensor terms) on the single-particle energies deduced from the PVC calculations. To do this, the interactions in the PVC vertex have been switched off accordingly.

Table 1. The energies of the neutron single-particle states around the Fermi surface in ^{208}Pb calculated in various approximations (see the text for details). The spectroscopic factors obtained in the full calculation are also shown in this Table. The results are obtained by using T44 parameter set. The experimental data are taken from Refs. 15,16.

	HF	PVC central		PVC central+S.O.		PVC full			Spectroscopic factors	
	$\varepsilon^{(0)}$	$\Delta\varepsilon_i$	ε_i	$\Delta\varepsilon_i$	ε_i	$\Delta\varepsilon_i$	ε_i	ε_i^{exp}	S_i^{th}	S_i^{exp}
$3d_{3/2}$	0.20	-0.55	-0.35	-0.44	-0.24	-0.44	-0.24	-1.40	0.895	1.09
$2g_{7/2}$	0.14	-0.85	-0.71	-0.53	-0.39	-0.61	-0.47	-1.44	0.832	1.05
$4s_{1/2}$	-0.35	-0.48	-0.83	-0.50	-0.85	-0.47	-0.82	-1.90	0.896	0.98
$3d_{5/2}$	-0.88	-0.72	-1.60	-0.81	-1.69	-0.81	-1.69	-2.37	0.855	0.98
$1j_{15/2}$	-0.30	-0.87	-1.17	-1.80	-2.10	-1.77	-2.07	-2.51	0.583	0.58
$1i_{11/2}$	-2.19	-0.39	-2.58	-0.51	-2.70	-0.57	-2.76	-3.16	0.884	0.86
$2g_{9/2}$	-3.28	-0.52	-3.80	-0.69	-3.97	-0.68	-3.96	-3.94	0.877	0.83
$3p_{1/2}$	-7.91	-0.08	-7.99	0.04	-7.87	0.03	-7.88	-7.37	0.905	0.90
$2f_{5/2}$	-8.92	0.03	-8.89	0.19	-8.72	0.18	-8.74	-7.94	0.888	0.60
$3p_{3/2}$	-9.14	0.11	-9.03	0.19	-8.95	0.21	-8.93	-8.26	0.844	0.88
$1i_{13/2}$	-9.18	0.17	-9.01	-0.01	-9.19	0.01	-9.17	-9.24	0.903	0.91
$2f_{7/2}$	-12.10	0.54	-11.56	0.49	-11.61	0.68	-11.42	-9.81	0.580	0.95
$1h_{9/2}$	-13.14	0.13	-13.01	0.43	-12.71	0.44	-12.70	-11.40	0.831	0.98

In Table 1 and Fig. 1 we show the results for the energies ε_i of neutron single-particle levels around the Fermi surface in ^{208}Pb calculated within various approximation. The results (denoted by $\varepsilon^{(0)}$) in the second column are obtained within the HF mean field approximation by including the contribution of tensor interaction. The results in the fourth, sixth and eighth columns in Table 1 correspond to the PVC calculations with the central Skyrme interaction, the central plus spin-orbit interaction and the central plus spin-orbit as well as tensor interactions in the PVC vertex, respectively. The values in the columns labelled by $\Delta\varepsilon_i$ are the difference between the PVC results and the original HF values. The results are also compared with the available experimental data.

From Table 1 and Fig. 1, we can see that the PVC calculations give a small repulsive contribution to the energy for most of the hole states below the Fermi surface. One noticeable exception is the $2f_{7/2}$ hole state

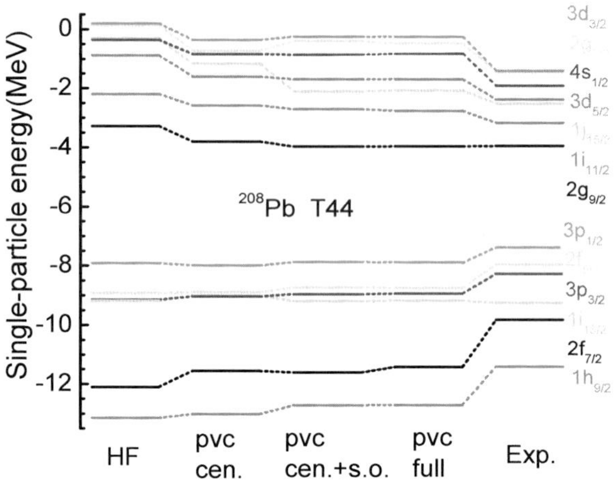

Fig. 1. (Color online) Neutron states in ^{208}Pb calculated with the parameter set T44. The various columns, from left to right, correspond to the HF calculation and to HF + PVC with only the central interaction in the vertex, to HF + PVC with central and spin-orbit interaction in the vertex, and to HF + PVC with central plus spin-orbit and tensor interactions in the vertex. The last column displays the experimental data.

which is shifted up in energy by 0.70 MeV by using T44 parameter set. On the other hand, for its spin-orbit partner state $2f_{5/2}$ the energy shift is rather small. This goes against the prejudice that spin-orbit partner states should be affected more or less in the same way by the PVC effects: in this case, the special role of the coupling with the low-lying 3^- breaks this rule. For the particle states, the energy shift $\Delta\varepsilon_i$ is always negative but its magnitude depends on the state chosen and on the approximation scheme. In particular, the spin-orbit and tensor terms of the force do not give a systematic effect: for some states they give a repulsive contribution to the energy shift (with respect to the shift obtained by retaining only the central part of the Skyrme force at the PVC vertex), whereas for some other states they give attractive contributions. We have also calculated the r.m.s. deviation σ between theoretical and experimental single-particle states. The values of σ are 1.421, 1.002, 0.907, 0.873 for T44 Skyrme interaction in the case of HF, PVC with central interactions, PVC with central plus spin-orbit interactions, and full PVC calculations, respectively. In this case the inclusion of all terms in the PVC vertex produces an improvement of the results.

In Table 1 we also show the calculated spectroscopic factors of the single-particle states and the corresponding experimental data. The calculations are performed within full PVC model. For the particle states, the agreement between the calculated and the measured spectroscopic factors is generally satisfied. For the hole states, the largest disagreement between theoretical and experimental data is found in the case of the $2f_{5/2}$ and $2f_{7/2}$ states. For the $2f_{5/2}$ state, we can see that it is rather fragmented from the experimental side, whereas the calculated fragmentation is rather small. For its spin-orbit partner state $2f_{7/2}$, the situation is opposite.

Table 2. The calculated effective mass around the Fermi surface for neutrons in ^{208}Pb in various approximations. The results are obtained by using the T44 parameter set.

HF	PVC central		PVC central+S.O.		PVC full	
$\tilde{m}/m$	$\overline{m}/m$	m^*/m	$\overline{m}/m$	m^*/m	$\overline{m}/m$	m^*/m
0.841	1.157	0.973	1.200	1.009	1.235	1.038

In Table 2, we show the effective k-mass, E-mass and total effective mass in ^{208}Pb. We have averaged the effective masses of the states that we have considered (above and below the Fermi surface). The effective k-mass is about 0.84 around the Fermi surface in the pure HF calculation with T44 parameter sets. When one goes beyond the mean field calculation, we see that the calculated E-mass is approximately between 1.16 and 1.24 around the Fermi surface. The total effective mass, that is, the product of the k-mass and E-mass, of the states around the Fermi surface is about one, which is comparable to the empirical value.

3. Summary

In summary, the properties of the single-particle states, in particular the energies, the spectroscopic factors and the effective masses, in the magic nucleus ^{208}Pb, have been studied in a fully self-consistent particle-vibration coupling approach by using T44 Skyrme parameter set. All the vibrations (phonons) are produced within a fully self-consistent RPA approach.

We have paid a specific attention to the effect produced on the single-particle properties by the non-central part of the Skyrme interaction. It has been found, in the case of the single-particle energies, that the contributions to their energy shift induced by the tensor and spin-orbit terms are smaller than those coming from the central Skyrme terms. The results are improved with respect to the experimental findings by the contributions of spin-orbit and tensor forces. The calculated single-particle energies and spectroscopic factors show an overall good agreement with data. This is reflected in the enhancement of the single-particle level density around the Fermi surface, due to the PVC correlations. The effective mass becomes indeed close to one around the Fermi energy, and this is consistent with the empirical information.

Acknowledgments

This work is supported by the National Science Foundation of China under Grant Nos. 11175216, 11435014.

References

1. M. Bender, P.-H. Heenen, P.-G. Reinhard, Rev. Mod. Phys. **75** (2003) 121.
2. A. Bohr and B. R. Mottelson, Nuclear structure. Vol. II (W. A. Benjamin, New York, 1975).
3. C. Mahaux, P. F. Bortignon, R. A. Broglia, and C. H. Dasso, Phys. Rep. **120** (1985) 1.
4. V. Bernard and N. V. Giai, Nucl. Phys. **A348** (1980) 75.
5. G. Colò, H. Sagawa, and P. F. Bortignon, Phys. Rev. **C82** (2010) 064307.
6. K. Mizuyama, G. Colò, and E. Vigezzi, Phys. Rev. **C86** (2012) 034318.
7. E. Litvinova and P. Ring, Phys. Rev. **C73** (2006) 044328.
8. H. Sagawa, G. Colò, Prog. Part. Nucl. Phys. **76** (2014) 76.
9. L. G. Cao, H. Sagawa, and G. Colò, Phys. Rev. **C83** (2011) 034324; L. G. Cao, G. Colò, and H. Sagawa,Phys. Rev. **C81** (2010) 044302; L. G. Cao, G. Colò, and H. Sagawa, *et al.*, Phys. Rev. **C80** (2009) 064304.
10. C. L. Bai, H. Q. Zhang, H. Sagawa, *et al.*, Phys. Rev. **C83** (2011) 054316; C. L. Bai, H. Q. Zhang, H. Sagawa, *et al.*, Phys. Rev. Lett. **105** (2010) 072501.
11. A. Pastore, M. Martini, V. Buridon, D. Davesne, K. Bennaceur, and J. Meyer, Phys. Rev. **C86**, 044308 (2012); A. Pastore, D. Davesne, Y. Lallouet, M. Martini, K. Bennaceur, and J. Meyer, Phys. Rev. **C85**, 054317 (2012); D. Davesne, M. Martini, K. Bennaceur, and J.Meyer, Phys. Rev. **C80**, 024314 (2009); **C84**, 059904(E) (2011).
12. L. G. Cao, G. Colò, H. Sagawa, P. F. Bortignon, Phys. Rev. **C89**, 044314 (2014).
13. T. Lesinski, M. Bender, K. Bennaceur, T. Duguet, J. Meyer, Phys. Rev. **C76**, 014312 (2007).
14. G. Colò, L.G. Cao, N. Van Giai, L. Capelli, Comp. Phys. Comm. **184**, 142 (2013).
15. N. Schwierz, I. Wiedenhöver, and A. Volya, arXiv:0709.3525.
16. A. Oros, Ph.D. thesis, University of Köln, 1996.

COLLECTIVE HAMILTONIAN FOR CHIRAL AND WOBBLING MODES

Q. B. Chen, S. Q. Zhang†, P. W. Zhao

State Key Laboratory of Nuclear Physics and Technology, School of Physics, Peking University, Beijing 100871, China
†E-mail: sqzhang@pku.edu.cn

R. V. Jolos

Joint Institute for Nuclear Research, Dubna 141980, Russia
Institut für Kernphysik der Universität zu Köln, Köln 50937, Germany

J. Meng (孟杰)*

State Key Laboratory of Nuclear Physics and Technology, School of Physics, Peking University, Beijing 100871, China
School of Physics and Nuclear Energy Engineering, Beihang University, Beijing 100191, China
Department of Physics, University of Stellenbosch, Stellenbosch, South Africa
**E-mail: mengj@pku.edu.cn*

The recent progresses of the collective Hamiltonian for chiral and wobbling modes are briefly introduced. The collective Hamiltonian is constructed from the collective potential and mass parameter obtained in the tilted axis cranking approach. The collective Hamiltonian can reproduce the exact solutions by the particle rotor model very well for both chiral and wobbling modes.

Keywords: Chirality; wobbling; collective Hamiltonian.

1. Introduction

Due to being regarded as the unique fingerprint of a stable triaxial nuclei, the chirality[1] and wobbling[2] have attracted lots of attentions in nuclear physics. More than thirty candidate chiral doublet bands have been reported in the $A \approx 80$, 100, 130 and 190 mass regions; see, e.g., Refs. 3–5. The wobbling bands have been extensively identified in the triaxial strongly deformed (TSD) region around $N = 94$, such as in ^{161,163,165,167}Lu and ^{167}Ta; see, e.g., Refs. 6,7. Very recently, a *transverse wobbling* band[6] was reported in ^{135}Pr,[8] which sheds new light on the study of wobbling modes.

Theoretically, the chirality and wobbling are mainly investigated by the triaxial particle rotor model (PRM),[9–13] the tilted axis cranking (TAC) model,[14,15] and the random phase approximation (RPA).[16–19] Recently, a new approach with collective Hamiltonian was proposed.[7,20] This approach goes beyond the mean-filed approximation and could microscopically describe not only the yrast sequence but also the highly excited bands.

In this proceeding, the recent progresses of the collective Hamiltonian for chiral and wobbling modes[7,20] will be briefly reviewed.

2. Collective Hamiltonian

The detailed framework of collective Hamiltonian has been formulated in Refs. 7,20. If in the case of the triaxial nuclei the numeration of the axes corresponds to the following relation between the components of the inertia tensor $\Im_3 \geq \Im_2 \geq \Im_1$ the angle θ is a dynamical variable of the wobbling motion. If the Hamiltonian is invariant with respect to reflection $\varphi \to -\varphi$ the corresponding parity will be a good quantum number and the azimuth angle φ describes the chiral mode. Taking φ as the collective variable, the quantized form of the collective Hamiltonian is written as

$$\hat{H}_{\text{coll}} = \hat{T}_{\text{kin}}(\varphi) + V(\varphi) = -\frac{\hbar^2}{2\sqrt{B(\varphi)}}\frac{\partial}{\partial\varphi}\frac{1}{\sqrt{B(\varphi)}}\frac{\partial}{\partial\varphi} + V(\varphi), \tag{1}$$

in which both the collective potential $V(\varphi)$ and mass parameter $B(\varphi)$ are obtained by TAC solutions. Solving the collective Hamiltonian on the basis states with appropriate boundary condition, e.g., box boundary condition, the collective levels and corresponding wave functions can be obtained.

3. Chiral and Wobbling Modes

The chiral modes were studied based on the tilted axis solutions of the TAC model. Considering the quantal fluctuations along the chiral degree of freedom, the collective Hamiltonian with the collective potential and mass parameter obtained by TAC was constructed. Solving the collective Hamiltonian, the collective energy levels and the wave functions for the chiral modes were obtained. In order to examine the accuracy of the collective Hamiltonian, the chiral doublet bands calculated by the collective Hamiltonian were compared with the the exact solutions by the PRM. It was found that the collective Hamiltonian well reproduces the PRM results, which motivates future applications for realistic nuclei. For details, see Ref. 20.

The collective Hamiltonian was also applied to investigate the wobbling modes, including the simple, longitudinal, and transverse wobbling modes.[7] Similar as the case of the chiral modes, the collective potential is calculated by the TAC model. From the TAC model, the wobbling modes were obtained based on the principal axis solutions. With the considerations of the quantal fluctuations along the wobbling degree of freedom, the wobbling excitation bands were obtained by the collective Hamiltonian.

A triaxial rotor can be considered as a simple wobbler. Its wobbling frequency increases with increasing spin.[2] Its angular momentum geometry has been investigated very recently.[21] For a system of a high-j particle coupled to a triaxial rotor, the longitudinal (transverse) wobbler is achieved by choosing the shortest axis to be (be perpendicular to) the axis with largest moments of inertia by setting the rigid (irrotational flow) type.

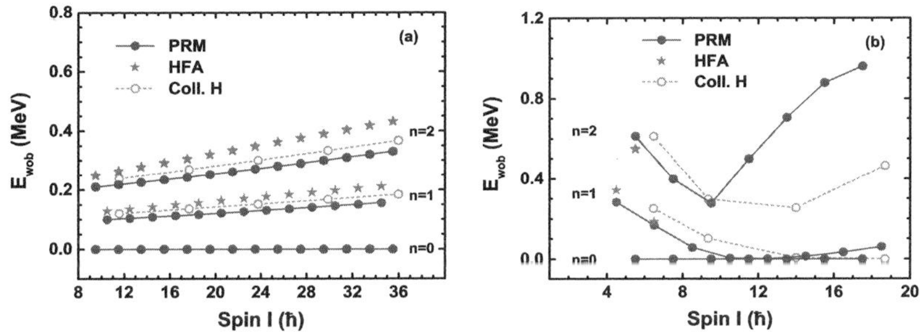

Fig. 1. (Color online) Energy spectra of the longitudinal (a) and transverse (b) wobbling bands $n = 1, 2$ relative to the $n = 0$ yrast sequence obtained by collective Hamiltonian in comparison with PRM and HFA. Taken from Ref. 7.

In Fig. 1, the energies of the longitudinal (a) and transverse (b) wobbling bands $n = 1, 2$ relative to the $n = 0$ yrast sequence obtained by collective Hamiltonian are shown in comparison with the PRM solutions and the analytic harmonic frozen approximation (HFA) results. The numerical details can be found in Ref. 7. It is seen that the collective Hamiltonian can reproduce the PRM results well at the region of wobbling motions. Furthermore, the wobbling frequency increases with increasing spin for the longitudinal wobbling mode, while decreases for the transverse wobbling mode. These variation trends of the wobbling frequency are revealed to be in association with the stiffness of the collective potentials. For details, see Ref. 7.

4. Summary

In summary, the recent progresses of the collective Hamiltonian for chiral[20] and wobbling modes[7] are briefly introduced. The collective Hamiltonian is constructed from the potential energy and mass parameter obtained in the tilted axis cranking approach. The collective Hamiltonian can reproduce the exact solutions by the particle rotor model very well for both chiral and wobbling modes. The success of the collective Hamiltonian here guarantees its applications for realistic nuclei.

Acknowledgments

This work was supported in part by the Major State 973 Program of China (Grant No. 2013CB834400), the National Natural Science Foundation of China (Grants Nos. 11175002, 11335002, 11375015, 11345004, 11461141002), Research Fund for the Doctoral Program of Higher Education (Grant No. 20110001110087).

References

1. S. Frauendorf and J. Meng, *Nucl. Phys. A* **617**, p. 131 (1997).
2. A. Bohr and B. R. Mottelson, *Nuclear structure* (Benjamin, New York, 1975).
3. J. Meng and S. Q. Zhang, *J. Phys. G: Nucl. Part. Phys.* **37**, p. 064025 (2010).
4. J. Meng, *Int. J. Mod. Phys. E* **20**, 341 (2011).
5. J. Meng, Q. B. Chen and S. Q. Zhang, *Int. J. Mod. Phys. E* **23**, p. 1430016 (2014).
6. S. Frauendorf and F. Dönau, *Phys. Rev. C* **89**, p. 014322 (2014).
7. Q. B. Chen, S. Q. Zhang, P. W. Zhao and J. Meng, *Phys. Rev. C* **90**, p. 044306 (2014).
8. J. T. Matta *et al.*, *Phys. Rev. Lett. (in press)* (2015).
9. J. Peng, J. Meng and S. Q. Zhang, *Phys. Rev. C* **68**, p. 044324 (2003).
10. S. Q. Zhang, B. Qi, S. Y. Wang and J. Meng, *Phys. Rev. C* **75**, p. 044307 (2007).
11. B. Qi, S. Q. Zhang, J. Meng, S. Y. Wang and S. Frauendorf, *Phys. Lett. B* **675**, p. 175 (2009).
12. I. Hamamoto, *Phys. Rev. C* **65**, p. 044305 (2002).
13. K. Tanabe and K. Sugawara-Tanabe, *Phys. Rev. C* **73**, p. 034305 (2006).
14. V. I. Dimitrov, S. Frauendorf and F. Dönau, *Phys. Rev. Lett.* **84**, 5732 (2000).
15. P. Olbratowski, J. Dobaczewski, J. Dudek and W. Plociennik, *Phys. Rev. Lett.* **93**, p. 052501 (2004).
16. S. Mukhopadhyay *et al.*, *Phys. Rev. Lett.* **99**, p. 172501 (2007).
17. D. Almehed, F. Dönau and S. Frauendorf, *Phys. Rev. C* **83**, p. 054308 (2011).
18. Y. R. Shimizu and M. Matsuzaki, *Nucl. Phys. A* **588**, p. 559 (1995).
19. M. Matsuzaki, Y. R. Shimizu and K. Matsuyanagi, *Phys. Rev. C* **65**, p. 041303 (2002).

20. Q. B. Chen, S. Q. Zhang, P. W. Zhao, R. V. Jolos and J. Meng, *Phys. Rev. C* **87**, p. 024314 (2013).
21. W. X. Shi and Q. B. Chen, *Chin. Phys. C (in press)* (2015).

THE ALTERNATIVE SCHEME TO DESCRIBE NUCLEI WITH $X(5)$ SYMMETRY

L. R. Dai,* F. Pan, R. L. Liu, H. Yu and W. L. Zhang

Department of Physics, Liaoning Normal University, Dalian 116029, China
**E-mail: dailr@lnnu.edu.cn*

The possible nuclei with $X(5)$ symmetry are investigated in the Interacting Boson Model (IBM), in which the traditional scheme and a new alternative scheme from the spherical to the axially deformed limit of the IBM with a schematic Hamiltonian are studied by using the SU(3) quadrupole-quadrupole term and O(6) cubic interaction, respectively. The low-lying energy levels and $E2$ transition rates from the new scheme are calculated and compared with the experimental data and those of the traditional $U(5) - SU(3)$ description. It is shown that the results from this new scheme seem better than those of the traditional description.

Keywords: Interacting boson model; quantum phase transitions; higher-order term; $X(5)$ symmetry.

1. Introduction

The nature of shape phase transitions in finite many-body systems is a fundamental issue, and it is of great interest in many areas of physics.[1–3] Transitional nuclei experienced renewed interest over the last decade and were extensively studied in the Interacting Boson Model (IBM).[4–8] It is now widely accepted that the three limiting cases[9–11] of the theory correspond to three different geometric shapes of nuclei, referred to as spherical (vibrational limit with U(5) symmetry), axially deformed (rotational limit with SU(3) symmetry), and γ-soft (triaxial with O(6) symmetry), respectively. The full range of the model can be parameterized in terms of the Casten triangle[12] shown in Fig. 1. Interesting phenomena occur when a system falls between two limits of the theory, in which case a quantum phase transition occurs at a critical point[4,13,14] where the dominance of one of the symmetries yields to the dominance of the other. Indeed, the $X(5)$ symmetry at the critical point of this transition was studied,[15] in which Iachello proposed analytical solutions of a Bohr Hamiltonian in the

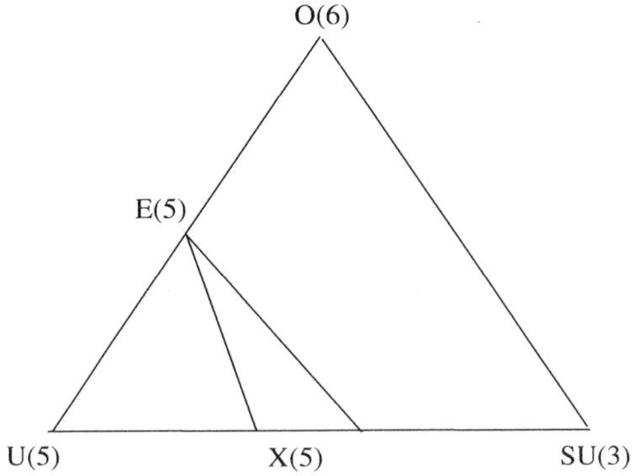

Fig. 1. Casten tirangle in the Interacting Boson Model.

situation appropriate for the description of nuclei near the critical point of the spherical to axially deformed shape-phase transition. Casten and Zamfir showed in 16 that ^{152}Sm and other $N = 90$ isotones demonstrate these characteristics. Transitional patterns from the spherical, U(5), to the axially deformed, SU(3), limit of the IBM with a schematic Hamiltonian were studied in Refs. 17–19, in particular the transitional behavior of some physically relevant quantities across the entire span of the $U(5)$-$SU(3)$ transitional region were explored. Recently, a new idea was firstly suggested by van Isacker[20] to investigate axially deformed nuclei, in which the well-known $SU(3)$ quadrupole-quadrupole interaction was replaced by the $O(6)$ cubic interaction, where $\hat{Q}_\mu(0) = s^\dagger \tilde{d}_\mu + d^\dagger_\mu s$ are generators of the $O(6)$ algebra. This idea was then further confirmed by Thiamova and Cejnar.[21] Inspired by this, in the present work we use two schemes, one is traditional scheme called $U(5)$-$SU(3)$, another is an new alternative scheme called UQ, to study the possible $X(5)$ candidates. In these two schemes, the well-known $SU(3)$ quadrupole-quadrupole interaction in the schematic Hamiltonian is replaced by the $O(6)$ cubic interaction. By using this idea, we have dynamically made an systematic investigation of this new alternative scheme, in which the transitional behavior of the low-lying energy levels, eigenstates, isomer shifts, $E2$ transition rates, and expectation values of shape variables across the entire transitional region are all examined. The results show that the new scheme can also describe the spherical to axially deformed shape-phase transition, especially near the critical point. And this new scheme was also successfully applied to ^{152}Sm nucleus with $X(5)$ symmetry.[22]

As we know, many other possible $X(5)$ nuclei have been experimentally found, such as ^{150}Nd, ^{154}Gd, ^{176}Os, ^{178}Os and ^{98}Sr, therefore, the main purpose of this work is to further study these possible $X(5)$ candidates, the low-lying spectrum and $E2$ transition rates will be investigated within the usual $U(5)$-$SU(3)$ scheme and the new UQ scheme.

2. Hamiltonian

The consistent-Q is defined as

$$\hat{Q}_\mu(\chi) = (d^\dagger s + s^\dagger \tilde{d})^{(2)}_\mu + \chi(d^\dagger \tilde{d})^{(2)}_\mu, \tag{1}$$

2.1. *For $U(5)$-$SU(3)$ scheme*

The schematic Hamiltonian for $U(5)$-$SU(3)$ scheme[22] is given by

$$\hat{H}_1 = c_1 \left[x\hat{n}_d + \frac{x-1}{f_1(N)} \hat{Q}(-\frac{\sqrt{7}}{2}) \cdot \hat{Q}(-\frac{\sqrt{7}}{2}) \right], \tag{2}$$

where N is the total number of bosons, $\hat{n}_d$ is the number operator for counting d-bosons, x is the control parameter with $0 \leq x \leq 1$, $f_1(N)$ is a linear function of N with $f_1(N) = 4N$ and $c_1 > 0$ is the scaling factor.

2.2. *For UQ scheme*

Alternatively, the schematic Hamiltonian for UQ scheme[22] is given by

$$\hat{H}_2 = c_2 \left[y\hat{n}_d + \frac{(1-y)}{f_2(N)} [Q(0) \times Q(0) \times Q(0)]^0 \right]. \tag{3}$$

where y is the control parameter $0 \leq y \leq 1$, $f_2(N)$ is a quadratic function of N with $f_2(N) = 0.8N^2$ which puts the critical point y_c of the UQ scheme close to x_c of the $U(5)$-$SU(3)$ scheme, $c_2 > 0$ is the scaling factor.

The $E2$ transition is a very sensitive signature of the structure. The $B(E2)$ reduced transition probability is defined by

$$B(E2; L_i \to L_f) = \frac{|\langle L_f \| T(E2) \| L_i \rangle|^2}{2L_i + 1} \tag{4}$$

here the $T(E2)$ operator in the $U(5)$-$SU(3)$ scheme is chosen as

$$T_\mu(E2) = q_1 \hat{Q}_\mu(-\sqrt{7}/2), \tag{5}$$

the $\hat{Q}_\mu(-\sqrt{7}/2)$ is the SU(3) quadrupole generator, q_1 is the effective charge. In the UQ scheme it is chosen as

$$T_\mu(E2) = q_2 \hat{Q}_\mu(0), \tag{6}$$

here the $\hat{Q}_\mu(0)$ is the $O(6)$ quadrupole operator and q_2 is the effective charge.

2.3. *How to solve equation*

In order to diagonalize Hamiltonians (2) and (3), we expand the corresponding eigenstates in terms of the $U(6) \supset SU(3) \supset SO(3)$ basis vectors $|N(\lambda\mu)KL\rangle$ as

$$|NL_\xi\rangle = \sum_{(\lambda\mu)K} C^{L_\xi}_{(\lambda\mu)K} |N(\lambda\mu)KL\rangle, \tag{7}$$

where ξ is used to denote the ξ-th level with angular momentum quantum number L, and the $C^{L_\xi}_{(\lambda\mu)K}$ are expansion coefficients. Since the total number of bosons N is fixed for a given nucleus, the eigenstates given in (7) will also be denoted as $|L_\xi; z\rangle$ with $z = x$ for the $U(5)$-$SU(3)$ scheme or $z = y$ for the UQ scheme in the following, where the value of the control parameter z is explicitly shown.

Table 1. The energy ratios of experimental data[23–30] and calculated results.

	$\frac{E(J_i^+)}{E(2_1^+)}$	4_1^+	6_1^+	8_1^+	10_1^+	0_2^+	2_2^+	4_2^+	6_2^+	2_3^+	4_3^+
^{152}Sm	Exp	3.00	5.80	9.22	13.19	5.61	6.64	8.39	10.75	8.93	11.25
	$U(5)$-$SU(3)$	2.59	4.71	7.32	10.39	3.11	4.75	6.98	9.66	6.06	8.70
	UQ	2.79	5.29	8.36	11.84	3.02	5.54	8.29	11.26	7.63	10.03
^{150}Nd	Exp	2.93	5.33	8.68	12.28	5.19	6.53	8.74		8.16	10.37
	$U(5)$-$SU(3)$	2.93	5.69	9.20	13.41	4.69	6.76	9.56		8.38	11.60
	UQ	2.93	5.66	9.00	12.75	3.53	6.07	8.95		8.13	10.59
^{154}Gd	Exp	3.01	5.83	9.30	13.30	5.53	6.62	8.51		8.03	10.05
	$U(5)$-$SU(3)$	3.01	5.92	9.66	14.18	4.79	6.90	8.29		8.78	11.98
	UQ	2.94	5.74	9.25	13.36	3.23	5.83	8.81		8.29	10.81
^{178}Os	Exp	3.02	5.76	9.03	12.72	4.92	5.83	7.74	10.56	6.54	9.18
	$U(5)$-$SU(3)$	2.21	3.76	5.46	7.34	3.01	4.51	6.42	8.44	5.90	8.20
	UQ	2.84	5.43	8.69	12.50	3.27	6.08	9.13	12.53	9.20	11.78
^{176}Os	Exp	2.93	5.50	8.57	12.10	4.45	5.50	7.59	10.60	6.39	9.06
	$U(5)$-$SU(3)$	2.25	3.73	5.41	7.25	3.00	4.40	6.29	8.27	5.75	8.01
	UQ	2.82	5.39	8.59	12.30	3.22	5.88	8.85	12.10	8.65	11.17
^{98}Sr	Exp	3.00	6.00	9.93		1.49					
	$U(5)$-$SU(3)$	3.01	5.98	9.87		2.13					
	UQ	3.00	5.98	9.87		1.46					

Table 2. The B(E2) transition ratios of experimental data[23–30] and calculated results.

	$\frac{B(E2;J_i^+\to J_f^+)}{B(E2;2_1^+\to 0_1^+)}$	$4_1^+\to 2_1^+$	$6_1^+\to 4_1^+$	$8_1^+\to 6_1^+$	$0_2^+\to 2_1^+$	$2_2^+\to 0_2^+$	$2_3^+\to 2_1^+$
^{152}Sm	Exp	1.45	1.70	1.98	0.23		0.07
	$U(5)$-$SU(3)$	1.64	1.90	2.00	0.90		0.02
	UQ	1.57	1.77	1.83	0.46		0.13
^{150}Nd	Exp	1.56	1.78	1.86	0.37	0.98	> 0.03
	$U(5)$-$SU(3)$	1.49	1.64	1.66	0.38	0.61	0.02
	UQ	1.50	1.66	1.68	0.28	0.49	0.12
^{154}Gd	Exp	1.55	1.68	1.97	0.28	0.34	0.08
	$U(5)$-$SU(3)$	1.49	1.67	1.73	0.34	0.63	0.02
	UQ	1.53	1.72	1.79	0.35	0.48	0.10
^{98}Sr	Exp	1.32			0.59		
	$U(5)$-$SU(3)$	1.56			0.61		
	UQ	1.61			0.58		

3. Results and Discussion

In order to demonstrate that the UQ scheme indeed serves as an alternative description of the spherical to axially deformed phase transition, we have systematically investigated the properties of some $X(5)$ candidates, including some low-lying energy levels and E2 transition rates, which are shown in Table 1 and Table 2.

It is shown that both the $U(5)$-$SU(3)$ scheme and the UQ scheme can reasonably describe the experimental data.[23–30] However, the UQ scheme seems better in describing the energy levels in higher excited levels and γ band, yet the 0_2^+ level is lower than those of $U(5)$-$SU(3)$ scheme and the experimental data. As for the $E2$ transition, for the intraband transition rate within the ground band seems better than those $U(5)$-$SU(3)$ scheme. It is also shown that this new scheme can greatly improve the interband $E2$ transition relating γ and ground bands.

The present work is very meaningful in helping us to deeply understand the new characteristics of symmetry due to the higher order $O(6)$ cubic interaction.

Acknowledgments

This work is partly supported by the National Natural Science Foundation of China under Grant Nos. 11375080, and the Natural Science Foundation of Liaoning Scientific Committee under Grant No 2013020091.

References

1. D. J. Rowe, G. Thiamova, Nucl. Phys. A **760**, 59 (2005).
2. G. Thiamova, Eur. Phys. J. A **45**, 81 (2010).
3. L. Fortunato, C. E. Alonso, J. M. Arias, J. E. García-Ramos, and A. Vitturi, Phys. Rev. C **84**, 014326 (2011).
4. D. H. Feng, R. Gilmore, and S. R. Deans, Phys. Rev. C **23**,1254 (1981).
5. J. Jolie, P. Cejnar, R. F. Casten, S. Heinze, A. Linnemann and V. Werner, Phys. Rev. Lett. **89**, 182502 (2002).
6. F. Iachello and N. V. Zamfir, Phys. Rev. Lett. **92**, 212501 (2004).
7. N. V. Zamfir, P. von Brentano, R. F. Casten and J. Jolie, Phys. Rev. C **66**, 021304(R) (2002).
8. V. Werner, P. von Brentano, R. F. Casten and J. Jolie, Phys. Lett. B **527**, 55 (2002).
9. A. Arima and F. Iachello, Ann. Phys. (N.Y.) **99**, 253 (1976).
10. A. Arima and F. Iachello, Ann. Phys. (N.Y.) **111**, 201 (1978).
11. A. Arima and F. Iachello, Ann. Phys. (N.Y.) **123**, 468 (1979).
12. R. F. Casten, in: F. Iachello (Ed.), Interacting Bose-Fermi System (Plenum Press, New York, 1981).
13. A. E. L. Dieperink, O. Scholten, and F. Iachello, Phys. Rev. Lett. **44**, 1747 (1980).
14. A. E. L. Dieperink and O. Scholten, Nucl. Phys. A **346**, 125 (1980).
15. F. Iachello, Phys. Rev. Lett. **87**, 052502 (2001).
16. R. F. Casten and N. V. Zamfir, Phys. Rev. Lett. **87**, 052503 (2001).
17. F. Pan, J. P. Draayer, and Y. Luo, Phys. Lett. B **576**, 297 (2003).
18. G. Rosensteela, D. J. Rowe, Nucl. Phys. A **759**, 92 (2005).
19. R. Fossion, C. E. Alonso, J. M. Arias, L. Fortunato, and A. Vitturi, Phys. Rev. **C 76**, 014316 (2007).
20. P. van Isacker, Phys. Rev. Lett. **83**, 4269 (1999).
21. G. Thiamova and P. Cejinar, Nucl. Phys. A **765**, 97 (2006).
22. L. R. Dai, F. Pan, L. Liu, L. X. Wang and J. P. Draayer, Phys. Rev. C **86** 034316 (2012).
23. R. Krücken, B. Albanna, C. Bialik, R. F. Casten, J. R. Cooper, et al. Phys. Rev. Lett. **88**, 232501 (2002).
24. R. V. Jolos and P. von Brentano,Phys. Rev. C **80**, 034308 (2009).
25. Y. Yamazaki, E. B. Shera, M. V. Hoehn, and R. M. Steffen, Phys. Rev. C **18**, 1474 (1978).
26. S. K. Basu, A. A. Sonzogni, Nuclear Data Sheets **114**, 435 (2013).
27. B. Harmatz, Nuclear Data Sheets **26**, 281 (1979).
28. B. Singh, Nuclear Data Sheets **84**, 565 (1998).
29. M. S. Basunia, Nuclear Data Sheets **107**, 791 (2006).
30. E. Acherberg, O. A. Capurro, G. V. Marti, Nuclear Data Sheets **110**, 1473 (2009).

THE EXPERIMENTAL STUDY OF NUCLEAR ASTROPHYSICS REACTION RATE OF $^{93}Zr(n,\gamma)^{94}Zr$

L. Gan, Z. H. Li*, J. Su, S. Q. Yan, B. Guo, X. C. Du, Z. D. Wu, S. Zeng, S. J. Jin, Y. B. Wang, X. X. Bai, W. J. Zhang

China Institute of Atomic Energy
Beijing, China
**E-mail: zhli@ciae.ac.cn*
http://www.ciae.ac.cn

H. B. Sun, E. T. Li

College of Physics Science and Technology, ShenZhen University
ShenZhen, China

The slow neutron capture (s-) process plays a very important role in the nucleosynthesis, which produces about half of the elements heavier than iron. ^{94}Zr is mainly from $^{93}Zr(n,\gamma)^{94}Zr$ in the s-process, and the direct component of the $^{93}Zr(n,\gamma)^{94}Zr$ capture reaction can be derived from the neutron spectroscopic factor of ^{94}Zr. As the existing neutron spectroscopic factors of ^{94}Zr vary from each other up to 60%, a new work should be adopted to measure it exactly. In the present work, the angular distributions of $^{94}Zr(^{13}C,^{13}C)^{94}Zr$, $^{94}Zr(^{12}C,^{12}C)^{94}Zr$ and $^{94}Zr(^{12}C,^{13}C)^{93}Zr$ were obtained using the high-precision Q3D magnetic spectrograph. In addition, distorted-wave Born approximation (DWBA) calculations of the transfer differential cross sections were performed. The calculated result displays a good agreement with the experiment data, and a value of 2.60±0.20 for the neutron spectroscopic factor of ^{94}Zr was extracted, and the direct capture cross section versus neutron energy of $^{93}Zr(n,\gamma)^{94}Zr$ for the ground state of ^{94}Zr was obtained too.

Keywords: Single neutron transfer; DWBA; neutron spectroscopic factor; direct capture cross section.

1. Background

About half of the elements heavier than iron are produced via slow neutron capture reactions and competing β-decays (the s-process).[1] Therefore, the research on the s-process is important to solve the mystery of nucleosynthesis of elements heavier than iron. ^{94}Zr is mainly made by the neutron capture reaction of ^{93}Zr. As the $^{93}Zr(n,\gamma)^{94}Zr$ reaction is on the s-process route, it can help us to understand the s-process nucleosynthesis through studying this reaction. Otherwise, Zr alloys are largely used in nuclear reactors, and ^{93}Zr is one of the major long-lived fission products. Because the long-term radioactive hazard can be reduced via transforming ^{93}Zr into steady ^{94}Zr , the study of ^{93}Zr(n,γ)^{94}Zr reaction is also helpful to reduce the radioactive hazard of reactor materials.[2] According to the radiation capture theory, the direct capture reaction section of ^{93}Zr(n,γ)^{94}Zr can be deduced with the spectroscopic factor of ^{94}Zr. Several works had been performed to study the spectroscopic factor of ^{94}Zr, but these results are different. As shown in Fig. 1, it can be seen that the values of the spectroscopic factor of ^{94}Zr from five different works vary from each other up to 60%.[3–7] Such a huge difference will affect the research of neutron capture reaction section. Therefore, a high-precision experiment should be proposed to deal with this problem.

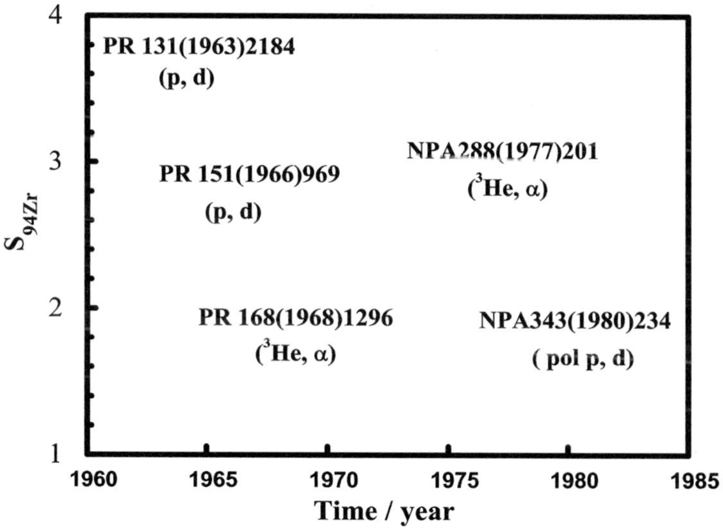

Fig. 1. The research state of the neutron spectroscopic factor of ^{94}Zr.

2. Experiment

The experiment data were obtained at the HI-13 tandem accelerator of the China Institute of Atomic Energy in Beijing. The incident ions were ^{12}C and ^{13}C, with the energy of 66 MeV and 64 MeV respectively. The ^{12}C support $^{94}ZrO_2$ target was made by ion sputtering, and the abundance of ^{94}Zr was 98.5%. The thickness of $^{94}ZrO_2$ and ^{12}C were 41 $\mu g/cm^2$ and 46 $\mu g/cm^2$ respectively.

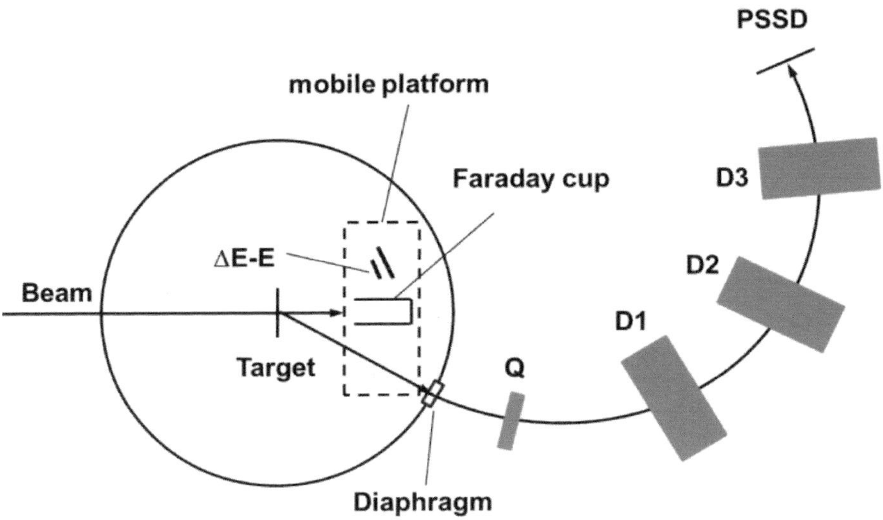

Fig. 2. Experimental settings.

The experimental settings are shown in Fig. 2. The target was in the target chamber center, and a faraday cup aimed to collect the ion steam through the target was on the backward of target, so that the reaction cross section can be absolutely normalized. The faraday cup was set in a remote controlled electric mobile platform. When the measuring angular was within 6^0, in order to avoid the outgoing ions being shielded, the faraday cup was removed. On the backward of the target, a ΔE-E detector telescopic system was set on the left side of 25^0, so that the differential cross section for small angles can be relatively normalized. A diaphragm was set at the entrance of Q3D magnetic spectrograph to limit the angular distinguish within 0.4^0, and the receiving angular of Q3D magnetic spectrograph was 0.34 msr.

On the Q3D focal plane, there was a 50 mm 50 mm position sensitive silicon detector (PSSD) to measure the products. The PSSD consists of 16 silicon strip detectors, and the width of each silicon strip is 3 mm, the vertical position of particles in detectors can be given by the vertical strips. The impedance of silicon strip detectors is uniformity, so the horizon position can be calculated via charge division. The Q3D magnetic spectrograph cannot distinguish the particles with same magnetic rigidity, but the energies of these particles are different, so they can be identified with the energy signals. The position-energy ion differential spectrum was shown in Fig. 3. It can be inferred from Fig. 3 that the target products can be easily picked out.

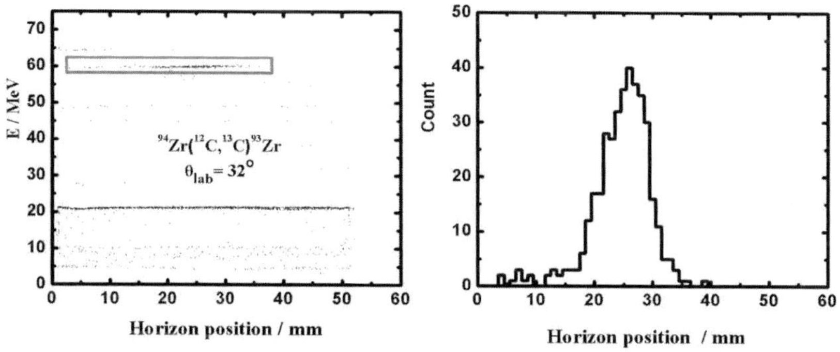

Fig. 3. Ion differential spectrum.

The angular distributions of $^{94}Zr(^{12}C,^{12}C)^{94}Zr$, $^{94}Zr(^{13}C,^{13}C)^{94}Zr$,and $^{94}Zr(^{12}C,^{13}C)^{93}Zr$ from 0-60 degree had been obtained. The ratio of the differential cross sections of elastic scattering to Rutherford cross sections had been drawn in Fig. 4, and the differential cross section of single neutron transfer reaction was shown in Fig. 5. The experimental error mainly came from statistical (3%) and target thickness (5%).

3. Data Analysis

The Distorted-wave Born approximation (DWBA) was applied to the analysis of experimental angular distributions using PTLOMY.[8] The calculation of the angular distribution of $^{94}Zr(^{12}C,^{13}C)^{93}Zr$ reaction needs the optical potential parameters of entrance and outgoing reaction channels, and these parameters can be extracted by fitting the angular distributions of elastic scattering reactions. Therefore, the angular distributions of $^{12}C+^{94}Zr$

and $^{13}C+^{94}Zr$ were measured too. As the radioactive of ^{93}Zr, the outgoing channel $^{13}C+^{93}Zr$ was replaced by $^{13}C+^{94}Zr$. Woods-Saxon potential was adopt when fitting, and the expression is as follow:

$$U(r) = \frac{V}{1+e^{(r-R_v)/a_v}} + \frac{W}{1+e^{(r-R_w)/a_W}} + V_c(r) \tag{1}$$

In the above formula, the first and second terms are the real and imaginary potentials. The third term is the coulomb part, and its expression is as follow:

$$V_c(r) = \begin{cases} \dfrac{Z_1Z_2e^2}{2R_c}\left[3-\left(\dfrac{r}{R_c}\right)^2\right], & r < R_c \\ \dfrac{Z_1Z_2e^2}{r}, & r > R_c \end{cases} \tag{2}$$

In Eq. (1) and Eq. (2), R_i refers to nuclear radius, and the general expression is$R_i = r_i(A_1^{1/3} + A_2^{1/3})$, $i = V, W, C$.

As is known that the contributions of the imaginary potential and coulomb part on the angular distributions of elastic scattering reactions are relative small, we fixed W=35 MeV, r_w=1.15 fm, a_w=0.56 fm, r_c=1.0 fm. The angular distributions of elastic scattering reactions are computed with different combinations of V, r_v and a_v, and then compared with the experimental cross sections. The most suitable combination was selected according to χ^2. The final parameters of optical potential of entrance and outgoing channels were list in Table 1.

Table 1. The parameters of Woods-Saxon potential extracted from the experiments. E_{lab} is the energy of the incident particles. V and W are the real and imaginary potential of well depth respectively, r_i and a_i are radius and diffusion respectively, i=V,W.

reaction	E_{lab} (MeV)	V (MeV)	r_v (fm)	a_v (fm)	W (MeV)	r_w (fm)	a_w (fm)	r_c (fm)	χ^2/point
$^{94}Zr(^{12}C,^{12}C)^{94}Zr$	66	189	1.124	0.55	35	1.15	0.56	1.00	1.67
$^{94}Zr(^{13}C,^{12}C)^{94}Zr$	64	197	1.131	0.55	35	1.15	0.56	1.00	1.45

It is easy to be seen in Fig. 4 that, the calculated angular distributions with the extracted optical potential parameters can fit the experimental data well. The geometrical parameters of bound state optical potential were fixed throughout the calculation of transfer angular distribution (r_0 = 1.25 fm, a=0.65 fm). The experimental and DWBA differential cross section of $^{94}Zr(^{12}C,^{13}C)^{93}Zr$ have the relationship of that,

$$(d\sigma/d\Omega)_{exp} = S_{13C}S_{94Zr}(d\sigma/d\Omega)_{DWBA} \tag{3}$$

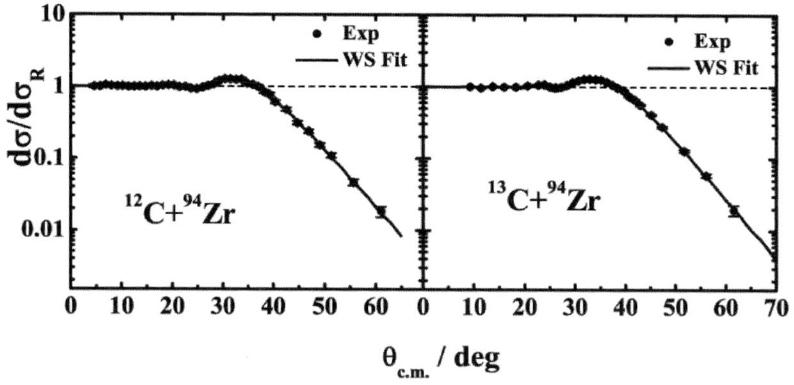

Fig. 4. The Woods-Saxon fit for the angular distributions of elastic scattering reactions.

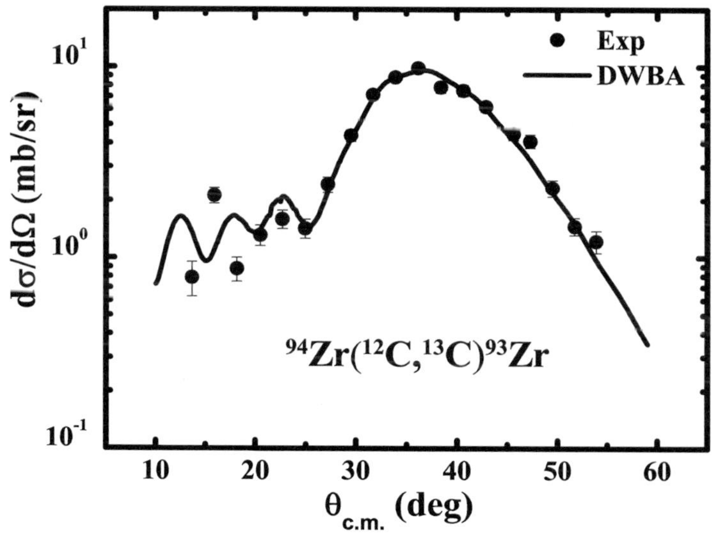

Fig. 5. Comparison between experimental angular distributions (dots) and calculations with DWBA (line) for $^{94}Zr(^{12}C,^{13}C)^{93}Zr$.

$(d\sigma/d\Omega)_{exp}$ and $(d\sigma/d\Omega)_{DWBA}$ refer to the experimental and DWBA differential cross section of ^{94}Zr(^{12}C,^{13}C)^{93}Zr respectively. S_{13C} and S_{94Zr} are the neutron spectroscopic factors of ^{13}Cand ^{94}Zr respectively. The neutron spectroscopic factors of ^{13}C employed the result of the work in 2010 by T. Al-Abdullah et al, using ^{13}C(^{12}C,^{13}C)^{12}C reaction, and the value is 0.65.[9] According to formula (3), the neutron spectroscopic factors of ^{94}Zr was gained, 2.60±0.20. As shown in Fig. 5, the normalized DWBA angular distribution can reproduce the experimental data. The error of spectroscopic factor was determined by experimental error.

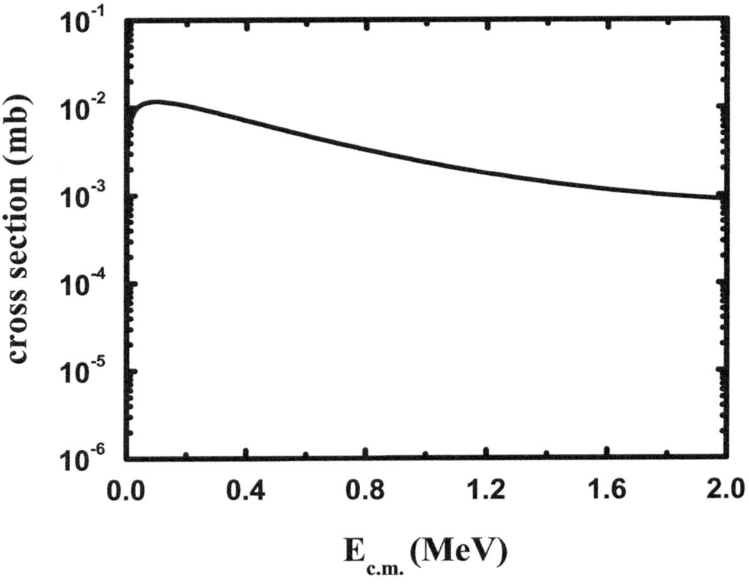

Fig. 6. The exciting function curve of ^{93}Zr(n,γ)^{94}Zr.

Table 2. The comparison of the neutron spectroscopic factors of ^{94}Zr.

Ref	3	4	5	6	7	presnt
^{94}Zr	3.41	2.5	2.16	2.61	2.18	2.06±0.20
Reaction	(d,t)	(p,d)	(^{3}He,α)	(^{3}He,α)	(pol p,d)	(^{12}C,^{13}C)
E/MeV	15	19.4	25	39	65	66

All the neutron spectroscopic factors of ^{94}Zr in former works and present work were list in Table 2. We can find out that the result in present work agrees with the values obtained from ^{94}Zr(p,d)^{93}Zr reaction in 1966 (Stautberg and Kraushaar)[4] and ^{94}Zr(^{3}He,α)^{93}Zr reaction in 1977 (Gales et al.) [6], but it is much different from the values gained by Cohen[3] and Rundquist [5]. In addition, it should be point out that, as the spin of ^{12}C and ^{13}C are 0+ and 1/2- respectively, just 1p1/2 orbit can contribute to the process of ^{12}C changing into ^{13}C by capturing a neutron. So the interference is smaller than other reactions, and the result is more believable.

With the neutron spectroscopic factor of ^{94}Zr obtained, the direct capture cross section of ^{93}Zr(n,γ)^{94}Zr for the ground state of ^{94}Zr can be calculated using REDCAP program.[10] The curve of direct capture cross section versus neutron Energy is shown in Fig. 6.

4. Conclusion

The differential cross sections of ^{94}Zr(^{12}C,^{13}C)^{93}Zr reaction and the elastic scattering reactions of its entrance and outgoing channels were measured in Q3D magnetic spectrograph of HI-13 tandem accelerator in Beijing. The optical parameters of ^{12}C + ^{94}Zr and ^{13}C + ^{94}Zr were obtained by fitting the corresponding elastic scattering angular distributions. The transfer reaction angular distribution was computed using PTLOMY with the extracted optical parameters, and then, the neutron spectroscopic factors of ^{94}Zr was gained as 2.60±0.20. It agrees with the values obtained from ^{94}Zr(p,d)^{93}Zr reaction in 1966 (Stautberg and Kraushaar) and ^{94}Zr(^{3}He,α)^{93}Zr reaction in 1977 (Gales et al). Because a better reaction system was adopted in the present work, a believable neutron spectroscopic factor of ^{94}Zr was obtained. And the direct capture cross section of ^{93}Zr(n,γ)^{94}Zr for the ground state of ^{94}Zr was obtained too.

This work was supported by the National Natural Science Foundation of China under grant Nos. 11321064, 11375269 and 11105228, and the National Basic Research Program of China (973 Program) under grant Nos. 2013CB834406.

References

1. B. Guo, Z. H. Li, M. Lugaro, et al., *J. Astrophysical Journal* **756**, 193 (2012).
2. G. Tagliente, P. M. Milazzo, K. Fujii, et al., *J. Physical Review C* **87**, 014622 (2013).
3. B. L. Cohen and O. V. Chubinsky, *J. Physical Review* **131**, 2184 (1963).
4. M. M. Stautberg and J. J. Kraushaar, *J. Physical Review* **151**, 969 (1966).

5. D. E. Rundquist, M. K. Brussel and A. I. Vavin, *J. Physical Review* **168**, 1296 (1968).
6. S. Gales, E. Hourani, S. Fortibr, et al., *J. Nuclear Physics A* **288**, 201 (1977).
7. K. Hosono, M. Kondo, T. Saito, et al., *J. Nuclear Physics A* **343**, 234 (1980).
8. I. T. Thompson, *J. Computer Physics Reports* **7**, 167 (1988).
9. T. A. Abdullah, F. Carstoiu, X. Chen, et al., *J. Physical Review C* **81**, 035802 (2010).
10. C. A. Bertulani, *J. Computer Physics Communications* **156**, 123 (2003).

A NEW POLYNOMIAL APPROACH WITH HEINE-STIELTJES CORRESPONDENCE TO THE STANDARD PAIRING PROBLEM

Xin Guan[1*], Haocheng Xu[1], Kristina D. Launey[2], Feng Pan[1,2] and J. P. Draayer[2]

[1]*Department of Physics, Liaoning Normal University, Dalian 116029, China*
[*]E-mail: ggguanxinnn@163.com

[2]*Department of Physics and Astronomy, Louisiana State University, Baton Rouge, LA 70803-4001, USA*

A new polynomial approach for solving the nuclear mean-field plus the standard pairing problem is proposed based on the Heine-Stieltjes correspondence. The advantage of the new polynomial approach lies in the fact that it offers exact numerical solutions with less CPU time, which makes it feasible to provide accurate results for systems with relatively larger configuration space.

Keywords: Heine-Stieltjes correspondence; the standard pairing problem.

1. Introduction

An exact solution of the standard pairing problem was first obtained by Richardson and is now referred to as the Richardson-Gaudin method.[1,2] Its advantage lies in the fact that the huge matrix in the Fock subspace is reduced to a set of equations, called Bethe ansatz equations (BAEs), such that the number of these equations equals exactly the number of pairs of the valence particles involved.[3,4] However, less attention has been paid to the Richardson solutions of the pairing problem in realistic applications mainly because the non-linear BAEs involved are very difficult to be solved numerically, especially for large-size systems.

In this study, we present a numerical method based on the new approach we proposed recently for solving the standard pairing problem.[5] In particular, we derive polynomial solutions of the second order Fuchsian equation. This makes exact pairing solutions feasible even when more energy levels or heavy nuclei are considered.

2. Polynomial Solutions to the Exact Pairing Problem

The Hamiltonian of the standard pairing model is given by

$$\hat{H} = \sum_{j=1}^{n} \epsilon_j \hat{n}_j - G \sum_{jj'} S_j^+ S_{j'}^-, \tag{1}$$

where the sums run over given j-levels of total number n, $G > 0$ is the overall pairing strength, $\{\epsilon_j\}$ are nondegenerate single-particle energies obtained from any spherical mean-field theory, $\hat{n}_j = \sum_m a_{jm}^\dagger a_{jm}$ is the number operator for valence particles in the j-th level, and $S_j^+ = \sum_{m>0}(-)^{j-m} a_{jm}^\dagger a_{j\,-m}^\dagger$ ($S_j^- = (S_j^+)^\dagger$) are pair creation (annihilation) operators.

According to the Richardson-Gaudin method, k-pair eigenstates of (1) can be written as

$$|k;x\rangle = S^+(x_1)S^+(x_2)\cdots S^+(x_k)|0\rangle, \tag{2}$$

where $|0\rangle$ is the pairing vacuum state satisfying $S_j^-|0\rangle = 0$ for all j, and $S^+(x_i) = \sum_j S_j^+/(x_i - 2\epsilon_j)$, in which x_i ($i = 1, 2, \cdots, k$) are spectral parameters satisfy the following set of BAEs:

$$1 - 2G \sum_j \frac{\rho_j}{x_i - 2\epsilon_j} - 2G \sum_{i'=1(\neq i)}^{k} \frac{1}{x_i - x_{i'}} = 0, \tag{3}$$

where the first sum runs over all j-levels and $\rho_j = -\Omega_j/2$ with $\Omega_j = j + 1/2$. For each solution, the corresponding eigen-energy is given by $E_k = \sum_{i=1}^k x_i$.

Through the Heine-Stieltjes correspondence, one can find the solutions[5] of the corresponding BAEs (3) by solving the second-order Fuchsian equation:

$$A(x)y''(x) + B(x)y'(x) - V(x)y(x) = 0, \tag{4}$$

where $A(x) = \prod_{j=1}^n (x - 2\epsilon_j)$ is a polynomial of degree n, the polynomial $B(x)$ is given as

$$B(x)/A(x) = \sum_{j=1}^{n} \frac{2\rho_j}{x - 2\epsilon_j} - \frac{1}{G}, \tag{5}$$

and $V(x)$ are called Van Vleck polynomials of degree $n-1$, which are determined according to Eq. (4). In search for polynomial solutions of Eq. (4), we write

$$y(x) = \sum_{j=0}^{k} a_j x^j, \quad V(x) = \sum_{j=0}^{n-1} b_j x^j, \tag{6}$$

where $\{a_j\}$ and $\{b_j\}$ are the expansion coefficients to be determined. Substitution of (6) into Eq. (4) yields two matrix equations. By solving these two matrix equations, we can obtain[5] the solutions of $\{a_j\}$ and $\{b_j\}$.

Furthermore, if we set $a_k = 1$ in $y(x)$, the coefficient a_{k-1} becomes equal to the negative sum of the $y(x)$ zeros, $a_{k-1} = -\sum_{i=1}^{k} x_i$, and hence, yields the corresponding eigen-energy $E_k = -a_{k-1}$.

3. Pairing Gaps for Sm Isotopes

The empirical like-particle pairing gap can be estimated by the third derivative of binding energies, BE, with respect to the number of valence like-particles,[6] which for neutrons is $\Delta \equiv \frac{1}{4}(BE(Z,N-2) - 3BE(Z,N-1) + 3BE(Z,N) - BE(Z,N+1))$.

Calculations for Sm isotopes are performed for a ^{144}Sm core and a model space of 6 j-levels. The single-particle energies used for (1) are $\epsilon_{7/2} = -1.056$ MeV, $\epsilon_{3/2} = -0.56$ MeV, $\epsilon_{3/2} = -0.162$ MeV, $\epsilon_{13/2} = 0.049$ MeV, $\epsilon_{9/2} = 0.368$ MeV, $\epsilon_{1/2} = 0.552$ MeV, $\epsilon_{5/2} = 0.603$ MeV. The average single-particle energy is $\epsilon_{avg} = -5.80$ MeV. We use $G = 24/A$ MeV.

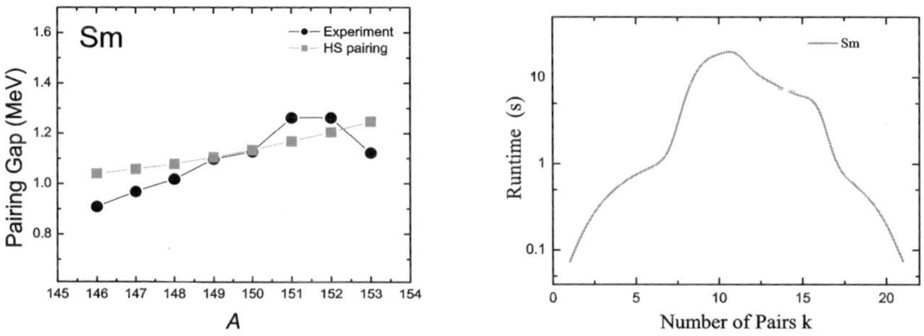

Fig. 1. Theoretical pairing gaps and the corresponding experimental values (in MeV) and the runtime (in seconds) to calculate the pairing energy for the ground state.

For the odd-A systems, the level blocked by an odd particle according to the experiment is $f_{7/2}$ for ^{145,147,149}Sm. For ^{151,153}Sm, the calculation shows that the lowest energy is obtained when the odd particle occupies the $f_{7/2}$ level.

The calculations for a large number of pairs are also possible when the polynomial approach is employed. We find that using $G = 24/A$ MeV yields a very close agreement of the theoretical paring gaps with the experimental counterparts. The runtime is in seconds for the polynomial approach to compute the pairing energy of the ground state shown in Fig. 1. This example shows that the polynomial approach can be straightforwardly applied for heavy nuclear systems, and hence, can provide exact pairing solutions in regions where the BAEs are impossible to be solved directly.

4. Conclusion

In summary, we have established a new approach for solving the standard pairing problem based on a robust mathematical foundation — the extended Heine-Stieltjes polynomials and the corresponding Van Vleck polynomials satisfying the polynomial solutions of the second-order Fuchsian equation. Thus, we reach the goal of the Richardson-Gaudin theory via the Heine-Stieltjes correspondence, which provides an exact solution to the pairing problem by solving only two matrix equations, which makes exact pairing solutions feasible even when more single-particle levels are involved or when it is applied to heavy nuclei. The approach can also easily be extended and applied to solve a large class of Gaudin-type quantum many-body problems.

5. Acknowledgments

Support from the U.S. National Science Foundation (OCI-0904874), U.S. Department of Energy (DE-SC0005248), the Southeastern Universities Research Association, the China-U.S. Theory Institute for Physics with Exotic Nuclei (CUSTIPEN) (DE-FG02-13ER42025), the NSFC (11405080, 11175078, and 11375080), and the LSU–LNNU joint research program (9961) is acknowledged.

References

1. R. W. Richardson, Phys. Lett. **3**, 277 (1963); **5**, 82 (1963).
2. M. Gaudin, J. Physique **37**, 1087 (1976).
3. Feng Pan and J. P. Draayer, Phys. Rev. C **66**, 044314 (2002).
4. J. Dukelsky, C. Esebbag, and P. Schuck, Phys. Rev. Lett. **87**, 066403 (2001);
5. X. Guan, K. D. Launey, M. Xie, L. Bao, F. Pan, and J. P. Draayer, Phys. Rev. C **86**, 024313 (2012); Comp. Phys. Comm. **185**, 2714 (2014).
6. A. Bohr and B. R. Mottelson, *Nuclear Structure*, Vol. 1: *Single-Particle Motion*, (Benjamin, New York, 1969).

THE ROLE OF SPIN-ORBIT AND TENSOR FORCE IN HEAVY-ION COLLISIONS

Lu Guo[1,2*], Gao-Feng Dai[2], En-Guang Zhao[2,3], Shan-Gui Zhou[2,3]

[1] *School of Physics, University of Chinese Academy of Sciences, Beijing 100049, China*

[2] *State Key Laboratory of Theoretical Physics, Institute of Theoretical Physics, Chinese Academy of Sciences, Beijing 100190, China*

[3] *Center of Theoretical Nuclear Physics, National Laboratory of Heavy Ion Accelerator, Lanzhou 730000, China*

In this contribution we present the one-body dissipation dynamics in the fully three-dimensional time-dependent Hartree-Fock (TDHF) theory. The calculations were performed with modern Skyrme energy functionals plus tensor terms and without any symmetry restrictions. The energy dissipation was revealed to decrease in deep-inelastic collisions of the light systems as the bombarding energy increases owing to the competition between collective motion and single-particle degrees of freedom. The role of spin-orbit and tensor force was given particular emphasis. The spin-orbit force causes a significant enhancement of the dissipation. About 40%∼65% of the total dissipation depending on the different parameter sets was predicted to arise from the spin-orbit force. The fusion cross section without tensor force overestimates the experimental value by about 25%, while the calculation with tensor force T11 has a good agreement with experimental cross section.

Keywords: Dissipation dynamics; TDHF; fusion cross section.

1. Introduction

The time-dependent Hartree-Fock (TDHF) theory has been applied in nuclear dynamics for more than 30 years;[1] for reviews see Refs. 2,3. It provides, from a fully microscopic point of view, the dynamical foundation of large amplitude collective motion. The extensive studies on the fusion excitation functions, deep inelastic collisions, fission, collective excitation, and nuclear molecular resonances revealed many useful insights on the basic

*E-mail: luguo@ucas.ac.cn

mechanisms of heavy-ion collisions and resonance dynamics. However, it is soon realized that these earlier TDHF calculations systematically underestimated the dissipation of energy from the relative collective motion into internal degrees of freedom due to the simplified effective interaction and symmetry restrictions.[4–6] For example, there has been the puzzle of fusion window anomaly, which was later solved by including the time-even terms of spin-orbit force in TDHF calculations.[7] The fusion excitation function was enhanced by as much as 20% for the ^{40}Ca+^{40}Ca system by removing the isospin symmetry restriction on the nuclear wave functions.[4]

With the development of computer ability, a fully three-dimensional TDHF calculation with modern energy functionals and without any symmetry restrictions has been realized in recent years. Accordingly, there is a renewed interest of the TDHF studies in heavy-ion collisions[8–18] and resonance dynamics.[19–23] In this contribution, we will present the dissipation dynamics in deep-inelastic collisions and the role of spin-orbit and tensor force by using modern three-dimensional TDHF calculations.

2. Theoretical Framework

TDHF calculations usually employ the Skyrme effective interaction. For the sake of later discussions, the time-even and time-odd spin-orbit contribution in the Skyrme energy functional

$$\mathcal{H}_{\rm ls}^{\rm even} = -\frac{1}{2}t_4\left(\rho\nabla\cdot\vec{J}+\sum_q \rho_q\nabla\cdot\vec{J}_q\right), \tag{1}$$

$$\mathcal{H}_{\rm ls}^{\rm odd} = -\frac{1}{2}t_4\left(\vec{s}\cdot\nabla\times\vec{j}+\sum_q \vec{s}_q\cdot\nabla\times\vec{j}_q\right), \tag{2}$$

are explicitly expressed by the density ρ, spin density $\vec{s}$, current $\vec{j}$, and spin-orbit density $\vec{J}$. The inclusion of both time-even and time-odd terms guarantees the Galilean invariance which should be met in a meaningful theory of heavy-ion collisions.

Most TDHF calculations do not include the tensor contribution, although the tensor component is present in the original Skyrme force. The time-even contribution of tensor terms

$$\mathcal{H}_{\rm t}^{\rm even} = \frac{1}{2}\alpha_T\left(\vec{J}_n^2+\vec{J}_p^2\right)+\beta_T\vec{J}_n\vec{J}_p, \tag{3}$$

is expressed by the strengths α_T and β_T of tensor force. For the details of theoretical formalism and numerical calculations, please refer to Refs. 17,18.

TDHF fusion cross section is calculated by the sharp-cutoff approximation[24]

$$\sigma_{\rm fus} = \pi(b_{\rm max}^2 - b_{\rm min}^2), \tag{4}$$

where the quantities $b_{\rm max}$ and $b_{\rm min}$ denote the maximum and minimum impact parameters at which fusion occurs.

3. Results and Discussions

3.1. *The effect of spin-orbit force on dissipation dynamics*

In Ref. 17, we have investigated the influence of the spin-orbit force on one-body dissipation dynamics in heavy-ion collisions of ^{16}O+^{16}O. In order to illustrate the energy dependence of dissipation dynamics, a quantity which measures the extent of energy dissipation in deep-inelastic collisions was defined as $P_{\rm dis} = 1 - E_{\rm fin}/E_{\rm c.m.}$, where $E_{\rm fin}$ and $E_{\rm c.m.}$ denote the final and initial kinetic energy, respectively. The percentage of energy dissipation is plotted in the left panel of Fig. 1 for head-on collisions of ^{16}O+^{16}O. As can be seen, the inclusion of spin-orbit (l*s) force results in a significant enhancement of energy dissipation in deep-inelastic collisions. For all the three cases the energy dissipation decreases as the incident energy increases. By observing the density distribution of separating ions, we found the nucleon density around neck region becomes lower as the increase of incident energy. At high energy of 170 MeV the nuclei are essentially spherical without a neck formation, indicating that less nucleons have been transferred between the colliding partners. This dynamical mechanism in the heavy-ion scattering may be interpreted by the concept of dissipative diabatic dynamics. As the increase of incident energy, the collective motion is so fast that the mean-field does not have enough time to rearrange itself and has to keep its identity as much as possible.

A quantity which expresses the proportion of dissipated energy coming from l*s force was defined as $P_{\rm s.o.} = 1 - P_{\rm dis}^{(\rm no\ l*s)}/P_{\rm dis}^{(\rm full\ l*s)}$, where $P_{\rm dis}^{(\rm no\ l*s)}$ and $P_{\rm dis}^{(\rm full\ l*s)}$ correspond to the blue and black lines in the left panel of Fig. 1. The results are displayed in the right panel of Fig. 1. The percentage of dissipated energy coming from l*s force has similar trend for SkM* and SLy4, but evident distinction appears for UNEDF1. As the increase of incident energy, the dissipation from l*s force increases and then starts to decrease after reaching the maximum value of 65% for SkM*, 63% for SLy4, and 48% for UNEDF1. The UNEDF1 calculations predict an overall smaller dissipation arising from l*s force and lower incident energy at the peak of

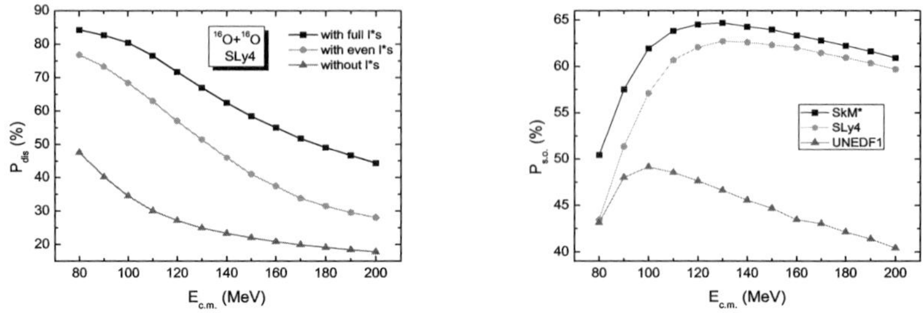

Fig. 1. (Color online) (Left panel) Percentage of energy dissipation as a function of center-of-mass energy for head-on collisions of $^{16}O+^{16}O$ with parametrization SLy4. The black, red, and blue lines represent the TDHF calculations involving full spin-orbit (l*s), time-even l*s, and no l*s force in the Skyrme energy functional. (Right panel) Percentage of dissipated energy arising from l*s force as a function of center-of-mass energy for head-on collisions of $^{16}O+^{16}O$ with Skyrme parametrizations SkM*, SLy4, and UNEDF1. Taken from Ref. 17.

dissipation due to the largest total dissipation and the smallest dissipated energy from l*s force. In spite of the sensitivity of force parameters, it is impressed that more than half of the total dissipation for SkM* and SLy4, and slightly less than half for UNEDF1 are attributed to the l*s force.

3.2. *The role of tensor force*

In Ref. 18, we investigated the role of tensor force in heavy-ion collisions. The fusion calculations for the $^{16}O+^{16}O$ system were carried out at $E_{\rm c.m.} =$ 70.5 MeV, where the experimental data is available. At high collision energy, there usually appears central transparency predicted in TDHF calculations and the lower limit of orbital angular momentum will be nonzero. Hence, both the maximum and minimum impact parameter $b_{\rm max}$ and $b_{\rm min}$ at which fusion occurs are searched within an interval of 0.1 fm. The fusion cross section shown in Table 1 without the tensor force (SLy5) overestimates the experimental value by about 25%. Since the strong dissipation results in an enhanced cross section, as we expected, the cross section is enhanced for SLy5+T, T31 and T33 and reduced for T11 and T13 when the tensor correlation has been taken into account. The good agreement between the result with parameter set T11 and experimental data suggests that T11 might be a good choice among the large number of tensor parameter sets to study the heavy-ion collision dynamics.

Table 1. Calculations of fusion cross section for $^{16}O+^{16}O$ at $E_{c.m.} = 70.5$ MeV using six Skyrme parameter sets, and the experimental data with errors.[25]

Force	σ_{fus} (mb)
SLy5	1307
SLy5+T	1327
T11	1161
T13	1265
T31	1326
T33	1327
Expt.	1056 ± 125

4. Summary

In this contribution, we present our recent work on the study of the dissipation dynamics and fusion cross section in heavy-ion collisions of $^{16}O+^{16}O$ with the microscopic TDHF theory. The calculations have been implemented in a fully three-dimensional coordinate space with modern Skyrme energy functionals plus tensor terms. We revealed that the percentage of energy dissipation in deep-inelastic collisions of the light systems decreases as the increase of incident energy due to the interplay between the collective motion and the single particle degrees of freedom. More than half of the total dissipation for SkM* and SLy4, and slightly less than half for UNEDF1 are predicted to come from spin-orbit force. The cross section without the tensor force using SLy5 overestimated the experimental value by about 25%, while the calculations with tensor force T11 had a good agreement with experiment.

Acknowledgments

This work was supported by Natural Science Foundation of China (Grants No. 11175252, No. 11121403, No. 11120101005, No. 11211120152, No. 11275248, and No. 11475115), the National Key Basic Research Program of China (Grant No. 2013CB834400), the Knowledge Innovation Project of the Chinese Academy of Sciences (Grant No. KJCX2-EW-N01), the President Fund of UCAS, the Scientific Research Foundation for the Returned Overseas Chinese Scholars, Ministry of Education of China, and the Open Project Program of State Key Laboratory of Theoretical Physics, Institute of Theoretical Physics, Chinese Academy of Sciences, China (No.Y4KF041CJ1). The results described in this paper were obtained on the High-performance Computing Clusters of SKLTP/ITP-CAS and the ScGrid of the Supercomputing Center, Computer Network Information Center of the Chinese Academy of Sciences.

References

1. P. Bonche, S. Koonin and J. W. Negele, *Phys. Rev. C* **13**, 1226(Mar 1976).
2. J. W. Negele, *Rev. Mod. Phys.* **54**, 913(Oct 1982).
3. C. Simenel, *Eur. Phys. J. A* **48**, p. 152 (2012).
4. S. J. Krieger and K. T. R. Davies, *Phys. Rev. C* **18**, 2567(Dec 1978).
5. K. T. R. Davies, H. T. Feldmeier, H. Flocard and M. S. Weiss, *Phys. Rev. C* **18**, 2631(Dec 1978).
6. J. A. Maruhn, K. T. R. Davies and M. R. Strayer, *Phys. Rev. C* **31**, 1289(Apr 1985).
7. A. S. Umar, M. R. Strayer and P. G. Reinhard, *Phys. Rev. Lett.* **56**, 2793(Jun 1986).
8. A. S. Umar and V. E. Oberacker, *Phys. Rev. C* **73**, p. 054607(May 2006).
9. L. Guo, J. A. Maruhn and P.-G. Reinhard, *Phys. Rev. C* **76**, p. 014601(Jul 2007).
10. L. Guo, J. A. Maruhn, P.-G. Reinhard and Y. Hashimoto, *Phys. Rev. C* **77**, p. 041301(R)(Apr 2008).
11. K. Washiyama, D. Lacroix and S. Ayik, *Phys. Rev. C* **79**, p. 024609(Feb 2009).
12. C. Simenel, *Phys. Rev. Lett.* **106**, p. 112502(Mar 2011).
13. N. Loebl, A. S. Umar, J. A. Maruhn, P.-G. Reinhard, P. D. Stevenson and V. E. Oberacker, *Phys. Rev. C* **86**, p. 024608(Aug 2012).
14. L. Guo and T. Nakatsukasa, *EPJ Web Conf.* **38**, p. 09003 (2012).
15. K. Sekizawa and K. Yabana, *Phys. Rev. C* **88**, p. 014614(Jul 2013).
16. C. Simenel and A. S. Umar, *Phys. Rev. C* **89**, p. 031601(Mar 2014).
17. G.-F. Dai, L. Guo, E.-G. Zhao and S.-G. Zhou, *Phys. Rev. C* **90**, p. 044609(Oct 2014).
18. G.-F. Dai, L. Guo, E.-G. Zhao and S.-G. Zhou, *Sci. China-Phys. Mech. Astron.* **57**, p. 1618 (2014).
19. T. Nakatsukasa and K. Yabana, *Phys. Rev. C* **71**, p. 024301(Feb 2005).
20. J. A. Maruhn, P. G. Reinhard, P. D. Stevenson, J. R. Stone and M. R. Strayer, *Phys. Rev. C* **71**, p. 064328(Jun 2005).
21. A. S. Umar and V. E. Oberacker, *Phys. Rev. C* **71**, p. 034314(Mar 2005).
22. P.-G. Reinhard, L. Guo and J. Maruhn, *Eur. Phys. J. A* **32**, 19 (2007).
23. C. Simenel and P. Chomaz, *Phys. Rev. C* **80**, p. 064309(Dec 2009).
24. P. Bonche, B. Grammaticos and S. Koonin, *Phys. Rev. C* **17**, 1700(May 1978).
25. F. Saint-Laurent, M. Conjeaud, S. Harar, J. Loiseaux, J. Menet and J. Viano, *Nucl. Phys. A* **327**, 517 (1979).

EFFECT OF NUCLEAR DEFORMATION ON THE α-DECAY HALF-LIVES

Shu Qing Guo, Xiao Jun Bao and Hong Fei Zhang*

School of Nuclear Science and Technology, Lanzhou University
Lanzhou, 730000, China
**Email: zhanghongfei@lzu.edu.cn*

The influence of nuclear deformation on the α-decay half-lives of even-even nuclei from ground states to ground states is systematically investigated within the framework of a generalized liquid drop model (GLDM) by taking into account the deformation of the daughter nucleus. The preformation factors of α particle are extracted from the experimental α-decay half-lives of 158 even-even nuclei for $62 \leq Z \leq 118$, and an analytic expression for calculating α clustering factor is proposed. The experimental half-lives are accurately reproduced and some predictions are reasonably made by employing the proposed preformation formula.

Keywords: α-decay; preformation factor; nuclear deformation; GLDM.

1. Introduction

The investigation of the α-decay mechanism of the heavy and superheavy nuclei is an important work both theoretically and experimentally. On the experimental side, the α-decay is used as a suitable tool to identify the new superheavy elements produced at accelerators.[1,2] On the theoretical side, the α-decay is the quantum mechanical tunneling process and it can give useful information about the structure of the parent nuclei.[3] Recently the α-decay calculations taking into account the deformation has attracted a great deal of interest,[4] and the hindrance factors are very sensitive to the deformations and therefore may constitute a powerful tool to extract spectroscopic information.[5] Moreover, the problem of α preformation in a nucleus is very difficult to handle microscopically if one wants to treat the relative motion of the cluster versus the core nucleus correctly. In principle, this is a very complicated four-body problem. Despite this, various microscopical approaches have been developed expending the parent wave function over a basis containing shell-model as well as cluster-model type elements.[6–8] In the present work the preformation factors are extracted phenomenologically from the experimental half-lives to properly describe their behavior and the aim of work is to investigate the role of the deformation effects and the preformation factors explaining the α-decay half-lives of the deformed nuclei in the case of favored transition in the frame of GLDM.[9]

2. Results and Discussion

First of all, we systematically calculate the α-decay half-lives of even-even isotopes of Th, U, Pu, Cm, Fm with and without considering the deformation of the daughter nuclei with the fixed assault frequency $\nu_0 = 10^{20} s^{-1}$, respectively. The calculated half-lives with considering the deformations of the daughter nuclei decrease by a factor of 2.0-3.0 as compared with those of spherical ones. The conclusions are made in following aspects. One the one hand, the deformation does sensitively affect the penetration probability $P(\theta)$ in a certain direction θ, but can also average the values over all directions, thus making the final penetration probability to be close to the value of the case for the spherical daughter nucleus. On the other hand, although reasonably reproducing the experimental data, the fixed assault frequency can result in larger deviations around the magic and heavier nuclei and cannot describe the detailed features of the nuclei.[10] Based on this situation, the assault frequency $\nu_E = \nu_C P_0$ defined as the product of the classical assault frequency $\nu_C = 1/2\sqrt{2E_\alpha/M_\alpha}$, and the preformation factor P_0 is used in present work.

To shed light on the preformation factors, various microscopical approaches have been developed. However, the problem of alpha preformation in a nucleus is very difficult to handle microscopically because it is a very complicated four-body problem. In the present work we extracted the values from the experimental half-lives to properly describe their behavior, as shown in Fig. 1 with open circles. It is obvious that the extracted preformation factors decrease with increasing neutron number up to the spherical shell closure N=126, and then increase quickly, indicating the shell effects in α particle preformation. To theoretically outline the regularity, a formula of α particle preformation is proposed in the following form

$$\begin{aligned}\log_{10} P_0 =& a + b(Z - Z_1)(Z_2 - Z) + c(N - N_1)(N_2 - N) \\ &+ dA + e(Z - Z_1)(N - N_1),\end{aligned} \tag{1}$$

where Z, N, and A are the charge, neutron, and mass number of the parent nucleus. Z_1 (N_1) and Z_2 (Z_2) are the proton (neutron) magic numbers around Z (N), respectively. The first and the fourth term can macroscopically describe the position and the trend of the preformation with the decreasing mass number of the parent nuclei. The second and the third term show a parabolic dependence in $\log P_0$ as a function of the N (or Z), which can properly describe their behavior with respect to

Table 1. The Parameters for Eq.(1) in different region represented as I: $50 \leq Z \leq 82, 82 \leq N \leq 126$, II: $82 \leq Z \leq 126, 82 \leq N \leq 126$, III: $82 \leq Z \leq 126, 126 \leq N \leq 152$ and IV: $82 \leq Z \leq 126, 152 \leq N \leq 184$, respectively.

	I	II	III	IV
a	7.118509	0.451318	-22.08835	34.90593
b	0.000345	0.006894	-0.002675	0.003011
c	0.001372	0.000721	0.001292	0.003717
d	-0.053566	-0.012391	0.098519	-0.151216
e	0.000944	-0.004534	-0.007509	0.006681
$\sqrt{\bar{\sigma}^2}$	0.160	0.130	0.167	0.229

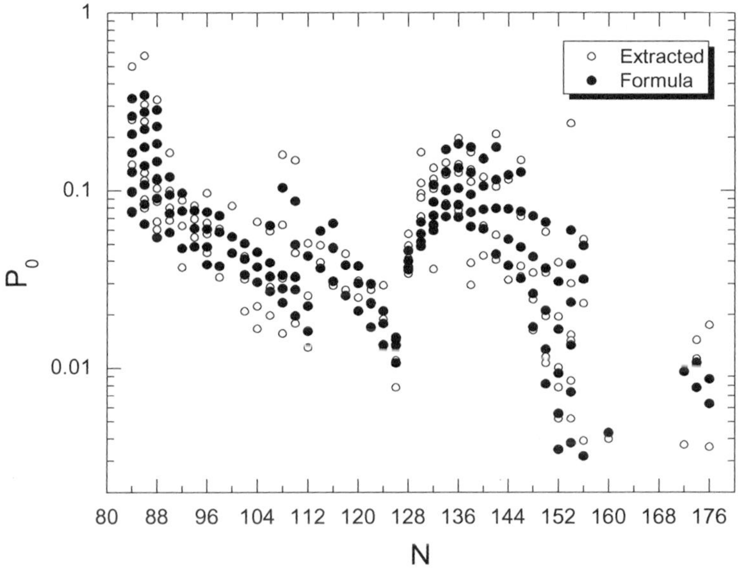

Fig. 1. Comparisons of the α particle preformation factor calculated by Eq. (1) and the values extracted from the experimental half-lives.

the proton and neutron.[11] The last term is related to the integrated neutron–proton interaction strength parameterized by the valence protons and neutrons.[12]

Here we would like to point out that the fitted parameters presented in Table 1 usually vary significantly, due to the different behaviors of the extracted preformation factors in respective region. However they can reproduce experimental half-lives excellently and predicate some unknown very neutron-deficient nuclei and SHN which are not far from the known ones we have used to fit the parameters.

3. Summary

In the present paper, the half-lives is investigated within a generalized liquid drop model where the deformation of the daughter nuclei is considered. The preformation factors of α particle are extracted from the experimental α-decay half-lives and an analytic expression for α particle preformation is proposed. The proposed formula associated with the introduction of the nuclear deformation can better produce accurate results and show shell structure. Predictions of some very neutron-deficient nuclei and SHN nuclei, are consistent with experimental values from other references, confirming that appropriate extrapolation can well predict those unknown nucleus.

4. Acknowledgments

The work is supported by the National Natural Science Foundation of China (Grants 11175074, 11475050, 11120101005 and 11105035), the Fundamental Research Funds for the Central Uni-versities (Grantlzujbky-2014-230).

References

1. S. Hofmann, F. Heßberger *et al.*, *The Eur. Phys. J. A* **10**, 5 (2001).
2. Y. T. Oganessian, V. K. Utyonkov *et al.*, *Phys. Rev. C* **74**, 044602 (2006).
3. R. G. Lovas, R. J. Liotta *et al.*, *Phys. Rep* **294**, 265 (1998).
4. D. Ni and Z. Ren, *Phys. Rev. C* **83**, 067302 (2011).
5. D. S. Delion, A. Insolia and R. J. Liotta, *Phys. Rev. C* **46**, 1346 (1992).
6. T. Fliessbach, H. J. Mang and J. O. Rasmussen, *Phys. Rev. C* **13**, 1318 (1976).
7. D. S. Delion, A. Sandulescu and W. Greiner, *Phys. Rev. C* **69**, 044318 (2004).
8. K. Varga, R. G. Lovas and R. J. Liotta, *Phys. Rev. Lett.* **69**, 37 (1992).
9. G. Royer, *J. Phys. G: Nucl. Part. Phys.* **26**, 1149 (2000).
10. S. Guo, X. Bao, Y. Gao, J. Li and H. Zhang, *Nucl. Phys. A* **934**, 110 (2015).
11. H. Zhang, G. Royer *et al.*, *Phys. Rev. C* **80**, 057301 (2009).
12. G. Gangopadhyay, *J. Phys. G: Nucl. Part. Phys.* **36**, 095105 (2009).

HIGH SPIN STUDY IN ^{126}Te

L. C. He, L. H. Zhu*, B. H. Sun, F. Wang, G. X. Zhang
School of Physics and Nuclear Energy Engineering, Beihang University, Beijing, 100191, China
**E-mail: zhulh@buaa.edu.cn*
www.buaa.edu.cn

Y. Zheng†, C. Y. He, X. G. Wu
China Institute of Atomic Energy, Beijing, 102413, China
†E-mail: zhengyun@ciae.ac.cn

The high-spin states in the even-even nucleus ^{126}Te have been investigated with in-beam γ-ray spectroscopy methods using the heavy-ion fusion-evaporation reaction ^{124}Sn(^{7}Li,4n1p)^{126}Te at 48 MeV beam. On the basis of the γ-γ coincidence relations and DCO ratios of γ transitions, a new level scheme includes 30 more γ transitions and one more negative-parity band was established. The deformation of the nucleus was discussed according to the plot of E-GOS curves.

Keywords: Heavy ion fusion-evaporation reaction; high spin states; negative-parity band; deformation.

1. Introduction

Discovery of the magic number is the most important evidence of shell structure theory. Based on previously experimental and theoretical study, nuclear shape tends spherical when its number of protons (Z) or neutrons (N) is near to the magic number and a lot of peculiar structures like magnetic rotation, chiral rotation show up when N or Z is becoming different from the magic number[1]. For ^{126}Te, its Z is near to the magic number 50 while its N is between two magic numbers 50 and 82 which will exhibit a rich variety of nuclear structures. Although there already have several excellent researches in ^{126}Te[2–4], heavy-ion fusion-evaporation reaction has never been used because of the low cross section of such nuclei during the region of stability. More study in ^{126}Te is worth to do.

2. Experiment and Results

High spin states in ^{126}Te were populated through the heavy-ion fusion-evaporation reaction ^{124}Sn(^{7}Li,4n1p)^{126}Te at a beam energy of 48 MeV provided by the HI-13 tandem accelerator of China Institute of Atomic Energy (CIAE), and studied by in-beam γ-ray spectroscopic method with a detector array consisting of 12 BGO-Compton-suppressed HPGe detectors and 2 planar detectors. The Ge detectors in the array were placed at 90°,

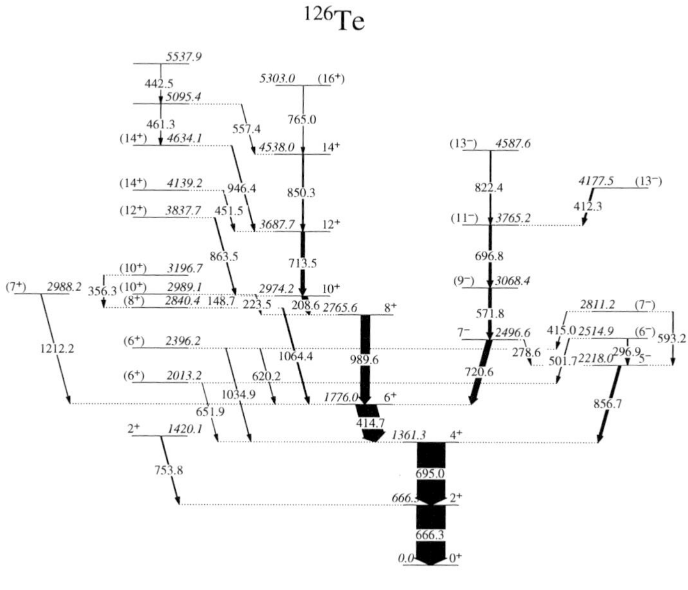

Fig. 1. The new level schemes of ^{126}Te from present experiment.

±37° , ±30° and ±60° relative to the beam direction. Each detector had an energy resolution of about 2 keV for 1332.5 keV γ rays. The target was an enriched self-supporting ^{124}Sn (97.5±0.2%) metallic foil of 4.6 mg/cm^2 thickness. Total of 9.42×10^7 γ-γ coincidence events was accumulated. Energy and efficiency calibrations of the Ge detectors were done with standard sources ^{60}Co and ^{152}Eu. After careful energy calibration and gain matching of each detector, the γ-γ coincidence data were sorted offline into conventional E_γ-E_γ matrices and asymmetry DCO (Directional Correlation ratios of Oriented states) matrices. The DCO matrices were created by sorting the detectors at $\pm 37°$ on one axis and the detectors at $\sim 90°$ on the other. These matrices were analyzed using the RADWARE package based on Linux-PC system. According to the γ-γ coincidence relations and DCO ratios of γ transitions, a new level scheme of ^{126}Te has been established, as shown in Fig. 1. Almost 30 new γ transitions were added. The level of the positive-parity yrast band has been extended to the 18^+ and an negative-patity band was constructed from the present data.

3. Discussion

The positive-parity yrast band could be obviously separated to three parts due to the life-times of levels in spin 6^+ and 10^+ measured by Zhang et al.[2] According to the research of Iachello, the ratio of the level energy of 4^+ and 2^+ could exhibit the shape of the nucleus[5,6]. ^{126}Te tends a near spherical shape and a typical vibration state in the first part consists of 2^+ and 4^+ because the ratio is near 2 which represents a spherical shape. This also corresponds to the Z near the magic number of ^{126}Te. The plot of E-GOS

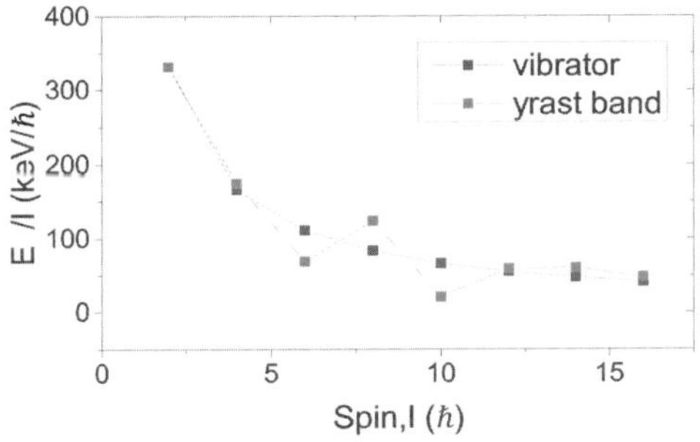

Fig. 2. The E-GOS plot of the positive-parity yrast band together with an ideal vibrator.

for the positive-parity yrast band is shown in Fig. 2 together with a vibrator of an ideal rigid rotor by fitting the first 2+ and 4+ states[7]. It is noted from Fig. 2 that the E-GOS curve of the band exhibits a drastic zigzag shape in the spin region from 6^+ to 10^+, and a smooth increasing after spin of 12^+. This phenomena indicates that the nucleus may undergo a shape evolution from spherical at low spin states to triaxial at medium spin states and further to prolate or oblate at high spin states. A complete understanding of the structure in ^{126}Te with a help of theoretical calculations is underway.

References

1. S. Frauendorf and J. Meng, Nucl. Phys. A 617(1997), 131-147.
2. C. T. Zhang et al., Nucl. Phys. A, 628(1998), 386-402.
3. T. von Egidy et al,. Nucl. Phys. A, 714(2003), 355-390.
4. J.R. Vanhoy et al, Phys. Rev. C 69(2004), 064323.
5. F. Iachello, Phys. Rev. Lett. 85(2000) 3580.
6. F. Iachello, Phys. Rev. Lett. 87(2001) 052502.
7. P. H. Regan, Phys. Rev. Lett., 90(2003) 152502.

NUCLEAR SHAPE PHASE TRANSITIONS AT THE NUCLEON LEVEL WITHIN SD-PAIR SHELL MODEL

Lei Li, Yanan Luo*

School of Physics, Nankai University, Tianjin, 300071, P.R. China
**E-mail: luoya@nankai.edu.cn*

Yu Zhang

Department of Physics, Liaoning Normal University, Dalian 116029, P. R. China

Feng Pan[1,2]

[1] *Department of Physics, Liaoning Normal University, Dalian 116029, P. R. China*

[2] *Department of Physics and Astronomy, Louisiana State University, Baton Rouge, LA 70803, USA*

Jerry P. Draayer

Department of Physics and Astronomy, Louisiana State University, Baton Rouge, LA 70803, USA

The effects of strength of each interaction on the nuclear shape phase and their transitional patterns are studied in the SD-pair shell model for both identical nucleon system. We found that the nuclear shape phase transitional patterns (vibraion↔ rotation) can be produced by changing the interactional strengths.

1. Introduction

In the last ten years, a number of theoretical developments have provided new insights on understanding the evolution of nuclear structure in transitional regions through the shape phase transition(SPT) analysis.[1–3]

Nuclei, as a mesoscopic system, have been found to possess interesting geometric shapes. Theoretical study of shape phase transitions and critical point symmetries in nuclei has mainly been carried out[2–21] in the interacting boson model.[2] The IBM is a phenomenological model of nuclear structure which has a deep connection with the microscopic shell model.[22,23] Recently there have been studies on nuclear shape phase transitions and their critical point symmetries in the framework of shell model,[24–30] density functional approach[31] and relativistic mean field approach.[32]

The investigations on nuclear shape phase transition and critical point symmetry for identical nucleon system have also been carried out with fermionic degrees of freedom in.[33–38]

In Ref.,[39] a correspondence between the strength of each of the interactions and the nuclear shape phases is obtained with the Dyson boson mapping approach SD-pair shell model(SDPSM) is built up from SD fermion pairs, and the Hamiltonian is diaganolized in the fermion space directly.[41] Our previous work in the SDPSM show that the vibrational, rotational, triaxial and γ-soft spectra can be well reproduced.[42,43] The nuclear shape phase transitional patterns and properties of the critical point symmetry can also be produced within the framework of the SDPSM with fermionic degrees of freedom.[44,45] It is interesting to see if the similar conclusion as in Ref.[39] can be obtained in the SDPSM. This is the main objective of this paper.

2. Model

As in Ref.,[39] we take a general shell model Hamiltonian [28,30] to study the dependence of the shape phases on each of the interactions, which is a combination of the single particle energy, monopole pairing, quadrupole-pairing and quadrupole-quadrupole interaction with

$$H = H_0 - G_0 \mathcal{S}^\dagger \mathcal{S} - G_2 \mathcal{P}^\dagger \mathcal{P} - \kappa Q^{(2)} \cdot Q^{(2)} \tag{1}$$

$$H_0 = \sum_a \epsilon_a n_a; \qquad \mathcal{S}^\dagger = \sum_a \frac{\widehat{j}_a}{2} \left(C_a^\dagger \times C_a^\dagger \right)$$

$$\mathcal{P}^\dagger = \sum_{ab} q(ab2) \left(C_a^\dagger \times C_b^\dagger \right)^2; \qquad Q_\mu^{(2)} = \sqrt{\frac{16\pi}{5}} \sum_{i=1}^{n} r_i^2 Y_{2\mu}(\theta_i \phi_i)$$

where a denote all quantum number necessary to specify a state $[a \equiv (nlj)]$. ε_a and n_a are the single-particle energy and number operator of state a, $\widehat{j}_a = \sqrt{2j_a+1}$ respectively. G_0, G_2 and κ is the monopole-pairing, quadrupole-pairing and quadrupole-quadrupole interaction strength, respectively.

The $E2$ transition operator is simply

$$T(E2) = e_{\text{eff}} Q^{(2)}, \tag{2}$$

where e_{eff} is the effective charge.

For the convenience of expression, we denote the characteristic quantities as $R_{4_1} = E_{4_1^+}/E_{2_1^+}$, $R_{6_1} = E_{6_1^+}/E_{2_1^+}$, $R_{0_2} = E_{0_2^+}/E_{2_1^+}$, $R_{0_3} = E_{0_3^+}/E_{2_1^+}$, $R_{2_2} = E_{2_2^+}/E_{2_1^+}$, $B_{4_12_1} = \frac{B(E2;4_1^+->2_1^+)}{B(E2;2_1^+->0_1^+)}$, $B_{6_14_1} = \frac{B(E2;6_1^+->4_1^+)}{B(E2;2_1^+->0_1^+)}$, $B_{0_22_1} = \frac{B(E2;0_2^+->2_1^+)}{B(E2;2_1^+->0_1^+)}$, $B_{0_32_1} = \frac{B(E2;0_3^+->2_1^+)}{B(E2;2_1^+->0_1^+)}$ and $B_{2_22_1} = \frac{B(E2;2_2^+->2_1^+)}{B(E2;2_1^+->0_1^+)}$.

3. Numerical Results

To see if the SDPSM can produce the similar results as in Ref.,[39] the same major shell and single particle energy levels as in Ref.[39] are used, which is 1.3, 2.8, 0, 0.8 and 2.5 for $j = 1/2,\ 3/2,\ 5/2,\ 2/7$ and $11/2$, respectively. The single-particle wave functions are taken to be the harmonic oscillators with oscillation constant $b_2 = 1.0A^{1/3}fm^2$. In addition, for simplicity, protons and neutrons are not differentiated, and the single-particle orbits energies are set to be invariant in the whole calculations. The effective charge is set to be 1.0 in the electric properties calculations.

3.1. *The effect of the monopole pairing interaction*

As in Ref.,[39] to study the effect of the monopole-pairing interaction on the collectivity of low-lying states, the quadrupole-pairing and quadrupole-quadrupole interactional strengths are set to be 0. To study the effect of the monopole pairing strength on the low-lying states, the monopole-pairing strength G_0 is set to be as $0.05MeV \leq G_0 \leq$ 0.5MeV.

The calculated results of the dependence of the energy levels and $B(E2)$ values on the monopole-pair interaction strength G_0 are illustrated in Fig.1.

From the figures, it is seen that when $G_0 <$ 0.1MeV, the energy levels do not change with G_0 and the $B(E2)$ values are all close to be 0, i.e., the collectivity of low-lying states can not be produced for small G_0 values. The energy levels and $B(E2)$ values begin to increase with G_0 from $G_0 \sim$ 0.1MeV. This analysis seems show that a new phase transition between collective motion and no-collective motion occurs in the SDPSM calculation.

One can also notice that the degenerate level structure of the vibrational states (U(5) symmetric states in the IBM)are reproduced for large G_0. For example, when $G_0 =$ 0.2MeV, 0_2^+,2_2^+ and 4_1^+ are degenerate, 0_3^+, 2_3^+, 3_1^+, 4_2^+ and 6_1^+ are also degenerate, these are the typical feature of the vibrational limit. Therefore, a large G_0 favors a spherical phase.

Ref.[39] shows that as G_0 is very small, the calculated results show approximately the feature of the axially rotational phase. But from the above

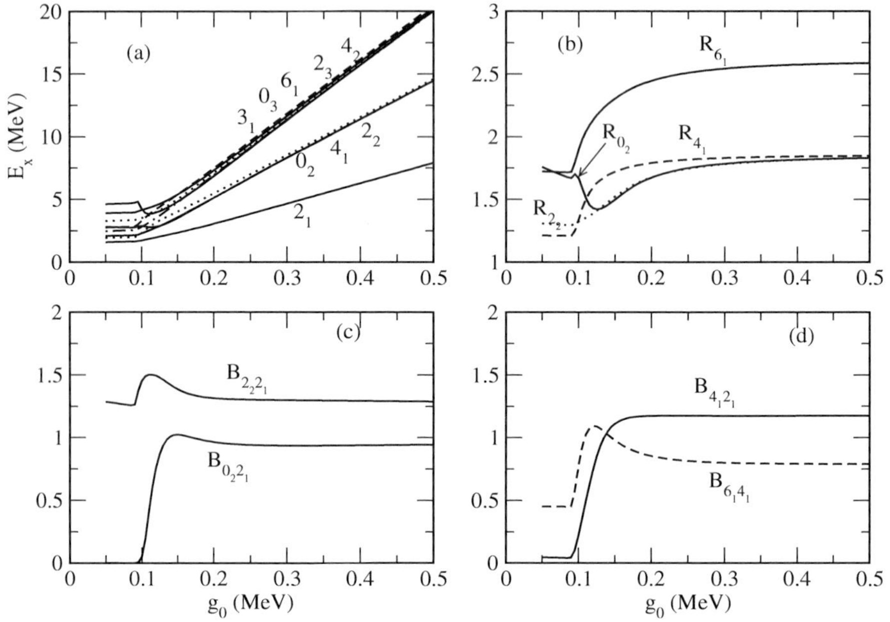

Fig. 1. Calculated result on the monopole pairing interaction G_0 when $G_2 = \kappa = 0$.

discussion one can see that the similar results can not be produced in the SDPSM.

3.2. *Effect of the quadrupole-pairing*

From Ref.[46] it is known that the quadrupole-pairing interaction is very important in producing the γ-soft spectrum for like-nucleon system. The effect of the quadrupole-pairing interaction on the collectivity of low-lying states were also be studied in Ref.[39]

To study the effect of the quadrupole pairing interaction on the collectivity of the low-lying states, we set G_0=0.15MeV and $\kappa = 0$ as in Ref.,[39] and $G_2 \in (0, 0.1)$. The results are given in Fig.2. Since $G_0 = 0.15$ and $G_2 = \kappa = 0$ corresponds to the spherical phase, our calculation here shows in fact the effect of the quadrupole-pair interaction on the spherical phase as in Ref.[39] The figures show that the general behaviors in Ref.[39] can all be produced in the SDPSM.

From above analysis one can see that considering the energy ratio and $B(E2)$ values together, the properties of the 0_2^+ state in the SDPSM

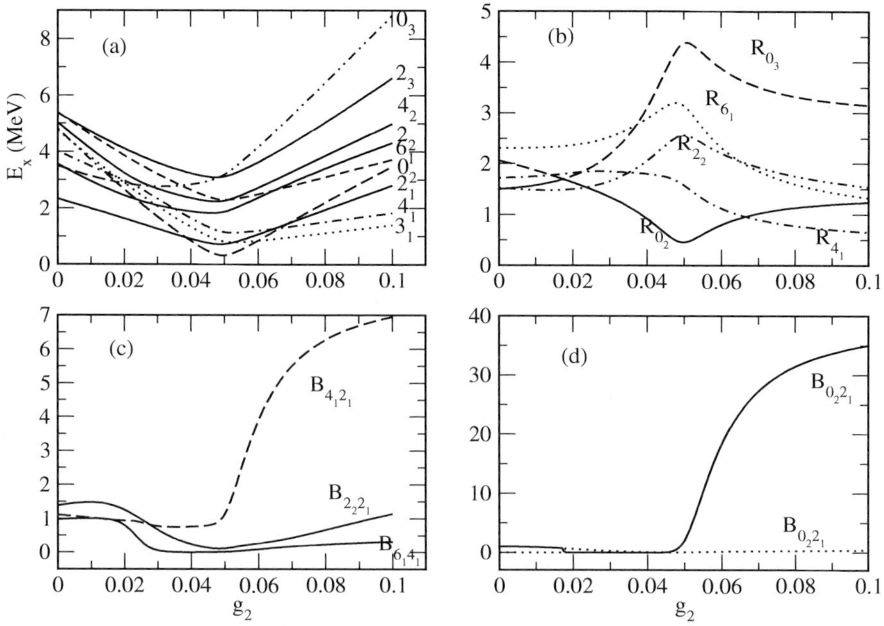

Fig. 2. Calculated result on the quadrupole pairing interaction G_2 when $G_0 = 0.15$MeV and $\kappa = 0$.

are also in consistent with that in the Ref.[39] It is also seen that for $G_2 \in (0, 0.02)\mathrm{MeV}/r_0^4$, the results are close to the vibrational phase, but for larger G_2 value, it is difficult to determine the shape phase of the nuclear system.

3.3. *Effect of the quadrupole-quadrupole interactional strength*

It is known that the mono-pole pairing and quadrupole-quadrupole interactions are dominant shell-model interactions. As in Ref.,[39] the effect of the quadrupole-quadrupole interactional strength are also studied in the SDPSM with G_0=0.15MeV, $G_2 = 0$ and $\leq 0\kappa \leq 0.1\mathrm{MeV}/r_0^4$. Since $G_0 = 0.15$MeV and $G_2 = 0$ corresponds to the spherical phase, our calculation here again shows in fact the effect of the quadrupole-quadrupole interaction on the spherical phase. The energies, energy ratios and $B(E2)$ ratios are given in Fig. 3.

From Fig. 3 one can notice that all the levels decrease with κ till $\kappa \sim 0.005\mathrm{MeV}/r_0^4$, and the degenerate level structure of the vibrational

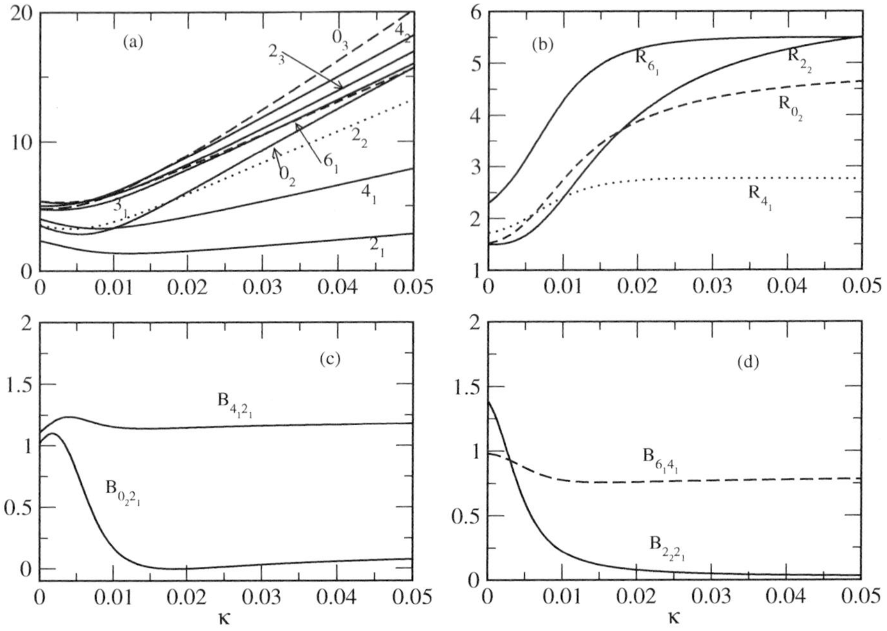

Fig. 3. Calculated result on the quadrupole-quadrupole interaction κ when $G_0 = 0.15\text{MeV}$ and $G_2 = 0$.

states (U(5) symmetric states in the IBM), while the level structure of the rotational states (SU(3) symmetric states in the IBM) can be reproduced for larger κ values. It is seen that the $B(E2)$ ratios are close to the vibrational limit when κ is small, and they are close to those values in the rotational limit when κ is large.

4. Discussion and Summary

From above analysis one can see that for the case of the vibrational limit and rotational limit, R_{41}s all smaller than the typical value of the vibrational limit 2.0 and rotational limit 3.3. The reason is the pauli-blocking effect play an important role in producing the collectivity of the low-lying states.

In summary, the effect of the interactional strength on the nuclear shape phase transition patterns have been studied within the framework of the SD-pair shell model for identical system. The results show that by changing the monopole pairing interactional strength, the nuclear phase from single-particle motion to collective motion can be produced. It is also shown that the shape phase transitional patterns as in the IBM case can also be produced in the SDPSM by changing the interactional strengths. This results also show that the results obtained in Ref.[39] about the validity of the boson mapping is reasonable if the general behavior of the vibration-rotation shape phase transitional patterns are considered.

Acknowledgments

One of the authors (Y.A. Luo) thanks Prof. Y.X. Liu for his valuable suggestions. This work was supported by the Natural Science Foundation of China (11475091,11005056,11175078,11075080,11075081,10935001,11075052) and the Natural Science Foundation for Talent Training in Basic Science (J1103208).

References

1. R. F. Casten, in: F. Iachello (Ed.), Interacting Bose-Fermi System, Plenum, New York, 1981.
2. F. Iachello and A. Arima, *The Interacting Boson Model*, Cambridge University Press, Cambridge New York, 1987.
3. J. N. Ginocchio and M. W. Kirson, Phys. Rev. Lett. **44**, 1744(1980).
4. D. H. Feng, R. Gilmore, and S. R. Deans, Phys. Rev. **23**, 1254(1981).
5. P. Van Isacker and J. Q. Chen, Phys. Rev. **C 24**, 684(1981).
6. F. Iachello, N. V. Zamfir and R. F. Casten, Phys. Rev. Lett. **81**, 1191(1998).
7. R. F. Casten and N. V Zamfir, Phys. Rev. Lett. **85**, 3584 (2000).
8. F. Iachello, Phys. Rev. Lett. **87**, 052502(2001).
9. J. Jolie, P. Cejnar, R. C. casten, S. Heinze, A. Linnemann and V. Werner, Phys. Rev. lett. **89**, 182502(2002).
10. J. Jolie, R. C. casten, P. von Brentano and V. Werner, Phys. Rev. lett. **87**, 162501(2001).
11. D. Warner, Nature **420**, 614(2002).
12. F. Iachello and N. V. Zamfir, Phys. Rev. Lett. **92**, 212501(2004).
13. D. J. Rowe, Phys. Rev. Lett. **93**, 122502(2004).
14. D. J. Rowe, P. S. Turner and G. Rosenstell, Phys. Rev. Lett. **93**, 232502(2004).
15. P. Cejnar, S. Heinze and J. Dobeš, Phys. Rev. **C 71**, 011304(R)(2005).
16. Y. X. Liu, L. Z. Mu and H. Wei, Phys. Lett. **B 633**, 49(2006); Y. Zhao, Y. Liu, L. Z. Mu and Y. X. Liu, Int. J. Mod. Phys. **E 15**, 1711(2006).
17. F. Pan, J. P. Draayer, and Y. A. Luo, Phys. Lett. **B 576**, 297(2003).
18. A. leviatan, Phys. Rev. Lett. **77**, 818(1996); **98**, 242502(2007); A. leviatan and P. Van Isacker, Phys. Rev. Lett. **89**, 222501(2002).
19. Y. Zhang, Z. F. Hou and Y. X. Liu, Phys. Rev. **C 76**, 011305(R)(2007).
20. J. M. Arias, J. E. García-Ramos and J. Dukelsky, Phys. Rev. Lett. **93**, 212501(2004) and the reference cited in this paper.
21. M. A. Caprio and F. Iachello, Phys. Rev. Lett. **93**, 242502(2004).

22. T. Otsuka, A. Arima, F. Iachello, Nucl. Phys. A **309** (1978) 1; T. Otsuka, A. Arima, F. Iachello, I. Talmi, Phys. Lett. B **76** (1978) 139; J.N. Ginocchio, Ann. Phys. **126** (1980) 234.
23. F. Iachello, I. Talmi, Rev. Mod. Phys. **59** (1987) 339.
24. D.J. Rowe, C. Bahri, W. Wijesundera, Phys. Rev. Lett. **80** (1998) 4394.
25. N. Shimizu, T. Otsuka, T. Mizusaki, M. Honma, Phys. Rev. Lett. **86** (2001) 1171; T. Otsuka, M. Honma, T. Mizusaki, N. Shimizu, Y. Utsuno, Prog. Part. Nucl. Phys. **47** (2001) 319; N. Shimizu, T. Otsuka, T. Mizusaki, M. Honma, Phys. Rev. C **70** (2004) 054313.
26. K. Kaneko, M. Hasegawa, T. Mizusaki, Phys. Rev. C **70** (2004) 051301(R); M. Hasegawa, K. Kaneko, T. Misusaki, Y. Sun, Phys. Lett. B **656** (2007) 51.
27. J.N. Ginocchio, Phys. Rev. C **71** (2005) 064325.
28. Y.A. Luo, F. Pan, T. Wang, P.Z. Ning, J.P. Draayer, Phys. Rev. C **73** (2006) 044323; Y.A. Luo, Y. Zhang, X.F. Meng, F. Pan, J.P. Draayer, Phys. Rev. C **80** (2009) 014311.
29. N. Marginean, et al., Phys. Lett. B **633** (2006) 696.
30. Y. Sun, P.M. Walker, F.R. Xu, Y.X. Liu, Phys. Lett. B **659** (2008) 165.
31. S. Cwiok, P.-H. Heenen, W. Nazarewicz, Nature **433** (2005) 705.
32. J. Meng, W. Zhang, S.Q. Zhang, H. Toki, L.S. Geng, Eur. Phys. J. A **25** (2005) 23; T. Niki c, D. Vretenar, G.A. Lalazissis, P. Ring, Phys. Rev. Lett. **99** (2007) 092502.
33. W. M. Zhang, D. H. Feng and J. N. Ginocchio, Phys. Rev. Lett. **59**, 2032(1987).
34. W. M. Zhang, D. H. Feng and J. N. Ginocchio, Phys. Rev. **C 37**, 1281(1988).
35. D. J. Rowe, C. Bahri and W. Wijesundera, Phys. Rev. Lett. **80**, 4394(1998).
36. C. Bahri, D. J. Rowe and W. Wijesundera, Phys. Rev. **C 58**, 1539(1998).
37. Y. X. Liu, Z. F. Hou, Y. Zhang abd H. Wei, arXiv:nucl-th/0611035v1.
38. J. N. Ginocchio, Phys. Rev. **C 71**, 064325(2005).
39. Z. F. Hou, Y. Zhang and Y. X. Liu, Physics Letters **B 688** (2010) 298?04.
40. D.Bonatsos, E.A.McCutchan,R. F.Casten, andR. J.Casperson, Phys. Rev. Lett. **100**, 142501 (2008).
41. J. Q. Chen, Nucl. Phys. **A 626**, 686(1997).
42. Y. A. Luo, F. Pan, C. Bahri, and J. P. Draayer, Phys. Rev. **C 71**, 044304(2005).
43. L.Li, Y. A. Luo, T. Wang, F. Pan and J. P. Draayer, J. Phys. G: Nucl. Part. Phys. **36**(2009)125107.
44. Y. A. Luo, F. Pan, T. Wang, P. Z. Ning and J. P. Draayer, Phys. Rev. **C 73**, 044323(2006).
45. Y. A. Luo, Y. Zhang, X. F. Meng, F. Pan and J. P. Draayer, Phys. Rev. **C 80**, 014311(2009).
46. Y. Wang, L. Li, Y. A. Luo, Y. Zhang, F. Pan and J. P. Draayer, Int. J. Mod. Phys. **E 20**, 2229 (2011).

EXPERIMENTAL PLAN OF THE ^{25}Mg(p,γ)^{26}Al RESONANCE CAPTURE REACTION AT JINPING UNDERGROUND LABORATORY

Z. H. Li*, J. Su, Y. J. Li, B. Guo, S. Q. Yan, Y. B. Wang, G. Lian, S. Zeng, Q. W. Zhang, G. Z. He, L. Gan, C. Zhou, and W. P. Liu

China Institute of Atomic Energy, Beijing 102413, China
** E-mail: zhli@ciae.ac.cn*
http://esic.ciae.ac.cn

K. A. Li, X. Q. Yu, X. D. Tang, and J. J. He

Institute of Modern Physics, Chinese Academy of Sciences, Lanzhou 730000, China

Y. Z. Qian

School of Physics and Astronomy, University of Minnesota, MN 55455, USA

The observation of ^{26}Al is an useful tool for γ-ray astronomy and in studies of galactic chemical evolution. The most likely mechanism for ^{26}Al nucleosynthesis is in the hydrogen burning MgAl cycle, and the ^{26}Al production from the ^{25}Mg(p,γ)^{26}Al reaction at the important temperature range below $T = 0.2$ is still not well known. We present a proposal to measure the resonance strength of 58 keV resonance level of the ^{25}Mg(p,γ)^{26}Al reaction, and the effective counting rate is estimated for the direct measurement at Jinping underground laboratory.

Keywords: γ-ray astronomy; astrophysical reaction rate; spectroscopic factor; resonance strength.

1. Introduction

In 1984, the 1.809 MeV γ-ray was first detected by Mahoney et al. with the spacecraft HEAO-C,[1] this evidences that ^{26}Al nucleosynthesis is still active in the Galaxy. Afterwards, the COMPTEL telescope on board of the CGRO satellite obtained the distribution of the 1.809 MeV γ-ray emission in the Galaxy.[2] The Galaxy is relatively transparent to such γ-rays, and emission has been found concentrated along its plane, as shown in Fig. 1. A total ^{26}Al mass of about 3 M$\odot$[3] were derived through the analysis of the measured gamma spectra. ^{26}Mg, the decay production of ^{26}Al, was also found excess in the Allende meteorite.[4] Nowadays, the origin of ^{26}Al in the galaxy was increasingly concerned.

The ^{25}Mg$(p,\gamma)^{26}$Al reaction is the main way to produce ^{26}Al in the galaxy and its cross section are dominated by the capture process of the isolated resonances in ^{26}Al. The temperature range of astrophysical interests is T = 0.02-2 GK, so the levels between 50 keV and 310 keV are more important in the study of galactic ^{26}Al. Many experiments have been performed to study the ^{25}Mg$(p,\gamma)^{26}$Al reaction sine 1970,[5–16] but the experiment on the surface of earth ground can only reach to 190 keV energy level due to the small cross section and large background effects of the cosmic rays. In 2012, the laboratory of underground nuclear astrophysics (LUNA) in Italy successfully measured the resonance strength at 92 keV with the help of high shielding conditions in the underground laboratory.[17,18] However, the ^{25}Mg$(p,\gamma)^{26}$Al cross section of 58 keV resonant capture is inaccessible for direct measurement in the shielding conditions of LUNA experiments. Jinping underground laboratory covered with the marble rock of 2400 meters, which makes the direct measurement of 58 keV resonance strength being possible with the new designed experimental setup.

In order to optimize the experimental setup for the ^{25}Mg$(p,\gamma)^{26}$Al reaction at the laboratory of Jinping underground nuclear astrophysics (JUNA), the resonance strength of 58 keV level is estimated by using the shell model calculation in the present work. The results show that the 58 keV resonance dominate the ^{25}Mg$(p,\gamma)^{26}$Al reaction rate at $T < 0.05$ GK, and the experiment can be achieved at JUNA.

2. The Astrophysical ^{25}Mg$(p,\gamma)^{26}$Al Reaction Rates

In energy region of nuclear resonances, the cross-section can be described with the Breit-Wigner formula:

$$\sigma_{BW}(E) = \frac{\lambda^2}{4\pi} \cdot \omega \cdot \frac{\Gamma_a \Gamma_b}{(E_r - E)^2 + \Gamma^2/4}, \tag{1}$$

where λ is the center-of-mass de Broglie wavelength of the projectile, ω the statistical factor contains all spin dependent measures; Γ, Γ_a, Γ_b, are the total energetic width, the partial energetic widths of the entrance and exit channel; E and E_r represent the center-of-mass energy and the resonant energy.

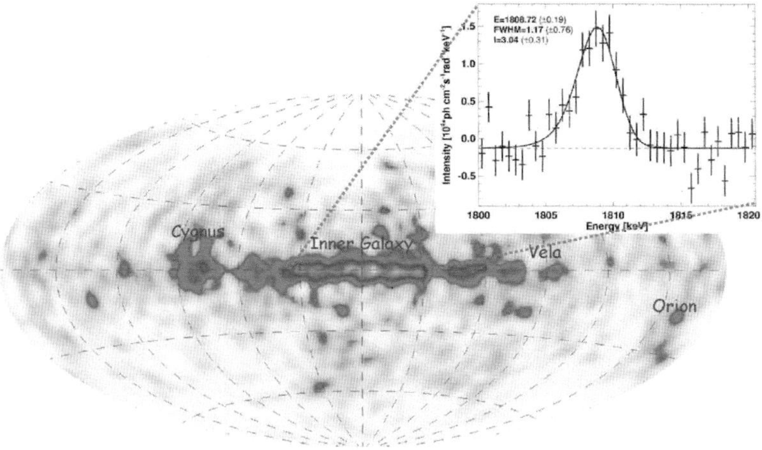

Fig. 1. (Color online) Skymap of ^{26}Al γ-rays from the 9-year CGRO mission with COMPTEL.[2]

The thermonuclear reaction rate for a narrow isolated resonance with the Breit-Wigner cross-section is introduced in the following:

$$N_A\langle\sigma v\rangle = N_A\left(\frac{2\pi}{\mu kT}\right)^{\frac{3}{2}}\hbar^2 e^{-E_r/kT}\cdot\omega\gamma, \tag{2}$$

where N_A and k denote the Avogadros and Boltzmanns constant, μ the reduced mass of the system, T the stellar temperature. $\omega\gamma$ is the so called resonance strength corresponding to the product of the spin and width, which is expressed as

$$\omega\gamma = \frac{(2J+1)(1+\delta_{01})}{(2j_1+1)(2j_2+1)}\cdot\frac{\Gamma_a\Gamma_b}{\Gamma}. \tag{3}$$

One can see that the resonance strength $\omega\gamma$ and the resonance energy E_r are needed to deduce the thermonuclear reaction rate. The factor $(1+\delta_{01})$ is included since the cross section for identical particles in the entrance channel increases by a factor of 2. For the ^{25}Mg$(p,\gamma)^{26}$Al reaction at 58 keV resonance, its reaction rate is determined by the proton partial width Γ_p, the γ-ray partial width Γ_γ, and the total width $\Gamma = \Gamma_p + \Gamma_\gamma$. The latter is estimated to be 0.03 eV from the half-life of 6.364 MeV excited state in ^{26}Al.[19]

The proton partial width Γ_p can be calculated by using the relation[20]

$$\Gamma_p = 2\frac{\hbar^2}{\mu a^2}PC^2S\theta_{sp}^2, \tag{4}$$

with a the channel radius, P the penetration factor, C^2S the proton spectroscopic factor of the final state, and θ_{sp}^2 the dimensionless single-particle reduced width. The channel radius is defined as $1.25 \cdot A^{1/3}$ in the present calculation, and the numerical results presented in Ref. 20 is adopted to deduce the θ_{sp}^2 value.

Calculations of the proton spectroscopic factor C^2S were carried out in the sdpn model space with the usdbcdpn hamiltonian[21] using the code Nushell@MSU.[22] $2s1/2$, $1d3/2$ and $1d5/2$ orbits contribute to 58 keV 3^+ resonance state, the spectroscopic factor of these orbits are 0.0703, 0.1419 and 0.0234 respectively. The partial proton width of 58 keV is then deduced to be 3.67×10^{-13} eV. This value is much smaller than the total width 0.03 eV, which indicates that $\Gamma_\gamma \gg \Gamma_p$, and the resonance strength depends only on the charged-particle partial width. Equation (3) can then be rewritten as

$$\omega\gamma = \omega\frac{\Gamma_p\Gamma_\gamma}{\Gamma} \approx \omega\frac{\Gamma_p\Gamma_\gamma}{\Gamma_\gamma} = \omega\Gamma_p. \tag{5}$$

The final resonance strength of 58 keV 3^+ state is then computed to be 2.14×10^{-13} eV.

3. Experiments with Thick-Target Method

In order to estimate the contribution of 58 keV level resonant capture to the the $^{25}\mathrm{Mg}(p,\gamma)^{26}\mathrm{Al}$ reaction, the reaction rate components of direct proton capture to the ground state in ^{26}Al, the 58 keV, 92 keV, 190 keV, 304 keV resonant capture and their total reaction rate are all computed and shown in Fig. 2. The direct capture calculation is similar with our previous work[23–27] by using the shell model spectroscopic factor 0.52 of ^{26}Al ground state. For the narrow resonance capture reaction, we use a deformation of Eq. (2) to calculate the temperature dependence of the reaction rate. The formula is expressed as

$$N_A\langle\sigma v\rangle = \frac{1.5399 \times 10^{11}}{(\mu T_9)^{3/2}} \cdot \omega\gamma \cdot \exp\left(-11.605\frac{E_r}{T_9}\right), \tag{6}$$

where $\omega\gamma$ and E_r are in units of MeV, T_9 denotes the temperature in GK. The resonance strength for 92 keV, 190 keV and 304 keV used in the present calculation are the experimental data taken from Ref. 10 and Ref. 17. As can be seen from Fig. 2, the 58 keV resonance dominates the $^{25}\mathrm{Mg}(p,\gamma)^{26}\mathrm{Al}$ reaction rate at $T_9 < 0.05$, the 92 keV resonance dominates at $0.05 < T_9 < 0.14$, and the 304 keV resonance dominates at $T_9 > 0.14$. The contribution of the direct capture to the ^{26}Al ground state is very small and can be neglected.

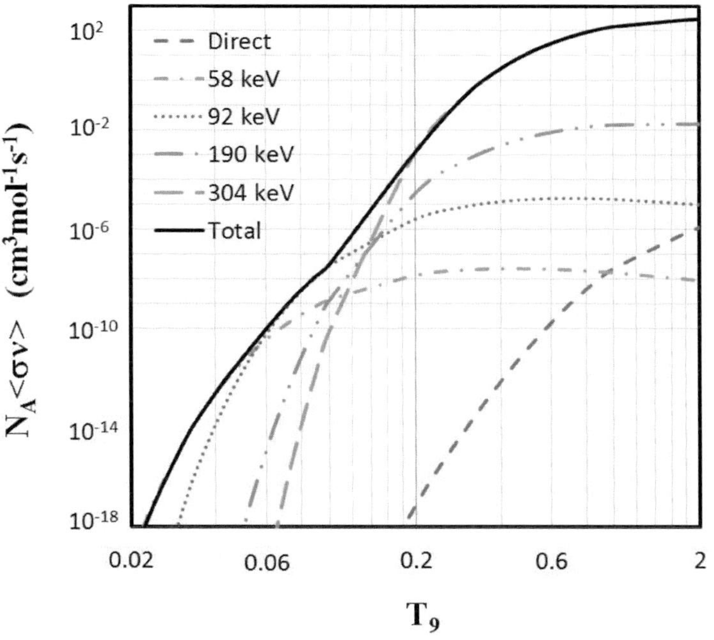

Fig. 2. The reaction rate of ^{25}Mg$(p,\gamma)^{26}$Al, which includes the direct proton capture to the ground state of ^{26}Al, the 58 keV, 92 keV, 190 keV, 304 keV resonance capture and the total reaction rate.

In a uniform density target, the yield of the reaction can be deduced with

$$Y = \int \sigma(E)\rho_T dx = \int \frac{\sigma(E)}{\epsilon(E)} dE. \tag{7}$$

Substituting a cross section of the Breit-Wigner form in Eq. (1), and integrating from the energy of the beam at a position x in the target, E_o - ΔE, to the initial beam energy, E_o, the maximum thick-target yield is expressed as

$$Y_{max}(\infty) = \frac{\lambda_r^2}{2}\omega\gamma\frac{M+m}{M\epsilon_r}, \tag{8}$$

where λ_r and ϵ_r are the de Broglie wavelength and stopping power at resonant energy, M and m are the mass of the target nucleus and projectile, respectively. According to the calculation with Eq. (8), the effect statistical counting rate of the ^{25}Mg$(p,\gamma)^{26}$Al at $E_{\rm lab} = 58$ keV is about 3.6 events per day by utilizing the high intensity proton beam of 10 mA and thick-target method. After the coincident measurement with the high efficiency 4π BGO gamma detector (as seen in Fig. 3), the 58 keV resonance strength of ^{25}Mg$(p,\gamma)^{26}$Al can be obtained at JUNA by a counting rate of 1.4 events per day. Thinking about the background counting rate is less than 0.2 events per day, we can thus accumulate the measurement about 36 reaction events in one month, the statistic error of the experiment will be around 17%.

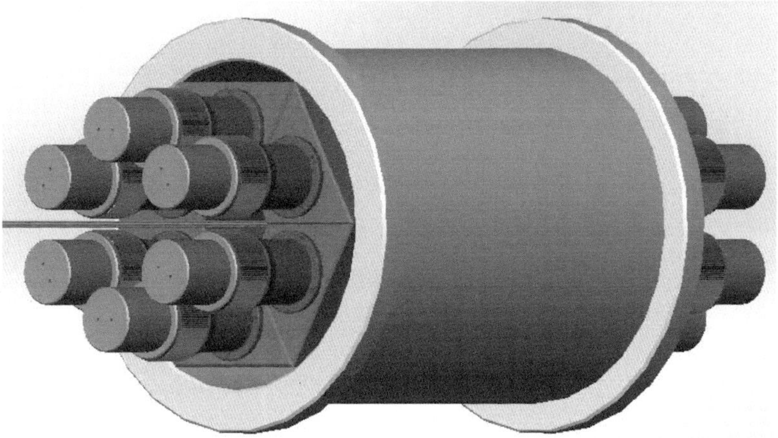

Fig. 3. (Color online) The 4π BGO γ-ray detector designed for JUNA (p,γ) experiments.

4. Summary

In summary, we have deduced the resonance strength for the 58 keV resonance level in ^{26}Al with the spectroscopic factor obtained by shell model calculation. The astrophysical ^{25}Mg$(p,\gamma)^{26}$Al reaction rates are then be deduced. The value is then used to estimate the counting rates of the Jinping underground experiments. The results show that the 58 keV resonance dominate the ^{25}Mg$(p,\gamma)^{26}$Al reaction rate at $T_9 < 0.05$, and the measurement of the 58 keV resonance strength determination can be achieved at JUNA with the new designed BGO detector.

5. Acknowledgments and Appendices

This work was supported by the National Natural Science Foundation of China (Grant Nos. 11490563, 11375269 and 11321064) and the 973 program of China under Grant No: 2013CB834406.

References

1. W. A. Mahoney et al., The Astrophysical Journal **286** (1984) 578.
2. S. Plüschke et al., ESA SP **459** (2001) 55.
3. R. Diehl et al., Nature **439**, (2006) 45.
4. C. M. Gray and W. Compston, Nature **251** (1974) 495.
5. R. R. Betts, H.T. Fortune, D.J. Pullen, Nucl. Phys. A **299** (1978) 412.
6. A. E. Champagne et al., Nucl. Phys. A **402** (1983) 159.
7. A. E. Champagne et al., Nucl. Phys. A **402** (1983) 179.
8. P. M. Endt, P. de Wit, C. Alderliesten, Nucl. Phys. A **459** (1986) 61.
9. A. E. Champagne et al., Nucl. Phys. A **451** (1986) 498.
10. P. M. Endt, C. Rolfs, Nucl. Phys. A **467** (1987) 261.
11. A.E. Champagne, A.J. Howard et al., Nucl. Phys. A **505** (1989) 384.
12. A.A. Rollefson, V. Wijekumar et al., Nucl. Phys. A **507** (1990) 413.
13. C. Iliadis, T. Schange et al., Nucl. Phys. A **512** (1990) 509.
14. C. Iliadis, L. Buchmann et al., Phys. Rev. C **53** (1996) 475.
15. D. C. Powell, C. Iliadis et al., Nucl. Phys. A **644** (1998) 263.
16. A. Arazi, T. Faestermann et al., Phys. Rev. C **74** (2006) 025802.
17. F. Strieder et al., Phys. Lett. B **707** (2012) 60.
18. O. Starniero et al., Astrophys. J. 763 (2013) 100.
19. A. G. Ledebuhr et al., Phys. Rev. C **22** 1723 (1980).
20. C. Iliadis, Nucl. Phys. A **618** (1997) 166.
21. W. E. Ormand and B. A. Brown, Nucl. Phys. A **491** (1989) 1.
22. B. A. Brown and W. D. M. Rae, MSU-NSCL report (2007).
23. Z.H. Li et al, Phys. Rev. C **71** (2005) 052801R.
24. B. Guo et al., Nucl. Phys. A **761** (2005) 162.
25. Z.H. Li et al., Phys. Rev. C **74** (2006) 035801.
26. B. Guo et al., Phys. Rev. C **73** (2006) 048801.
27. B. Guo et al., J. Phys. G: Nucl. Part. Phys. **34** (2007) 103.

THE $f_0(980)$ AND $f_0(500)$ PRODUCTION IN B^0 AND B_s^0 DECAYS INTO $J/\psi\pi^+\pi^-$

Wei-Hong Liang*

Department of Physics, Guangxi Normal University, Guilin 541004, China
**E-mail: liangwh@gxnu.edu.cn*

E. Oset

Departamento de Física Teórica and IFIC, Centro Mixto Universidad de Valencia-CSIC Institutos de Investigación de Paterna, Valencia 46071, Spain

We study $B^0(B_s^0) \to J/\psi\pi^+\pi^-$ decays, investigating the $f_0(980)$ and $f_0(500)$ production by taking into account the final state interaction between the pseudoscalar mesons in the framework of chiral unitary theory. We obtain the $\pi^+\pi^-$ invariant mass distribution for these decay processes, which allows us to compare directly to the experiments and test the nature of the scalar resonances.

Keywords: B decay; scalar resonance; the final state interaction.

1. Introduction

The nature of the scalar mesons is a topic of long-standing debate. The light scalars $f_0(500)$ and $f_0(980)$ have been associated to $q\bar{q}$ structures, a molecular state, and a $(q)^2(\bar{q})^2$ tetraquark among others.[1,2] In the chiral unitary theory,[3] which achieves success describing meson-meson and meson-baryon interactions, the $f_0(500)$ and $f_0(980)$ resonances are dynamically generated from the multiple scattering of the pseudoscalar mesons via the Bethe-Salpeter equation, with $f_0(500)$ coupling mostly to $\pi\pi$ and $f_0(980)$ mostly to $K\bar{K}$.[4] They can be interpreted as a kind of meson-meson molecular states.

Recently, the LHCb, Belle, CDF and D0 Collaborations have measured the B_s^0 or B^0 decays into $J/\psi\pi^+\pi^-$,[5-12] with a pronounced peak for the $f_0(980)$ and no signal for the $f_0(500)$ in $B_s^0 \to J/\psi\pi^+\pi^-$ decay, while a clear contribution from the $f_0(500)$ and no signal (or a small peak) for the $f_0(980)$ in $B^0 \to J/\psi\pi^+\pi^-$ decay. The decays of $\bar{B}_s^0(B^0) \to J/\psi f_0$, $f_0 \to \pi^+\pi^-$ become a good place for testing the nature of the light scalars $f_0(500)$ and

$f_0(980)$. The experimental data have served as a basis of discussion on the possible nature of the scalar mesons as a $q\bar{q}$ or tetraquark.[13]

In the present work, we study the $\bar{B}^0_s(\bar{B}^0) \to J/\psi\pi^+\pi^-$ decays, by considering the final state interactions between the meson pairs in chiral unitary theory. The $f_0(500)$ and $f_0(980)$ resonances can be produced dynamically. If we look at the $\pi^+\pi^-$ invariant mass distribution, we can see the $f_0(500)$ and $f_0(980)$ production in these decay processes.

2. Formalism

Following Refs. 5–13, we take the dominant weak mechanism for $\bar{B}^0$ and $\bar{B}^0_s$ decays at the quark level which we depict in Fig. 1. In order to have

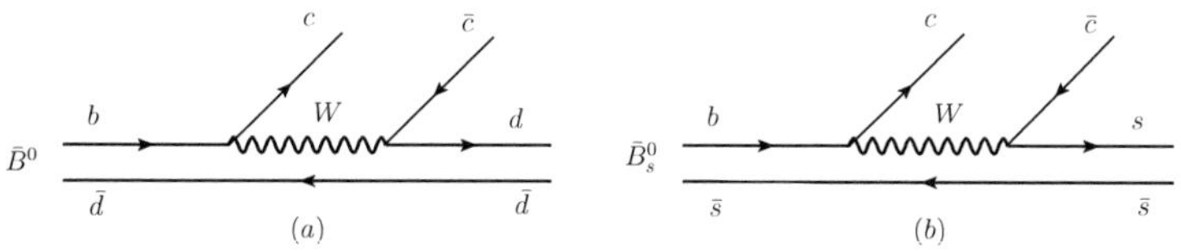

Fig. 1. Diagrams for $\bar{B}^0(\bar{B}^0_s) \to J/\psi + q\bar{q}$, $d\bar{d}$ for $\bar{B}^0$ and $s\bar{s}$ for $\bar{B}^0_s$.

$\pi^+\pi^-$ in the final state as in the experiments, we need the hadronization of the final $q\bar{q}$ pair, as shown in Fig. 2, adding an extra $q\bar{q}$ pair with the quantum numbers of the vacuum, $\bar{u}u+dd+\bar{s}s$. To write the $q\bar{q}(\bar{u}u+\bar{d}d+\bar{s}s)$

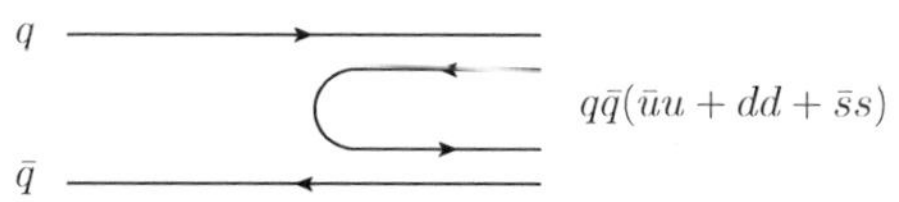

Fig. 2. Schematic representation of the hadronization $q\bar{q} \to q\bar{q}(\bar{u}u + \bar{d}d + \bar{s}s)$.

combination in terms of pairs of mesons, we define the $q\bar{q}$ matrix M,

$$M = \begin{pmatrix} u\bar{u} & u\bar{d} & u\bar{s} \\ d\bar{u} & d\bar{d} & d\bar{s} \\ s\bar{u} & s\bar{d} & s\bar{s} \end{pmatrix} = \begin{pmatrix} \frac{1}{\sqrt{2}}\pi^0 + \frac{1}{\sqrt{6}}\eta & \pi^+ & K^+ \\ \pi^- & -\frac{1}{\sqrt{2}}\pi^0 + \frac{1}{\sqrt{6}}\eta & K^0 \\ K^- & \bar{K}^0 & -\frac{2}{\sqrt{6}}\eta \end{pmatrix}, \tag{1}$$

having the property $M \cdot M = M \times (u\bar{u} + d\bar{d} + s\bar{s})$. Hence we have

$$d\bar{d}(\bar{u}u + \bar{d}d + \bar{s}s) \equiv (M \cdot M)_{22} = \pi^-\pi^+ + \frac{1}{2}\pi^0\pi^0 - \frac{1}{\sqrt{3}}\pi^0\eta + K^0\bar{K}^0 + \frac{1}{6}\eta\eta,$$

$$s\bar{s}(\bar{u}u + \bar{d}d + \bar{s}s) \equiv (M \cdot M)_{33} = K^-K^+ + K^0\bar{K}^0 + \frac{4}{6}\eta\eta. \tag{2}$$

Let us call V_P the production vertex which contains all dynamical factors common to both reactions. The $\pi^+\pi^-$ production will proceed via primary production or final state interaction as depicted in Fig. 3.

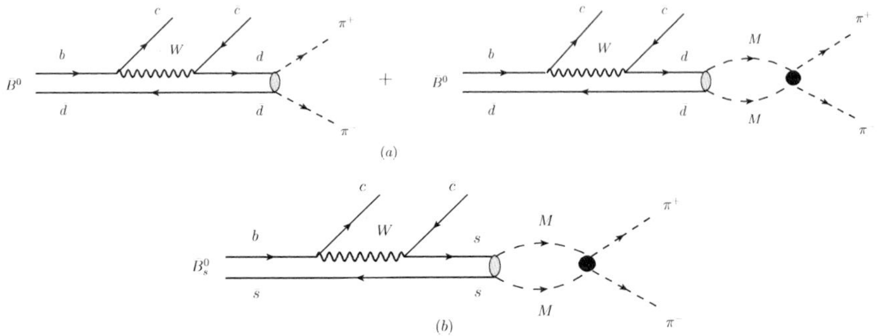

Fig. 3. Diagrammatic representation of $\pi^+\pi^-$, via direct plus rescattering mechanisms in $\bar{B}^0$ decay (a), and via rescattering for $\bar{B}_s^0$ decay (b).

The amplitudes for $\pi^+\pi^-$ production are given by

$$\begin{aligned}
t(\bar{B}^0 \to J/\psi\pi^+\pi^-) &= V_P V_{cd}(1 + G_{\pi^+\pi^-}t_{\pi^+\pi^-\to\pi^+\pi^-} + \frac{1}{4}G_{\pi^0\pi^0}t_{\pi^0\pi^0\to\pi^+\pi^-} \\
&\quad + G_{K^0\bar{K}^0}t_{K^0\bar{K}^0\to\pi^+\pi^-} + \frac{1}{12}G_{\eta\eta}t_{\eta\eta\to\pi^+\pi^-}), \\
t(\bar{B}_s^0 \to J/\psi\pi^+\pi^-) &= V_P V_{cs}(G_{K^+K^-}t_{K^+K^-\to\pi^+\pi^-} + G_{K^0\bar{K}^0}t_{K^0\bar{K}^0\to\pi^+\pi^-} \\
&\quad + \frac{1}{3}G_{\eta\eta}t_{\eta\eta\to\pi^+\pi^-}) \,, \qquad (3)
\end{aligned}$$

where V_{cd} and V_{cs} are CKM matrix elements, related to the Cabbibo angle,

$$V_{cd} = -\sin\theta_c = -0.22534, \qquad V_{cs} = \cos\theta_c = 0.97427. \qquad (4)$$

G_i are the loop functions of two meson propagators, given in ref.[14] The t_{ij} are elements of the scattering matrices for transitions of channel i to j, following the Bethe-Salpeter equation,

$$t = [1 - VG]^{-1}V, \qquad (5)$$

with the V matrix taken from refs. 4,15.

One final element of information is needed to complete the formula for $d\Gamma/dM_{inv}$, with M_{inv} the $\pi^+\pi^-$ invariant mass, which is the fact that in a $0^- \to 1^-0^+$ transition we shall need an $L' = 1$ for the J/ψ to match angular momentum conservation. Hence, $V_P = A\, p_{J/\psi}\cos\theta$, and we assume A to be constant (equal to 1 in the calculations). Thus,

$$\frac{d\Gamma}{dM_{inv}} = \frac{1}{(2\pi)^3}\frac{1}{4M_{\bar{B}_j}^2}\frac{1}{3}p_{J/\psi}^2 p_{J/\psi}\tilde{p}_\pi \sum\sum \left|\tilde{t}_{\bar{B}_j^0\to J/\psi\pi^+\pi^-}\right|^2, \quad (6)$$

where the factor 1/3 is coming from the integral of $\cos^2\theta$ and $\tilde{t}_{\bar{B}_j^0\to J/\psi\pi^+\pi^-}$ is $t_{\bar{B}_j^0\to J/\psi\pi^+\pi^-}/(p_{J/\psi}\cos\theta)$, which depends on the $\pi^+\pi^-$ invariant mass. In Eq. (6) $p_{J/\psi}$ is the J/ψ momentum in the global CM frame ($\bar{B}$ at rest) and $\tilde{p}_\pi$ is the pion momentum in the $\pi^+\pi^-$ rest frame.

3. Results and Conclusion

In Fig. 4 we show the $\pi^+\pi^-$ invariant mass distribution for the case of the $\bar{B}_s^0 \to J/\psi\pi^+\pi^-$ decay. We compare our results with the data of Ref.[11] The data points, given in events per bin in the experimental paper, are normalized here to match the theoretical strength at the peak of the distribution. We can see that the agreement is quantitatively good. We observe an appreciable peak for $f_0(980)$ production and basically no trace for $f_0(500)$ production. At lower invariant masses both the theory and experiment show a very small strength.

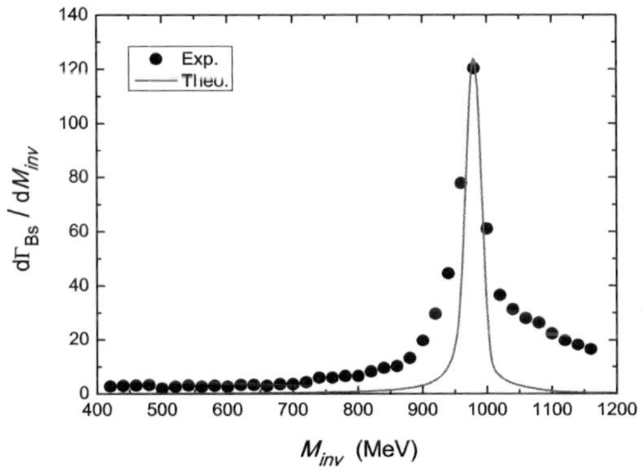

Fig. 4. $\pi^+\pi^-$ invariant mass distribution for the $\bar{B}_s^0 \to J/\psi\pi^+\pi^-$ decay, with arbitrary normalization. Data from.[11]

The second of Eqs. (3) tells us why the $f_0(500)$ contribution is so small. All intermediate states involved $K\bar{K}$, $\eta\eta$, have a mass in the 1 GeV region and the G_i functions are small at lower energies. Furthermore, the coupling of the $f_0(500)$ to both $K\bar{K}$ and $\eta\eta$ is also extremely small, such that the t matrices involved have also small magnitudes.

Note that in this decay we could have also J/ψ and vector meson production, but the $s\bar{s}$ component would give ϕ production which does not decay to $\pi\pi$. The case is quite different for the $\bar{B}^0 \to J/\psi\pi^+\pi^-$ decay, because now we can also produce $J/\psi\rho$ ($\rho \to \pi^+\pi^-$) decay and in fact this takes quite a large fraction of the $J/\psi\pi^+\pi^-$ decay, as seen in Ref.[12]

We plot our relative S-wave $\pi^+\pi^-$ production for the $\bar{B}^0 \to J/\psi\pi^+\pi^-$ decay in Fig. 5. We can see that the $f_0(500)$ production is clearly dominant. The $f_0(980)$ shows up as a small peak. The strength of the peak of the $f_0(980)$ is about 1/3 of that of the $f_0(500)$. A direct comparison with the extracted results for the $f_0(500)$ and $f_0(980)$ from the partial wave analysis of Ref.[12] (see Fig. 13 of Ref.[12]) is not possible because of the smallness of the $f_0(980)$ signal, which is of the order of the fluctuations of the data and is not shown in the figure.

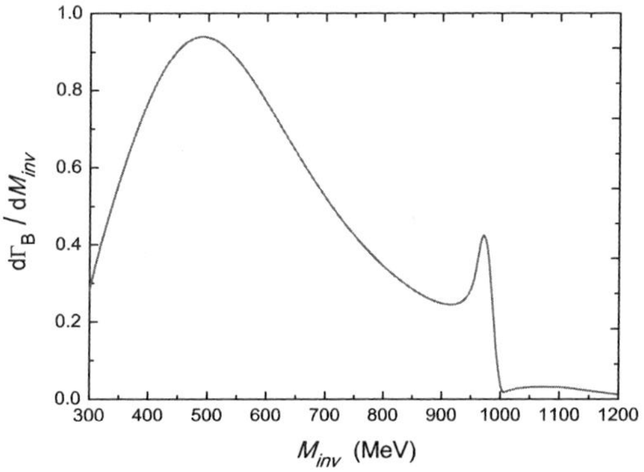

Fig. 5. $\pi^+\pi^-$ invariant mass distribution for the $\bar{B}^0 \to J/\psi\pi^+\pi^-$ decay, with arbitrary normalization.

There is another point in our calculations worth noting. The normalization of Fig. 4 and Fig. 5 is arbitrary but the relative size is what the theory predicts. It is easy to compute that

$$\frac{\Gamma(B^0 \to J/\psi f_0(500))}{\Gamma(B_s^0 \to J/\psi f_0(980))} \simeq (4.5 \pm 1.0) \times 10^{-2}. \tag{7}$$

This number is in agreement within errors with the band of $(2.08 \sim 4.13) \times 10^{-2}$ that one obtains from the branching fractions of $9.60^{+3.79}_{-1.20} \times 10^{-6}$ for $\bar{B}^0 \to J/\psi f_0(500)$ from Ref.[10] and $3.40^{+0.63}_{-0.16} \times 10^{-4}$ for $\bar{B}_s^0 \to J/\psi f_0(980)$ from Ref. 9.

We have addressed the B^0 and B_s^0 decays into J/ψ $f_0(980)$ and J/ψ $f_0(500)$ and have looked at the $f_0(980)$ and $f_0(500)$ production in both cases. We observe a pronounced $f_0(980)$ peak in the $\pi\pi$ invariant mass distribution for the B_s^0 decay and no visible trace of the $f_0(500)$ production, like in the experiment. On the other hand, in the B_s^0 decay, both the $f_0(500)$ and $f_0(980)$ excitations are visible, but the $f_0(980)$ production represents only a very small fraction of the $f_0(500)$, also in quantitative agreement with experiment.

The results obtained here add to the analysis of other reactions where both the $f_0(500)$ and $f_0(980)$ are also produced. The systematic and accurate agreement of the predictions of chiral unitary theory with experiment is remarkable and gives a strong support to the idea originated from chiral unitary theory that the $f_0(500)$ and $f_0(980)$ resonances are dynamically generated from the interaction of pseudoscalar mesons and could be interpreted as a kind of molecular states of meson-meson with the largest component $\pi\pi$ for the $f_0(500)$ and $K\bar{K}$ for the $f_0(980)$.

Acknowledgments

This work is supported by the National Natural Science Foundation of China (No. 11165005), the Spanish Ministerio de Economia y Competitividad and European FEDER funds (Nos. FIS2011-28853-C02-01 and FIS2011 28853 C02-02).

References

1. E. Klempt and A. Zaitsev, Phys. Rept. **454**, 1 (2007).
2. V. Crede and C. A. Meyer, Prog. Part. Nucl. Phys. **63**, 74 (2009).
3. J. A. Oller, E. Oset and A. Ramos, Prog. Part. Nucl. Phys. **45**, 157 (2000).
4. J. A. Oller and E. Oset, Nucl. Phys. A **620**, 438 (1997) [Erratum-ibid. A **652**, 407 (1999)].
5. R. Aaij *et al.* [LHCb Collaboration], Phys. Lett. B **698**, 115 (2011).
6. J. Li *et al.* [Belle Collaboration], Phys. Rev. Lett. **106**, 121802 (2011).
7. T. Aaltonen *et al.* [CDF Collaboration], Phys. Rev. D **84**, 052012 (2011).
8. V. M. Abazov *et al.* [D0 Collaboration], Phys. Rev. D **85**, 011103 (2012).
9. RAaij *et al.* [LHCb Collaboration], Phys. Rev. D **86**, 052006 (2012).
10. RAaij *et al.* [LHCb Collaboration], Phys. Rev. D **87**, no. 5, 052001 (2013).
11. R. Aaij *et al.* [LHCb Collaboration], Phys. Rev. D **89**, 092006 (2014).
12. R. Aaij *et al.* [LHCb Collaboration], Phys. Rev. D **90**, 012003 (2014).
13. S. Stone and L. Zhang, Phys. Rev. Lett. **111**, no. 6, 062001 (2013).
14. W. H. Liang and E. Oset, Phys. Lett. B **737**, 70 (2014).
15. D. Gamermann, E. Oset, D. Strottman and M. J. Vicente Vacas, Phys. Rev. D **76**, 074016 (2007).

DYNAMICALLY GENERATED $J^P = 1/2^-$ CHARMED AND BOTTOM HEAVY BARYONS

Jun-Xu Lu*, Yu Zhou, Hua-Xing Chen, and Li-Sheng Geng

School of Physics and Nuclear Energy Engineering and International Research Center for Nuclei and Particles in the Cosmos, Beihang University, Beijing, 100191, China
**E-mail: ljxwohool@buaa.edu.cn*
www.buaa.edu.cn

Ju-Jun Xie

Institute of Modern Physics, Chinese Academy of Sciences, Lanzhou 730000, China

Approximate heavy-quark spin and flavor symmetry and chiral symmetry play an important role in our understanding of the nonperturbative regime of the strong interactions. In this talk, we report on a recent study of the consequences of these symmetries in the description of the interactions between the ground-state singly charmed (bottom) baryons and the Nambu-Goldstone bosons in the unitarized chiral perturbation theory. In particular, by fixing the only parameter in the theory to reproduce the $\Lambda_b(5912)$ and the $\Lambda_c(2595)$, we predict a number of dynamically generated states, which are contrasted with those of other approaches and available experimental data.

Keywords: Chiral lagrangians; bound and unstable states; Bethe-Salpeter equations; bottom baryons; charmed baryons.

1. Introduction

In recent years, heavy-flavor hadron physics has yielded many surprising results and attracted a lot of attention due to the intensive worldwide experimental activities. Different from the case of heavy meson states, no similar exotic states have been firmly established in the heavy-flavor baryon sector. Up to now, there are only a few experimental observations of excited charmed and bottom baryons. In the present work, we choose the UChPT to study the interaction between the ground-state charmed (bottom) baryons and the pseudoscalar mesons using the leading order chiral Lagrangian.

2. Theoretical Framework

There are two building blocks in the UChPT: a kernel provided by chiral Lagrangians up to a certain order and a unitarization procedure. In the present work, we choose to work with relativistic chiral Lagrangians and in the Bethe-Salpeter equation framework.

The Bethe-Salpeter equation can be written schematically as

$$T = V + VGT, \tag{1}$$

where T is the unitarized amplitude, V the potential, and G the one-loop 2-point scalar function. In the context of the UChPT, the integral Bethe-Salpeter equation is often simplified and approximated to be an algebraic equation with the use of the on-shell approximation.[1,2]

The leading order interaction between a singly charmed baryon of the ground-state sexet and anti-triplet and a pseudoscalar meson of the pion octet is provided by the following chiral Lagrangian,[3,4]

$$\begin{aligned}\mathcal{L} = &\frac{i}{16f^2}\mathrm{Tr}(\overline{H}_{[\bar{3}]}(x)\gamma^\mu[H_{[\bar{3}]}(x),[\phi(x),(\partial_\mu\phi(x))]_-]_+)\\ &+\frac{i}{16f^2}\mathrm{Tr}(\overline{H}_{[6]}(x)\gamma^\mu[H_{[6]}(x),[\phi(x),(\partial_\mu\phi(x))]_-]_+),\end{aligned} \tag{2}$$

where ϕ collects the pseudoscalar octet, $H_{[\bar{3}]}$ and $H_{[6]}$ collect the charmed (bottom) baryons, respectively.

Expanding the Lagrangian of Eq. (2) up to two pseudoscalar fields, one obtains the interaction kernel needed to describe the $\phi(p_2)B(p_1) \to \phi(p_4)B(p_3)$ process, where p_i (i = 1, 2, 3, and 4) are the four momenta of the respective particles. In deriving the final form of V, we have assumed that the three momentum of a baryon is small compared to its mass.

The loop function is divergent and needs to be properly regularized. In principle, one can either adopt the dimensional regularization scheme or the cutoff scheme. In Ref. 5, a so-called heavy quark symmetry (HQS) inspired regularization scheme has been suggested, which manifestly satisfies both the chiral power counting and the heavy-quark spin and flavor symmetry up to $1/M_H$, where M_H is a generic heavy-hadron mass. In the present work, we adopt the HQS regularization scheme, which reads:

$$G_{HQS} = G_{\overline{MS}} - \frac{2\mathring{M}}{16\pi^2}\left(\log\left(\frac{\mathring{M}^2}{\mu^2}\right) - 2\right) + \frac{2m_{\mathrm{sub}}}{16\pi^2}\left(\log\left(\frac{\mathring{M}^2}{\mu^2}\right) + a\right), \tag{3}$$

$$G_{\overline{MS}}(s,M^2,m^2)=\frac{2M}{16\pi^2}\left[\frac{m^2-M^2+s}{2s}\log\left(\frac{m^2}{M^2}\right)\right.$$
$$-\frac{q}{\sqrt{s}}(\log[2q\sqrt{s}+m^2-M^2-s]+\log[2q\sqrt{s}-m^2+M^2-s]$$
$$-\log[2q\sqrt{s}+m^2-M^2+s]-\log[2q\sqrt{s}-m^2+M^2+s])$$
$$\left.+\left(\log\left(\frac{M^2}{\mu^2}\right)-2\right)\right]. \tag{4}$$

In the above equations, $m_{\rm sub}$ is a generic pseudoscalar meson mass, which can take the value of m_π in the u, d flavor case or an average of the pion, the kaon, and the eta masses in the u, d, and s three flavor case.

For the sake of comparison, we have also performed calculations using the cutoff regularization scheme,

$$G_{\rm cut}=\int_0^\Lambda \frac{q^2\,dq}{2\pi^2}\frac{E_M+E_m}{2E_M E_m}\frac{2M}{s-(E_M+E_m)^2+i\epsilon}, \tag{5}$$

with $E_M=\sqrt{q^2+M^2}$, and $E_m=\sqrt{q^2+m^2}$.

3. Results and Discussions

The only unknown parameter in the UChPT is related to the regularization of the loop function G, i.e., the subtraction constant a in the dimensional regularization scheme or the cutoff value Λ in the cutoff scheme. Conventionally, in the latter method one often chooses a cutoff of the order of 1 GeV (the chiral symmetry breaking scale). In our work, the subtraction constant and the cutoff value are both fixed in a way such that the $\Lambda_b(5912)$ or the $\Lambda_c(2595)$ mass is produced exactly.

Once the subtraction constant or the cutoff value is fixed in the HQS approach, one can use the same constant to predict the counterparts of the dynamically generated bottom (charmed) baryons. Broken SU(3) chiral symmetry then predicts a number of additional resonances or bound states as shown in Tables 1 and 2. The results obtained in the cutoff scheme can be found in Ref. 6.

In anticipation of future lattice chromodynamics simulations, we have calculated the corresponding scattering lengths and compared them with the existing predictions from a $\mathcal{O}(p^3)$ chiral perturbation theory study. For details, see Ref. 6.

Table 1. Dynamically generated bottom baryons of $J^P = 1/2^-$. The subtraction constant is fixed by reproducing the $\Lambda_b(5912)$ mass.

Pole positions(MeV)	SU(3) multiplet	(S,I)	Main channels (threshold)
$(6082, -i57)$	$[\bar{3}]$	$(0, 1)$	$\Xi_b K(6285.1)$
$(5922, 0)$	$[\bar{3}]$	$(0, 0)$	$\Xi_b K(6285.1)$
$(5895, 0)$	$[\bar{3}]$	$(-1, \frac{1}{2})$	$\Lambda_b \overline{K}(6115.0)$
$(6100, -i32)$	$[\bar{3}]$	$(-1, \frac{1}{2})$	$\Xi_b \pi(5928.5)$
$(6202, 0)$	$[\bar{3}]$	$(-2, 0)$	$\Xi_b \overline{K}(6285.1)$
$(6201, 0)$	$[6]$	$(1, \frac{1}{2})$	$\Sigma_b K(6308.6)$
$(5967, -i9)$	$[6]$	$(0, 1)$	$\Sigma_b \pi(5951.0)$
$(6223, -i14)$	$[6]$	$(0, 1)$	$\Sigma_b \eta(6360.9)$
$(5912, 0)^*$	$[6]$	$(0, 0)$	$\Sigma_b \pi(5951.0)$
$(6307, -i12)$	$[6]$	$(0, 0)$	$\Xi'_b K(6421.6)$
$(6213, -i25)$	$[6]$	$(-1, \frac{3}{2})$	$\Sigma_b \overline{K}(6308.6)$
$(5955, 0)$	$[6]$	$(-1, \frac{1}{2})$	$\Sigma_b \overline{K}(6308.6)$
$(6105, -i17)$	$[6]$	$(-1, \frac{1}{2})$	$\Xi'_b \pi(6064.0)$
$(6381, -i25)$	$[6]$	$(-1, \frac{1}{2})$	$\Omega_b K(6566.6)$
$(6377, -i58)$	$[6]$	$(-2, 1)$	$\Omega_b \pi(6209.0)$
$(6180, 0)$	$[6]$	$(-2, 0)$	$\Xi'_b \overline{K}(6421.6)$,$\Omega_b \eta(6618.9)$

Table 2. Dynamically generated charmed baryons of $J^P = 1/2^-$. The subtraction constant is fixed by reproducing the $\Lambda_c(2595)$ mass.

Pole positions(MeV)	SU(3) multiplet	(S,I)	Main channels (threshold)
$(2723, 0)$	$[\bar{3}]$	$(0, 0)$	$\Xi_c K(2965.1)$
$(2655, -i26)$	$[\bar{3}]$	$(-1, \frac{1}{2})$	$\Lambda_c \overline{K}(2782.1)$,$\Xi_c \pi(2607.5)$
$(2965, 0)$	$[\bar{3}]$	$(-2, 0)$	$\Xi_c \overline{K}(2965.1)$
$(2948, 0)$	$[6]$	$(1, \frac{1}{2})$	$\Sigma_c K(2949.1)$
$(2674, -i51)$	$[6]$	$(0, 1)$	$\Sigma_c \pi(2591.5)$
$(2999, -i16)$	$[6]$	$(0, 1)$	$\Sigma_c \eta(3001.4)$,$\Xi'_c \overline{K}(3072.4)$
$(2591, 0)^*$	$[6]$	$(0, 0)$	$\Sigma_c \pi(2591.5)$
$(3070, -i12)$	$[6]$	$(0, 0)$	$\Xi'_c K(3072.4)$
$(2694, 0)$	$[6]$	$(-1, \frac{1}{2})$	$\Sigma c \overline{K}(2949.1)$
$(2827, -i55)$	$[6]$	$(-1, \frac{1}{2})$	$\Xi'_c \pi(2714.7)$
$(3124, -i43)$	$[6]$	$(-1, \frac{1}{2})$	$\Omega c K(3190.8)$
$(2947, 0)$	$[6]$	$(-2, 0)$	$\Xi'_c \overline{K}(3072.4)$,$\Omega_c \eta(3243.1)$

References

1. J. A. Oller and E. Oset, *Nucl. Phys. A* **620** (1997).
2. E. Oset and A. Ramos, *Nucl. Phys. A* **635** (1998).
3. M. F. M. Lutz and E. E. Kolomeitsev, *Nucl. Phys. A* **730** (2004).
4. Z. W. Liu and S. L. Zhu, *Phys. Rev. D* **86** (2012).
5. M. Altenbuchinger, L. S. Geng and W. Weise, *Phys. Rev. D* **89** (2014).
6. J. X. Lu, Y. Zhou, H. X. Chen, J. J. Xie and L. S. Geng, *arXiv:1409.3133* .

n + BaF_2 REACTION STUDIED BY TALYS1.4 TOOLKIT

Cui-Juan Lü and Chun-Wang Ma*

Depart. Phys., Henan Normal University, Xinxiang, 453007, China
**E-mail: machunwang@126.com*

The neutron induced reactions on the ^{138}Ba and ^{19}F nuclei are investigated by using the optical model implanted in the Talys1.4 toolkit. The results of the n + ^{138}Ba and n + ^{19}F reactions within the incident energy ranging from 1 keV to 30 MeV are calculated and compared to the measured results.

Keywords: Neutron absorption; neutron activation; optical model; ^{138}Ba; ^{19}F.

1. Introduction

Fluorine is the strongest electro negativity and elemental oxidant in nature. It can be used in organic synthesis, enzyme inhibitor, inorganic material, fluorine-containing polymer, oral disease prevention and control of and bio-pharmaceutical. In nuclear applications, the study of fluoride is of great significance and the n + ^{19}F reaction has also been extensively reported. Meanwhile, in the 4th generation of nuclear reactor TMSR (Thorium-based Molten Salt Reactor), the compounds of fluorine (^{7}LiF-BeF_2-ThF_4-UF_4) are of the mainly concerned compositions in the salts. The natural fluorine has no other isotope but ^{19}F. The different compounds, such as the CaF_2, have been used in experiments to study the n + F reactions previously.[1] In this article, the compound ^{138}Ba^{19}F_2 is chosen as a target since ^{138}Ba has the greatest natural abundance, which facilitates the data analysis. Experimentally, the cross sections of n + ^{19}F can be extracted by subtracting the cross sections of n + ^{138}Ba from those of n + ^{138}Ba^{19}F_2. Therefore, both of the n + ^{138}Ba and the n + ^{19}F reactions should be calculated in theory. To compare the calculated results with the measured ones, the incident energy of neutron calculated covers a wide range from 1keV to 30MeV.

2. Model Description

The optical model can well describe the neutron induced reaction in a wide range of incident energies. The optical model code ECIS-06[2] is implanted as a subroutine to perform all the optical model calculations in Talys1.4. Due to the limited pages of the article, the readers are referred to the Talys1.4 manual[3] for the optical model and codes.

3. Results and Discussion

3.1. $n + {}^{19}F$

The fluorine element has only one stable isotope ^{19}F, thus only the n + ^{19}F reaction is calculated. The calculated and measured results are plotted in Fig. 1. The calculated reaction include the ^{19}F(n, γ)^{20}F, ^{19}F(n, α)^{16}N, ^{19}F(n, p)^{19}O, ^{19}F(n, np)^{18}O, ^{19}F(n, 2n)^{18}F and ^{19}F(n, tot) channels. The calculated threshold energies E_{th} for ^{19}F(n, α)^{16}N, ^{19}F(n, p)^{19}O, ^{19}F(n, np)^{18}O and ^{19}F(n, 2n)^{18}F are 1.61, 4.25, 6.08 and 10.99MeV, respectively. For the n + ^{19}F reaction, when E_n < 1.61MeV, only the ^{19}F(n, γ)^{20}F happens because of the limitation of threshold energy. The ^{19}F(n, γ)^{20}F channel happens in the whole neutron energy range but the cross sections are smaller than the other channels when E_n > 0.2MeV. The calculated results for ^{19}F(n, γ)^{20}F, which decreases fast with E_n when E_n is low, is relative smaller than the measured results by Uberseder *et al.*[4] The calculated results agree with the measured values when the optical potential parameters are adjusted. ^{19}F nucleus is in the intermediate zone between the region of validity of statistical theory and the region of light nuclei which requires detailed treatment, while the optical model is a type of the statistical theory. For the ^{19}F(n, α)^{16}N channel, which happens when E_n > 1.61MeV, the calculated results are smaller than the results measured by Smith *et al.*[1] and Picard *et al.*[5] in the range of 3.87MeV< E_n <6.16MeV and 14.18MeV < E_n < 21MeV. The calculated results agree with the results measured by Durusoy *et al.*[6] in the range of 13.6MeV< E_n < 14.86MeV. For the ^{19}F(n, p)^{19}O channel, which is opened when E_n > 4.25MeV, the calculated results are a little smaller than the measured results by Marion *et al.*,[7] but agree well with those measured by Smith *et al.*,[8] Bormann *et al.*[9] and Picard *et al.*[5] For the ^{19}F(n, np)^{18}O channel, which is opened when E_n > 6.08MeV, no measured data are found. The ^{19}F(n, 2n)^{18}F channel is opened when E_n > 10.99MeV. The ^{18}F nucleus is a positron emitter with a half-life 109.771(20) minutes, which is one of the important nucleus in the applications of tomography scanning in medicine. For ^{19}F(n, 2n)^{18}F, the calculated results are consistent with the results measured by Brill *et al.*,[10] Lapenas *et al.*[11] and Picard *et al.*,[5] but is a little larger than the results measured by Hartmann *et al.*[12]

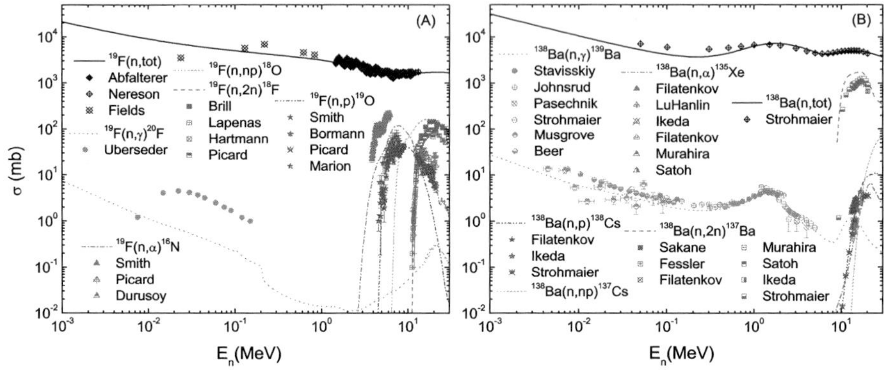

Fig. 1. (Color online) The results for the reactions n + ^{19}F in (A) and n + ^{138}Ba in (B). The calculated results are represented by lines and the measured results are denoted by symbols.

3.2. *n* + ^{138}*Ba*

The barium element has a stable isotope ^{138}Ba with 71.7% of natural abundance. For the n + ^{138}Ba reactions, the calculated channels include the ^{138}Ba(n, γ)^{139}Ba, ^{138}Ba(n, p)^{138}Cs, ^{138}Ba(n, α)^{135}Xe, ^{138}Ba(n, np)^{137}Cs, ^{138}Ba(n, 2n)^{137}Ba and ^{138}Ba(n, tot). The threshold energies for ^{138}Ba(n, α)^{135}Xe, ^{138}Ba(n, p)^{138}Cs, ^{138}Ba(n, np)^{137}Cs and ^{138}Ba(n, 2n)^{137}Ba are 0, 4.63, 6.83 and 8.67MeV, respectively. For ^{138}Ba(n, γ)^{139}Ba, which have no threshold energy, the calculated results agree with those results measured by Stavisskiy *et al.*,[13] Johnsrud *et al.*,[14] Pasechnik *et al.*,[15] Strohmaier *et al.*,[16] Musgrove *et al.*[17] and Beer *et al.*[18] For ^{138}Ba(n, 2n)^{137}Ba, which is opened when $E_n > 8.67$MeV, the calculated results are smaller than the results measured by Sakane *et al.*,[19] Fessler *et al.*,[20] Filatenkov *et al.*,[21] Murahira *et al.*,[22] Satoh *et al.*,[23] Ikeda *et al.*[24] and Strohmaier *et al.*[16] For ^{138}Ba(n, p)^{138}Cs, which is opened when $E_n > 4.63$MeV, the calculated results are larger than the results measured by Strohmaier *et al.*[16] when $E_n > 15$MeV, the calculated results are agree with the results measured by Filatenkov *et al.*,[21] Ikeda *et al.*[24] and Strohmaier *et al.*[16] when $E_n <$ 15MeV. While for ^{138}Ba(n, α)^{135}Xe, which can happen in the whole energy range, the calculated results are a little larger than the results measured by Murahira *et al.*,[22] Satoh *et al.*,[23] LuHanLin *et al.*,[25] Ikeda *et al.*[24] and Filatenkov *et al.*[21] For the ^{19}F(n, np)^{18}O channel, which is opened when $E_n > 6.83$MeV, no measured data are found.

4. Summary

In order to study the neutron activation reaction on ^{19}F, which is an important isotope in many materials and a widely existing isotope in nature, the $^{138}Ba^{19}F_2$ compound is adopted as the reaction target. The calculated results agree well with the measured results by using the default parameters in Talys1.4. Due to that ^{19}F is an intermediate mass isotope, the default parameters cannot well predict the results for the (n, γ) channel at low neutron energy, which need special treatment in calculation. The results are both useful for the neutron activation analysis and the TMSR physics.

References

1. D. M. Smith *et al*, *Phys. Rev.*, **117**, 514 (1960).
2. H. J. Xu, The 15th National Conference on Nuclear Physics, 2013 Shanghai.
3. A. J. Koning *et al*, *User Manual of Talys-1.4*.
4. E. Uberseder*et al*, *Phys. Rev. C*,**75**, 035801 (2007).
5. J. Picard, C. F. Williamson, *Nucl. Phys.*, **63**, 673 (1965).
6. A. Durusoy, I. A. Reyhancan, *Ann. Nucl. Energy.*, **33**, 159 (2006).
7. J. B. Marion, R. M. Brugger, *Phys. Rev.*, **100**, 69 (1955).
8. D. L. Smith *et al*, *Nucl. Sci. Eng.*, **77**, 256 (1981).
9. M. Bormann *et al*, *Nucl. Phys.*, **63**, 438 (1965).
10. O. D. Brill *et al*, *Dokl. Akad. Nauk+*, **136**, 55 (1961).
11. A. A. Lapenas *et al*, *Neutron Spect.Meas.by Activ.*, (1957).
12. C. L. Hartmann, P.M. DeLuca, *Nucl. Sci. Eng.*, **109**, 319 (1991).
13. Ju. Ya. Stavisskiy *et al*, *Atomnaya Energiya*, **10(5)**, 508 (1961).
14. A. E. Johnsrud *et al*, *Phys. Rev.*, **116**, 927 (1959).
15. M. V. Pasechnik *et al*, *Second Internat. At.En. Conf.*, **15**, 18 (1958).
16. B. Strohmaier *et al*, *Nucl. Sci. Eng.*, **65**, 368 (1978).
17. A. R. DeL. Musgrove *et al*, *Int.Conf. on Neutr.Phys.Nucl.Data*, 449 (1978).
18. H. Beer *et al*, *Astrophys. J.*, **474**, 843 (1997).
19. H. Sakane *et al*, *Ann. Nucl. Energy.*, **28**, 1175 (2001).
20. A. Fessler *et al*, *Nucl. Sci. Eng.*, **134(2)**, 171 (2000).
21. A. A. Filatenkov *et al*, *Khlopin Radiev.Inst., Leningrad Reports No.252*, (1999).
22. S. Murahira *et al*, *Japanese report to the I.N.D.C. No.175/U 171*, (1995).
23. Y. Satoh *et al*, *JAERI Conference proceedings*, **8(95)**, 189 (1994).
24. Y. Ikeda *et al*, *JAERI Reports No.1312*, (1988).
25. H. L. Lu *et al*, *Chinese J. of Nuclear Physics (Beijing)*, **14(3)**, 244 (1992).

SYSTEMATIC STUDY OF HYPERNUCLEAR MAGNETIC MOMENTS

H. F. Lü*

College of Science, China Agriculture University,
Beijing 100083, China
**E-mail: hongfeng@cau.edu.cn*

Magnetic moments of single Λ hypernuclei are systematically investigated using a perturbative mean-field approach. Under a perturb treatment, we improved an earlier expression of magnetic moment. With a self-consistent calculation, we found that large deviation in Λ-hypernucleus with $l_\Lambda \neq 0$ according to a perturbation treatment may suit medium-mass nuclei better than the other parts along the nuclear chart.

Keywords: Magntic moment; Lambda Hypernuclei; tensor coupling; polarization.

1. Introduction

Due to the strangeness, hyperon provides a highly sensitive probe of nuclear structure. Hypernuclear magnetic moment can show direct information about hyperon-nucleon interaction and the behavior of hadrons in nuclear medium, which provides a way to penetrate various theoretical models.[1,2] With technique development, promising measurements of hypernuclear magnetic moments are in process.[3,4]

Several relativistic studies show that $\omega\Lambda\Lambda$ tensor coupling and core response effect play an important role on hypernuclear magnetic moment.[5,6] Considering a tensor potential, these two effects have been further investigated in a simple relativistic approach with a Woods-Saxon form potential.[7] With a realistic scalar and vector potential for Dirac equation obtained self-consistently from meson-exchange, subsequent study following Ref.[7] is presented here. An improved core response effect including the contribution from $\omega\Lambda\Lambda$ tensor coupling term is shown in a linear response theory. With newly developed meson-baryon interaction PK1-Y1,[8] systematic hypernuclear magnetic moments are calculated in the relativistic mean field (RMF) theory with both a perturb treatment and a self-consistent investigation.

2. Theoretical Framework

For Λ hypernuclei, a standard effective Lagrangian density is generalized by including a tensor coupling term $\frac{f_{\omega BB}}{4m_B}\bar{\psi}_B\sigma^{\mu\nu}\Omega_{\mu\nu}\psi_B$ as,[9]

$$
\begin{aligned}
\mathcal{L} = &\bar{\psi}_B[i\gamma^\mu\partial_\mu - m_B - g_{\sigma B}\sigma - g_{\omega B}\gamma^\mu\omega_\mu - g_{\rho B}\gamma^\mu\vec{\tau}\cdot\vec{\rho}_\mu - Q_B\gamma^\mu A_\mu]\psi_B \\
&+\bar{\psi}_B\frac{f_{\omega BB}}{4m_B}\sigma^{\mu\nu}\Omega_{\mu\nu}\psi_B + \frac{1}{2}\partial_\mu\sigma\partial^\mu\sigma - \frac{1}{2}m_\sigma^2\sigma^2 - \frac{1}{3}g_2\sigma^3 - \frac{1}{4}g_3\sigma^4 - \frac{1}{4}\Omega_{\mu\nu}\Omega^{\mu\nu} \\
&+\frac{1}{2}m_\omega^2\omega_\mu\omega^\mu + \frac{1}{4}c_3(\omega_\mu\omega^\mu)^2 - \frac{1}{4}\vec{R}_{\mu\nu}\vec{R}^{\mu\nu} + \frac{1}{2}m_\rho^2\vec{\rho}_\mu\cdot\vec{\rho}^\mu - \frac{1}{4}F_{\mu\nu}F^{\mu\nu}, \quad (1)
\end{aligned}
$$

where $\sigma^{\mu\nu} = \frac{1}{2i}[\gamma^\mu,\gamma^\nu]$, m_B $(B = n, p, \Lambda)$ and $m_\phi(\phi = \sigma, \omega, \rho)$ are masses of baryons and mesons respectively. Field tensor for ω-meson is given as $\Omega_{\mu\nu} = \partial_\mu\omega_\nu - \partial_\nu\omega_\mu$ and by similar expressions for ρ-meson and photon. Q_B is charge of baryons. Dirac equation of baryons and Klein-Gordon equation for mesons can be derived from the Lagrangian density in Eq. (1) with variational principle. For odd-mass nuclei, time-odd (space-like) components of vector fields appear and it can modify the nuclear current and the single particle properties.[10,11] Thus, Dirac equations have to be solved separately in two subspaces with different simplex.

3. Results and Discussion

Staring from the Λ hyperon Dirac equation with a $\omega\Lambda\Lambda$ tensor term, there is

$$
\begin{aligned}
2(E_\Lambda - g_{\omega\Lambda}V_\Lambda^0)\psi_\Lambda^\dagger\vec{\alpha}\psi_\Lambda = &\psi_\Lambda^\dagger[i(\overleftarrow{\nabla} - \overrightarrow{\nabla})]\psi_\Lambda + \nabla\times[\psi_\Lambda^\dagger\vec{\Sigma}\psi_\Lambda] \\
&-2g_{\omega\Lambda}\vec{V}_\Lambda\psi_\Lambda^\dagger\psi_\Lambda + \frac{f_{\omega\Lambda\Lambda}}{m_\Lambda}\psi_\Lambda^\dagger\gamma_0(\nabla V_\Lambda^0\times\vec{\Sigma})\psi_\Lambda.
\end{aligned}
$$

The Λ hyperon current including the $\omega\Lambda\Lambda$ tensor coupling is,

$$
\begin{aligned}
\langle\vec{J}_\Lambda\rangle &= \psi_\Lambda^\dagger\vec{\alpha}\psi_\Lambda - \frac{f_{\omega\Lambda\Lambda}}{g_{\omega\Lambda}}\frac{1}{2m_\Lambda}\nabla\times(\bar{\psi}_\Lambda\vec{\Sigma}\psi_\Lambda) \\
&= \frac{1}{2(\varepsilon_\Lambda - g_{\omega\Lambda}\omega_0)}\{\psi_\Lambda^\dagger[i(\overleftarrow{\nabla} - \overrightarrow{\nabla})]\psi_\Lambda + \nabla\times[\psi_\Lambda^\dagger\vec{\Sigma}\psi_\Lambda]\} \\
&\quad -\frac{f_{\omega\Lambda\Lambda}}{g_{\omega\Lambda}}\frac{1}{2m_\Lambda}\nabla\times[\bar{\psi}_\Lambda\vec{\Sigma}\psi_\Lambda] \\
&\quad +\frac{1}{2(\varepsilon_\Lambda - g_{\omega\Lambda}\omega_0)}[\frac{f_{\omega\Lambda\Lambda}}{m_\Lambda}\psi_\Lambda^\dagger\gamma_0(\nabla\omega_0\times\vec{\Sigma})\psi_\Lambda - 2g_{\omega\Lambda}\vec{\omega}\psi_\Lambda^\dagger\psi_\Lambda], \quad (2)
\end{aligned}
$$

Thus, Λ convection current term $\psi_\Lambda^\dagger(\overleftarrow{\nabla}-\overrightarrow{\nabla})\psi_\Lambda$ vanished in $l_\Lambda = 0$ hypernucleus. Adopting the approximation $-\nabla^2+m_\omega^2 \simeq m_\omega^2$, the polarized Dirac magnetic moment becomes

$$\vec{\mu}_D^{\text{pol.}} \simeq \frac{1}{4}\frac{g_{\omega N}g_{\omega\Lambda}}{m_\omega^2}\int d^3r[1+(\frac{g_{\omega N}}{m_\omega})^2\frac{\langle\rho_V^N\rangle}{\bar{\varepsilon}_N - g_{\omega N}\omega_0}]^{-1}\frac{\langle\rho_V^N\rangle}{(\bar{\varepsilon}_N - g_{\omega N}\omega_0)}$$
$$\times\frac{1}{2(\varepsilon_\Lambda - g_{\omega\Lambda}\omega_0)}\vec{r}\times\{(\frac{f_{\omega\Lambda\Lambda}}{g_{\omega\Lambda}}\frac{m_\Lambda^*}{m_\Lambda}-1)\nabla\times[\psi_\Lambda^\dagger\vec{\Sigma}\psi_\Lambda]$$
$$+2g_{\omega\Lambda}\vec{\omega}\psi_\Lambda^\dagger\psi_\Lambda - \frac{f_{\omega\Lambda\Lambda}}{m_\Lambda}\psi_\Lambda^\dagger\gamma_0(\nabla\omega_0\times\vec{\Sigma})\psi_\Lambda\}, \qquad (3)$$

where $\bar{\varepsilon}_N$ is an average single particle energy of nucleons, m_Λ^* is the Dirac effective mass of Λ hyperon. Even for $f_{\omega\Lambda\Lambda}/g_{\omega\Lambda} = 1$ and $m_\Lambda^*/m_\Lambda = 1$ at the nuclear surface, the last two terms in big bracket are not necessarily to cancel each other. It shows that even with $\omega\Lambda\Lambda$ tensor coupling, the core-polarized magnetic moment in Λ-hypernucleus still exists and is proportional to the coupling strength $g_{\omega N}\cdot g_{\omega\Lambda}^2$ (or $g_{\omega N}\cdot f_{\omega\Lambda\Lambda}^2$) approximately. Therefore, a self-consistent calculation with the $\omega\Lambda\Lambda$ tensor coupling is necessary to investigate the dependence of polarized magnetic moment on the coupling strength $g_{\omega\Lambda}$.[12]

Table 1. Magnetic moment in excited hypernuclei with a double-closed core. Results are obtained with parameter set PK1-Y1 in time-odd triaxial RMF. Lambda single particle energy ε_Λ marked by star refers to the excited case when Λ occupies the lowest $l_\Lambda = 1$ orbit.[13]

Nuclei	ε_Λ	μ_D	μ_a^Λ	$\mu_{\text{tot.}}$	$\Delta\mu$
$^{17}_\Lambda$O	-2.924*	-0.030	-0.580	-0.610	0.003
$^{41}_\Lambda$Ca	-10.229*	-0.071	-0.608	-0.688	-0.075
$^{57}_\Lambda$Ni	-14.097*	-0.014	-0.576	-0.590	0.023
$^{133}_\Lambda$Sn	-19.514*	-0.016	-0.521	-0.537	0.076
$^{209}_\Lambda$Pb	-21.750*	-0.051	-0.571	-0.622	-0.009
$^{17}_\Lambda$O	-12.700	-0.009	-0.610	-0.619	-0.006
$^{41}_\Lambda$Ca	-18.739	-0.014	-0.610	-0.624	-0.011
$^{57}_\Lambda$Ni	-22.072	-0.018	-0.610	-0.628	-0.015
$^{133}_\Lambda$Sn	-24.556	-0.015	-0.611	-0.626	-0.013
$^{209}_\Lambda$Pb	-25.637	-0.016	-0.612	-0.628	-0.015

Note: The Schmidt value $\mu_{\text{Sch.}}$ is -0.613 μ_N. Deviation of hypernuclear magnetic moment from the Schmidt value is $\Delta\mu = \mu_{\text{tot.}} - \mu_{Sch.}$.

With non-vanishing Λ convection current term $\psi_\Lambda^\dagger(\overleftarrow{\nabla}-\overrightarrow{\nabla})\psi_\Lambda$, there may be a large deviation $\Delta\mu$ in excited Λ-hypernucleus with $l_\Lambda \neq 0$. A self-consistent calculation with $\omega\Lambda\Lambda$ tensor coupling may clear the uncertainty of perturbation treatment. By blocking the lowest $1p_{3/2}$ orbit, magnetic moment for $^{17}_{\Lambda}$O, $^{41}_{\Lambda}$Ca, $^{57}_{\Lambda}$Ni, $^{133}_{\Lambda}$Sn and $^{209}_{\Lambda}$Pb are obtained with parameter set PK1-Y1 in time-odd triaxial RMF, see Table 1. The maximum magnitude of $\Delta\mu$ for excited hypernuclei is 12% of the Schmidt value, see $^{41}_{\Lambda}$Ca and $^{133}_{\Lambda}$Sn. Hypernuclear magnetic moment is close to the Schmidt value by a 2% difference when Λ hyperon occupies the lowest $l_\Lambda = 0$ orbit, see the second data block in Table 1. Shift of Lambda hyperon from $l_\Lambda = 0$ to $l_\Lambda = 1$ increases the hypernuclear magnetic moment except $^{41}_{\Lambda}$Ca. If deviation $\Delta\mu$ changes from negative to positive, the magnitude of it may decrease. Magnitude of $\Delta\mu$ for excited state of $^{17}_{\Lambda}$O and $^{209}_{\Lambda}$Pb is smaller than corresponding ground state. Thus, the prediction of large deviation in Λ-hypernucleus with $l_\Lambda \neq 0$ according to a perturbation treatment suits medium-mass nuclei better than the other parts along the nuclear chart.[13]

4. Summary

Besides a modified expression of polarized Dirac magnetic moment, we further improved the result by a self-consistent calculation of the meson potentials and show a semi-linear relation between the polarized magnetic moment and the meson-Lambda coupling strength $g_{\omega\Lambda}$. Valence Lambda hyperon results in core polarized proton current and deviate magnetic moment from the Schmidt value as $\Delta\mu$. The deviation $\Delta\mu$ of single Λ-hypernuclei with p_Λ orbit occupied is large for medium mass region, but not for the others.

References

1. M. Ukai et al., Phys. Rev. Lett. 93, 232501 (2004).
2. O. Hashimoto, H. Tamura, Prog. Part. Nucl. Phys. **57**, 564 (2006).
3. T. Fukuda, T. Saito and HypHI collaboration, Nucl. Phys. A790, 161 (2007).
4. A. Gal, O. Hashimoto and J. Pochodzalla (Eds.) Nucl. Phys. **A881** (2012).
5. B. K. Jennings, Phys. Lett. **B246**, 325 (1990).
6. J. Cohen, Phys. Rev. C **48**, 1346 (1993).
7. J. M. Yao, H. F. Lü, G. Hillhouse, and J. Meng, Chin. Phys. Lett. **25**, 1629 (2008).
8. X. S. Wang et al., Commun. Theor. Phys. **60**, 479 (2013).
9. J. Cohen and H. J. Weber, Phys. Rev. C **44**, 1181 (1991).
10. R.J. Furnstahl and C.E. Price, Phys. Rev. C **40**, 1398 (1989).
11. J.M. Yao, H. Chen, and J. Meng, Phys. Rev. C **74**, 024307 (2006).
12. X.S. Wang et al., Eur. Phys. J. **A**49, 101 (2013).
13. H. Y. Sang et al., Phys. Rev. C **88**, 064304 (2013).

A DIRECT PROBE FOR CHEMICAL POTENTIALS DIFFERENCE BETWEEN NEUTRON AND PROTONS IN HEAVY-ION COLLISIONS

Chun-Wang Ma*, Yan-Li Zhang, and Shan-Shan Wang
Institute of Particle and Nuclear Physics, Henan Normal University, Xinxiang 453007, China
E-mail: machunwang@126.com

We briefly introduce the newly proposed probe to the neutron and proton chemical potential (and density) difference, which is called as the isobaric yield ratio difference (IBD). The IBD probe is related to the chemical potential difference of neutrons and protons between two reactions, at the same time, the nuclear density difference between two reactions. The relationship between the IBD probe and the isoscaling method has also been discussed.

Keywords: Isobaric yield difference; nuclear chemical potential; nuclear density; isoscaling.

1. Introduction

In the past fifteen years, the study of nuclear matters produced in heavy-ion collisions (HICs) have attracted much attentions. Due to the development of radioactive ion beam (RIB) facilities, which can provide nuclear matters ranging from sub-saturation to supra-saturation densities via HICs near or above the Fermi energies. The new generation of RIB techniques are hoped to greatly stimulate the very hot isospin physics newly born in nuclear physics, of which the main aims are to study the properties of nuclear matters with large isospin, especially the neutron-rich nuclear matters.[1]

In HICs, the nuclear matter evolves during the whole reaction process. With a certain incident energy, the projectile nucleus collides on the target nucleus, and the nucleons in the overlapping zone of the projectile and target nuclei are compressed in the initial stage of the collisions. Then the compressed system expands, which may make the density of the system be smaller than the saturation density. In the expansion of the system, the chemical freeze-out happens and forms the measurable fragments. In dynamical model, the whole HICs process are governed by the interaction potential and the information change can be traced theoretically.[a] One of the most important goal of the isospin physics is to understand the nuclear symmetry energy of neutron-rich matter. Various of probes have been proposed to study the nuclear symmetry energy in the framework of different theories as have been listed in Ref. 1. Since it is difficult to directly measure nuclear symmetry energy and nuclear density (typically the neutron density), the probes find different results of nuclear symmetry energy in both the sub-saturation densities and the supra-saturation densities. Besides, the probes depend on not only the theories but also the interaction potentials. The results have been compared by recent works.[2–5] It should be noted that the probes for the nuclear symmetry energy are indirect ones, which make the results be the "extracted" ones. For theoretical physicist, it is still an important to find new probes more direct to the nuclear symmetry energy. In this article, the newly proposed IBD probe, which denotes the isobaric yield ratio difference (between two reactions), will be introduced.

2. Description of IBD probe

The IBD probe was proposed in the framework of the thermal equilibrium models. Similar to the isoscaling method,[6] which extracts the nuclear symmetry energy from the isotopic or isotonic yield ratios of two reactions, the IBD probe is also constructed from the fragment yield ratio in reactions. Starting from the definition of the isobaric yield ratio (R) differing two units in neutron-excess I ($I \equiv N - Z$, with N and Z being the neutron and proton numbers of a fragment),[7,8]

$$\mathrm{R}(I+2, I, A) = \sigma(A, I+2)/\sigma(A, I), \tag{1}$$

with A being the mass number of the fragment and σ the yield of fragments.

Taking the canonical ensembles theory within the grand canonical limit as an example,[9,10] the yield of fragment can be written as,

$$\sigma(A, I) = CA^{\tau}\exp\{[F(A, I) + \mu_n N + \mu_p Z]/T\}, \tag{2}$$

where C is a constant. τ is also a constant in the grand canonical ensembles theory. μ_n and μ_p are chemical potentials of the neutron and proton, respectively; $F(A, I)$ is the free energy of the cluster (fragment), and T is

[a]Since there are a lot of works on this topic, we do not list the references due to the limited pages of this article. The readers are referred to the review article written by Bao-An Li *et al.*[1]

the temperature. τ can be nonuniform in different reaction systems[11] in the Modified Fisher model.[12] Inserting the expression of σ into Eq. (1), the following equation can be obtained,

$$\ln R(I+2, I, A) = (\Delta F + \Delta\mu)/T, \tag{3}$$

where $\Delta F = F(I+2, A) - F(I, A)$, and $\Delta\mu = \mu_n - \mu_p$. Since $F(A, I)$, μ_n, μ_p and T are all difficult to be measured, some approximations should be made to cancel out some of them. Following the method as isoscaling, we make the difference between $\ln R$ of two reactions, one is a symmetric system and the other is an asymmetric system (neutron-rich system), which yields[13,14]

$$\begin{aligned} \Delta\ln R_{21} &= \ln[R_2(I+2, I, A)] - \ln[R_1(I+2, I, A)] \\ &= \Delta\mu_{n21}/T - \Delta\mu_{p21}/T \equiv \Delta\mu_{21}/T = \alpha - \beta, \end{aligned} \tag{4}$$

where $\alpha \equiv \Delta\mu_{n21}/T$ and $\beta \equiv \Delta\mu_{p21}/T$ are the isoscaling parameters fit from the isotopic yield ratio and isotonic yield ratio, respectively. To obtain Eq. (4), T of fragments in the two reactions are assumed to be the same. This has been verified by using the isotopic thermometer.[15] With this assumption, the difference between the free energies of the isobaric fragments in the two reactions can be cancelled out [$F(A, I)$ can also be calculated considering the temperature effects or the deexcitation effects[16–18]].

3. Discussion

Eq. (4) shows the relationship between IBD and the chemical difference between the two reactions. Besides, Eq. (4) also illustrates the relationship between the IBD probe and the isoscaling parameters, which has been compared by using the 140A MeV ^{40,48}Ca + ^{9}Be and ^{58,64}Ni + ^{9}Be reactions,[19,20] and the 1A GeV ^{124,136}Xe and ^{112,124}Sn projectile fragmentation reactions,[21,22] which verify that the IBD and isoscaling results are consistent. It is also found out that the IBD results are sensitive to the shell effects.[14] The target effects in the IBD probe has also been studied by using the measured data of 140A MeV ^{40,48}Ca + ^{9}Be and ^{58,64}Ni + ^{9}Be reactions.[20]

Comparing to the isosclaing method, which fits the parameters α and β from the isotopic and isotonic ratio, the IBD probe does not fit any ratio but directly obtain the neutron and proton chemical potential difference between the two reactions from the isobaric yield ratio difference. The IBD probe is thus a direct probe to the chemical potential difference. Meanwhile, the IBD probe uses the yield of heavy fragments, which may produce different results comparing to the probes constructed from the the light particles like neutron, proton, α, etc.

The IBD probe can also be related to the density difference between the reactions, which is written as,[23–27]

$$\begin{aligned}\Delta\mu_{21}/T = \alpha - \beta &= (\ln\rho_{n,2} - \ln\rho_{n,1}) - (\ln\rho_{p,2} - \ln\rho_{p,1}) \\ &= (\ln\rho_{n,2} - \ln\rho_{p,2}) - (\ln\rho_{n,1} - \ln\rho_{p,1}) \end{aligned} \quad (5)$$

where ρ_n and ρ_p denote the neutron density and proton density, respectively; 1 and 2 denote the reaction systems. In Eq. (5), it points out the method to measure the neutron density in HICs. For example, if in reaction 1 the values of $(\ln\rho_{p,2} - \ln\rho_{p,1})$ can be omitted and $(\ln\rho_{n,1} \approx \ln\rho_{p,1})$ can be assumed, then $\rho_{n,2}$ can be obtained from the IBD result. And if in reaction 1 the values of $(\ln\rho_{n,1} - \ln\rho_{p,1})$ and $\rho_{p,2}$ is measurable, $\rho_{n,2}$ can also be obtained from the IBD result.[28] Equation (5) has been studied by using the calculated yields of fragments in the calcium reactions and the ^{48}Ca projectile reactions but with different neutron-skin thickness, which shows that the IBD probe is only sensitive to the nuclear density changes in the prefragments, but the sensitiveness is destroyed in the final fragments due to the simple deexcitation method in the modified statistical abrasion-ablation model.[23,24] The IBD probe has also been related to the information entropy uncertainty. With the definition of information entropy uncertainty as $U(A, I) = -\ln\sigma(A, I)$, the IBD probe and the isoscaling method can be explained in the information uncertainty theory.[29,30]

4. Conclusion

The IBD probe is a direct probe to the neutron and proton chemical potentials difference between two reaction systems. The present results indicate that the IBD probe still calls for more theoretical and experimental works, for example studying the IBD probe in the transport model like the quantum mechanism dynamics model etc. Since the IBD probe is sensitive to the nuclear density in HICs, it can be studied in a transport model to see the density evolution in the whole process of HICs.

Acknowledgment

This work is supported by the Program for Science & Technology Innovation Talents in Universities of Henan Province (No. 13HASTIT046).

References

1. B.-A. Li, L.-W. Chen, C. M. Ko, *Phys. Rep.* **464**, 113 (2008).
2. L. W. Chen, arXiv:1212.0284 [nucl-th].
3. B. A. Li, L. W. Chen, F. J. Fattoyev *et al.*, arXiv:1212.1178 [nucl-th].
4. C. J. Horowitz *et al.*, *J. Phys. G: Nucl. Part. Phys.* **41**, 093001 (2014); Z. Zhang, L. W. Chen, *Phys. Lett. B* **726**, 234 (2014).
5. (1) L.-W. Chen, *Systematics of the density dependence of nuclear symmetry energy.* and (2) B.-A. Li, *From Earth to Heaven: Probing Nuclear Symmetry Energy and its Astrophysical Impacts.* **The 4th International Workshop on Nuclear Dynamics in heavy-ion reactions (IWND2014)**, August 15-19, Lanzhou, China, 2014. Extracted on Dec. 27, 2014 from website http://iwnd2014.csp.escience.cn/dct/page/65580
6. H. S. Xu *et al.*, *Phys. Rev. Lett.* **85**, 716 (2000).
7. M. Huang *et al.*, *Phys. Rev. C* **81**, 044620 (2010).
8. C. W. Ma *et al.*, *Phys. Rev. C* **83**, 064620 (2011).
9. C. B. Das *et al.*, *Phys. Rev. C* **64**, 044608 (2001).
10. M. B. Tsang *et al.*, *Phys. Rev. C* **76**, 041302(R) (2007).
11. M. Huang *et al.*, *Phys. Rev. C* **82**, 054602(R) (2010).
12. A. S. Hirsch *et al.*, *Phys. Rev. C* **29**, 508 (1984).
13. C. W. Ma *et al.*, *Phys. Rev. C* **87**, 034618 (2013).
14. C. W. Ma *et al.*, *J. Phys. G: Nucl. Part. Phys.* **40**, 125106 (2013).
15. C. W. Ma *et al.*, *Commun. Theor. Phys.* **59**, 95 (2013).
16. C. W. Ma *et al.*, *Phys. Rev. C* **86**, 054611 (2012).
17. C. W. Ma *et al.*, *Phys. Rev. C* **88**, 014609 (2013).
18. C. W. Ma *et al Nucl. Sci. Tech.* **24**, 050510 (2013).
19. M. Mocko *et al.*, *Phys. Rev. C* **74**, 054612 (2006).
20. C. W. Ma *et al.*, *Phys. Rev. C* **91**, 014615 (2015).
21. D. Henzlova *et al.*, *Phys. Rev. C* **78** 044616 (2008).
22. V. Föhr *et al.*, *Phys. Rev. C* **84** 054605 (2011).
23. C.-W. Ma *et al.*, *Eur. Phys. J. A* **50**, 139 (2014).
24. C. W. Ma *et al.*, *Phys. Rev. C* **89**, 057602 (2014).
25. A. S. Botvina, O. V. Lozhkin, and W. Trautmann, *Phys. Rev. C* **65**, 044610 (2002).
26. M. B. Tsang *et al.*, *Phys. Rev. Lett.* **86**, 5023 (2001).
27. E. Geraci, M. Bruno, M. D'Agostino *et al.*, *Nucl. Phys. A* **732**, 173 (2004).
28. C.W. Ma, *New probe for nuclear density difference between reactions of projectile fragmentation.* **The 4th International Workshop on Nuclear Dynamics in heavy-ion reactions (IWND2014)**, August 15-19, Lanzhou, China, 2014. Extracted on Dec. 27, 2014 from weblink http://iwnd2014.csp.escience.cn/dct/attach/Y2xiOmNsYjpwZGY6MzQwNTc=
29. C. W. Ma, H.L. Wei, *Commun. Theor. Phys.* **62**, 717 (2014).
30. C. W. Ma *et al.*, *Phys. Lett. B* **742**, 19-22 (2015).

PAIRING CORRELATIONS AT HIGH SPINS

Hai-Liang Ma*, Bao-Guo Dong, Yan Zhang, Ping Fan, Da-Qing Yuan,
Shen-Yun Zhu, Huan-Qiao Zhang
China Institute of Atomic Energy, P.O. Box 275-50, Beijing 102413, China

C. M. Petrache
Centre de Sciences Nucléaires et de Sciences de la Matière, Université Paris-Sud and CNRS/IN2P3, Orsay, France

I. Ragnarsson and B. G. Carlsson
Division of Mathematical Physics, Lund University, Sweden

The pairing correcting energies at high spins in ^{161}Lu and ^{138}Nd are studied by comparing the results of the cranked-Nilsson-Strutinsky (CNS) and cranked-Nilsson-Strutinsky-Bogoliubov (CNSB) models. It is concluded that the Coriolis effect rather than the rotational alignment effect plays a major role in the reduction of the pairing correlations in the high spin region. Then we proposed an average pairing correction method which not only better reproduces the experimental data comparing with the CNS model but also enables a clean-cut tracing of the configurations thus the full-spin-range discussion on the various rotating bands.

Keywords: Pairing correlation; high spin; particle number projection; Cranked-Nilsson-Strutinsky-Bogoliubov model.

1. Introduction

The pairing correlation is the most important residual interaction beyond the independent single particle model. It shapes the nuclear rotational structure in the low-spin region of even-Z and even-N nuclei. For a rotating nucleus, there are two anti-pairing mechanisms, i.e. the Coriolis effect and rotational alignment effect. Their roles on the backbending were much debated in the 1970s and 1980s.[1,2] Now it is generally accepted that the alignment of two quasi-particles gives the most important push.[1,3] As the

*mhl624@ciae.ac.cn

nucleus rotates faster and achieves much higher spins approaching terminating states, the pairing correlations become weak, however the roles of the two anti-pairing mechanisms in the reduction of the pairing correlations have not been studied throughly.

In recent decades, the advances in experimental techniques and more efficient detectors result in a large number of rotational bands observed, possibly reaching dozens. It is formidable to interpret those bands within the overlapping spin range if they have the same parity and signature quantum numbers. The presence of pairing correlations which mix the wave functions of the orbitals near the Fermi level makes the interpretation even more difficult. By neglecting pairing correlations, the configuration-dependent cranked-Nilsson-Strutinsky (CNS) model is the only one capable of accurate tracing of the configurations using the number of particles in different "deformed" j shells as approximate quantum numbers.[4,5]

By varying the total energy with respect to the pairing gaps and Fermi levels, the new cranked-Nilsson-Strutinsky-Bogoliubov (CNSB) model[6,7] is capable of reaching the terminating states with the dynamic pairing correlations included. Since the macroscopic models are the same for the unpaired CNS and paired CNSB models, a direct comparison is possible and enables the study of the pairing effects at high spins.

2. The CNS and CNSB Models

The CNS model is a macroscopic-microscopic model, and has been highly successful in describing the high-spin rotational structures in the whole nuclear chart.[5,8] The orbitals are constructed diabatically regarding to rotating frequencies. After minimizing the total energies with respect to deformation parameters, the energies and deformations of a configuration can be obtained. The pairing correlations are not included in the CNS formalism. The neglecting of pairing has an advantage in that the tracing of a fixed configuration undergoing drastic deformation changes becomes possible over the considerable spin range up to the band termination.

The CNSB model[6,7] is based on the Ultimate Cranker formalism.[9] using the same modified oscillator potential as the CNS model, plus the monopole pairing term. The total energy after the particle number projection (PNP), is minimized not only in deformation space but also in a mesh of the pairing parameters, Fermi energy λ and paring gap Δ. However by introducing the pairing correlations, the wave functions of the orbitals near the Fermi levels are largely mixed. The nucleon number in the high-j shell is no longer conserved. That makes it impossible to tracing a configuration reliably in a deformation mesh, thus usually only the yrast bands were given.

3. Results and Discussions

We first present the results of high spin structure of ^{161}Lu. The $A = 150$ parameters[5,10] are used in both CNSB and CNS calculations. Because the parity π and signature α are conserved quantum numbers with the present Hamiltonian, in total 16 combinations designated with $(\pi_p, \alpha_p)(\pi_n, \alpha_n)$ can be formed. The comparison of unpaired and paired results of the 8 positive parity yrast bands in their full-spin-range are shown in Figs. 1(a)(b). In the figures the reference energies of a rotating liquid drop are subtracted. The energy differences between the unpaired and paired results, which can be seen as the pairing correcting energies, are displayed in Fig. 1(c). After the first backbending of the yrast bands $(+, \pm 1/2)(+, 0)$, the pairing correlations actually greatly reduced due to the alignment of two quasi-neutrons. At higher spins, the pairing correcting energies follow the same trend for the

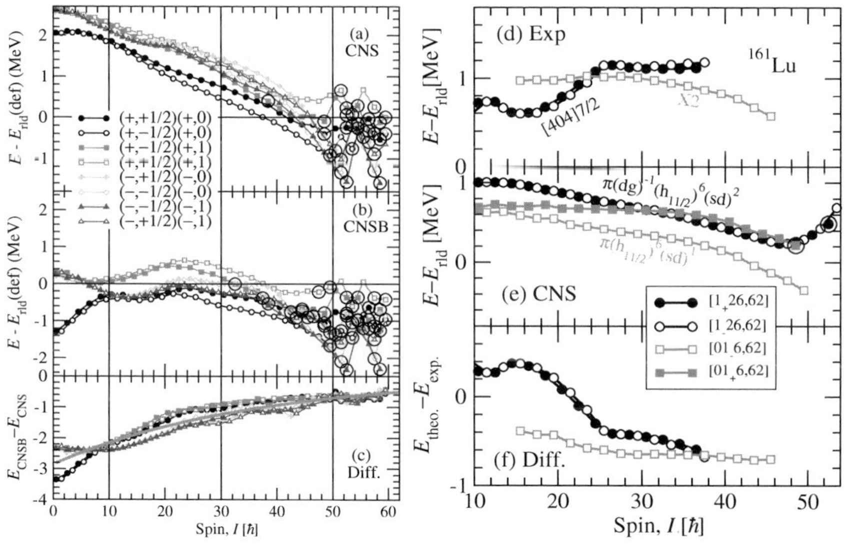

Fig. 1. (Left) Unpaired (a) and paired (b) calculation results and their differences (c) in ^{161}Lu for the positive parity configurations. The magenta line in (c) denote the average pairing correcting energies $\tilde{E}_{\text{pair}} = -2.85 \cdot \exp(-I/35.7)$. The aligned states at $\gamma = 60°$ are indicated by large open circles. (Right) The (d) (e) and (f) panels show the experimental, pairing corrected CNS energies and their differences, respectively. The detailed configurations relative to a $^{146}\text{Gd}_{82}$ core, $\pi(dg)^{-p_1}(sd)^{p_2}(h_{11/2})^{p_3}\nu(f_{7/2}h_{9/2})^{n_1}(i_{13/2})^{n_2}$, are given in the legend as $[p_1p_2p_3, n_1n_2]$. For an odd number of particles in a group, the subscript $+$ or $-$ defines the signature, $\alpha = 1/2$ or $\alpha = -1/2$. See Ref. 7 for details.

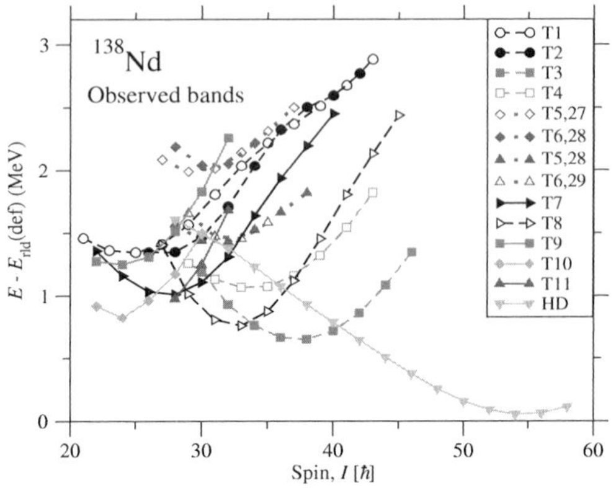

Fig. 2. The energy of the observed and linked high-spin bands in ^{138}Nd are drawn relative to a rotating liquid drop energy. Taken from Ref. 11.

different configurations. It indicates that the Coriolis effect plays a major role in the reduction of pairing correlations at high spins than the further alignment of quasi-particles. Our calculation in Ref. 7 shows that the negative parity bands also follow similar trend. A smooth function can be fitted to the curves of energy differences in Fig. 1(c) and will be taken as the average pairing correcting energies. Then it is straightforward to add this average paring correction to the CNS results.

As an example, experimental, pairing corrected CNS energies and their differences of the yrast and X2 bands are shown in Figs. 1(d)(e)(f), respectively. Here additional correction on the ground state energy due to the using of the PNP instead of the BCS procedure, the same as that in the CNSB approach, is also included, see Ref. 7 for details. The energy differences shown in Fig. 1(f) are comparable to the CNSB results, see Fig. 6 in Ref. 7. However, it is possible to fine distinguish the configurations in the CNS formalism,[5,8] therefore a more transparent discussion is achieved. In the case of the X2 band, the CNS model is even able to distinguish between the pseudo-spin partners in the proton $\mathcal{N} = 4$ shell, $(d_{5/2},g_{7/2})$ and $(d_{3/2},s_{1/2})$, which is crucial in the description of [404]7/2 and X2 bands simultaneously from low to high spins.

Another example presented in the article using the same strategy as the above is the description of the triaxial deformed bands in ^{138}Nd. Shown in Fig. 2 are the energies of observed high spin bands.[11] Several experimental bands, such as T1 and T3, T2 and T4, T5 and T6, have the same parity and signature quantum numbers. Without further distinguishing the valence

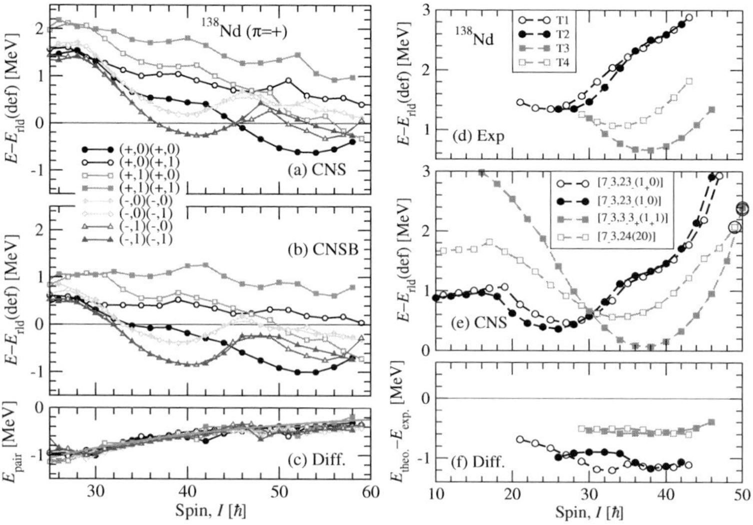

Fig. 3. (Left) The calculated energies with the (a) CNS and (b) CNSB models at triaxial shape, $(\varepsilon_2, \gamma) = (0.27, 35°)$, for 8 positive parity configurations are drawn relative to the rotating liquid drop energy. The energy differences between the CNS and CNSB results, seen as pairing corrections, are shown in (c). The bold magenta line denotes the average pairing correcting energies. (Right) The observed bands T1, T2, T3 and T4 in ^{138}Nd are shown in (d) with the calculated configurations (e) assigned to these bands. The panel (f) shows the difference between calculations and experiment.

orbitals, only those yrast bands within the same $(\pi_p, \alpha_p)(\pi_n, \alpha_n)$ can be concretely calculated in a deformation mesh. We have to turn to the CNS formalism again to interpret the observed bands.

We carried out CNS and CNSB calculations with the same constraints and searched the bands around the local minimum at $(\varepsilon_2, \gamma) = (0.27, 35°)$. The CNS and CNSB results for 8 positive parity configurations are compared in Figs. 3(a)(b). It is shown above in ^{161}Lu that the pairing correlations are lesser strongly dependent on the configurations at higher spins. Here for ^{138}Nd, it is evident that in the shown spin range, $I = 25 - 60$, the general features of the different configurations are essentially the same and pairing correcting energies vary smoothly as a function of spin. The main effect of pairing is an energy gain which decreases with increasing spin. The triaxial bands in ^{138}Nd are generally built upon the configurations with at least four quasi-particles. Thus it provides a better example of demonstration of the Coriolis anti-pairing effect. The fitting to the pairing correction energies can cover quite a large spin range, as shown in Fig. 3(c), the magenta line gives the fitting results $\tilde{E}_{\text{pair}} = -3.05 \cdot \exp(-I/23.5)$ MeV.

Shown in Fig. 3 (e), the new method much better reproduced the experimental data comparing with the ones of the CNS model in Ref. 12.

4. Summary

In summary, the pairing correcting energies at high spins in ^{161}Lu and ^{138}Nd are studied by comparing the results of the CNS and CNSB models. They are generally less dependent on the configurations at high spins where multiple quasi-particles are aligned. It is concluded that the Coriolis effect plays a major role in the reduction of the pairing correlations in the high spin region. Then we proposed an average pairing correction method. A smooth function is fitted to pairing correcting energies calculated by the CNS and CNSB models for the yrast bands. The average pairing correcting energies calculated by this function are added to the energies calculated by the configuration-dependent CNS model. The new method not only has better reproduced the experimental data comparing with the CNS model but also enables a clean-cut tracing of the configurations thus the full-spin-range discussion on the various rotating bands.

Acknowledgments

The authors acknowledge the support from the NSF of China under Contract Nos. 11205245, 11375266, 10975189 and 11405273.

References

1. M. Wakai and A. Faessler, *Nucl. Phys. A* **295**, 86 (1978).
2. R. Bengtsson and S. Frauendorf, *Nucl. Phys. A* **314**, 27 (1979).
3. M. J. A. de Voigt, J. Dudek and Z. Szymański, *Rev. Mod. Phys.* **55**, 949 (1983).
4. T. Bengtsson and I. Ragnarsson, *Nucl. Phys. A* **436**, 14 (1985).
5. A. V. Afanasjev, D. B. Fossan, G. J. Lane and I. Ragnarsson, *Phys. Rep.* **322**, 1 (1999).
6. B. G. Carlsson, I. Ragnarsson, R. Bengtsson, *et al.*, *Phys. Rev. C* **78**, 034316 (2008).
7. H.-L. Ma, B. G. Carlsson, I. Ragnarsson and H. Ryde, *Phys. Rev. C* **90**, 014316 (2014).
8. B. G. Carlsson and I. Ragnarsson, *Phys. Rev. C* **74**,011302 (2006).
9. T. Bengtsson, *Nucl. Phys. A* **496**, 56 (1989).
10. T. Bengtsson, *Nucl. Phys. A* **512**, 124 (1990).
11. C.M. Petrache, I. Ragnarsson, H.-L. Ma, *et al.*, Multiple triaxial bands in ^{138}Nd, (2014), submitted to *Phys. Rev. C.*
12. C. M. Petrache, I. Ragnarsson, H.-L. Ma, *et al.*, *Phys. Rev. C* **88**, 051303 (2013).

ELECTROMAGNETIC PROPERTIES OF ROTATIONAL BANDS IN ^{256,257}Rf WITH THE CRANKED-NILSSON-STRUTINSKY-BOGOLIUBOV MODEL

Hai-Liang Ma*, Bao-Guo Dong, Ping Fan, Yan Zhang, Da-Qing Yuan, Shen-Yun Zhu, Huan-Qiao Zhang

China Institute of Atomic Energy, P.O. Box 275-50, Beijing 102413, China
**E-mail: mhl624@ciae.ac.cn*

The rotational structures of ^{256,257}Rf are investigated by the particle-number-projected Cranked-Nilsson-Strutinsky-Bogoliubov (CNSB) model. The experimental energies of the yrast bands are well reproduced within 0.5 MeV. The electromagnetic properties with the CNSB model for ^{256}Rf generally agree with the projected shell model results. The calculated g-factor for ^{257}Rf increases with the increasing spin and approaches the value of ^{256}Rf for $I \gtrsim 20$. The resulted theoretical $|(g_K - g_R)/Q_0|$ values in ^{257}Rf are in agreement of available experimental data.

Keywords: Super heavy element; g-factor; particle number projection; Cranked-Nilsson-Strutinsky-Bogoliubov model.

1. Introduction

Under liquid drop model limit, the superheavy elements (SHEs) with $Z \geq 104$ are supposed not to exist because of the strong Coulomb repulsive force. The shell-correction energy which resulted from reduced level density around the Fermi level creates a barrier and prevents the SHE nuclei from fission. However, because the level density is very high in the SHE region, a successful description or accurate prediction of shell gaps depend sensitively on the accuracy of the single-particle energies. The advances in experimental techniques and more efficient detectors enabled the detailed spectroscopic study on the heavy transfermium and then doorstep SHE nuclei.[1,2] The study of these nuclei provides a critical test on the energies of single-particle orbitals, particularly of those with large j.

The ^{256,257}Rf are the heaviest ones for which excited states can be studied using in-beam spectroscopy.[2–4] The rotational energies of ^{256}Rf have been studied in the framework of the particle-number-conserving cranked shell model,[5] the total-Routhian-surface method[6,7] and the projected shell model (PSM).[8] The electromagnetic properties are invaluable information in the determination of nuclear configuration and deformation. Here we try to understand these features of the rotational bands in ^{256,257}Rf with the recently developed Cranked-Nilsson-Strutinsky-Bogoliubov (CNSB) model.[9,10]

2. The CNSB Model

The CNSB model[9,10] is based on the Ultimate Cranker formalism.[11] The total energy for the correct number of particles with pairing included ('particle number projection') is minimized not only in deformation space but also in a mesh of the pairing parameters, Fermi energy λ and paring gap Δ. The quasi-particles are connected diabatically between the ω mesh points removing virtual crossings[9,11] thus diabatic configurations are followed. This becomes relatively straightforward because it is done at constant values of the deformation and pairing parameters. The same rotating liquid drop parameters as the Cranked-Nilsson-Strutinsky (CNS) model are used in the CNSB calculations.

3. Results and Discussions

S. G. Nilsson originally employed the mass dependent κ and μ parameters,[13] then these parameters are made dependent on the main oscillator quantum number to facilitate the study of a wide range of nuclei with the same set of parameters.[14] However, these so called standard parameters are set to reproduce essentially the original Nilsson diagram in the rare-earth and actinide nuclei. The parameters of the highlying orbitals with $N \geqslant 7$ are set to be the same which obviously poorly reproduce the single particle levels for $Z \approx 100$.[5] In this paper, we have combined the standard parameters (those with $N \leqslant 5$ and $N \geqslant 8$ are kept) with the original $A \approx 250$ parameters and require that the κ and μ values vary smoothly with N.

We first carried out the CNS calculations without pairing in a quadrupole and hexadecapole deformation mesh $(\varepsilon_2, \gamma, \varepsilon_4)$ with the modified Nilsson parameters. The macroscopic energy and rigid rotor moment of inertia are calculated by the Lublin-Strasbourg drop (LSD) model. It is found that the nuclear deformation changes very little and the triaxial deformations are small for $I \leqslant 30$, so an average deformation, $(\varepsilon_2, \gamma, \varepsilon_4) = (0.236, 0, 0.016)$ for ^{256}Rf and $(\varepsilon_2, \gamma, \varepsilon_4) = (0.231, 0, 0.020)$ for

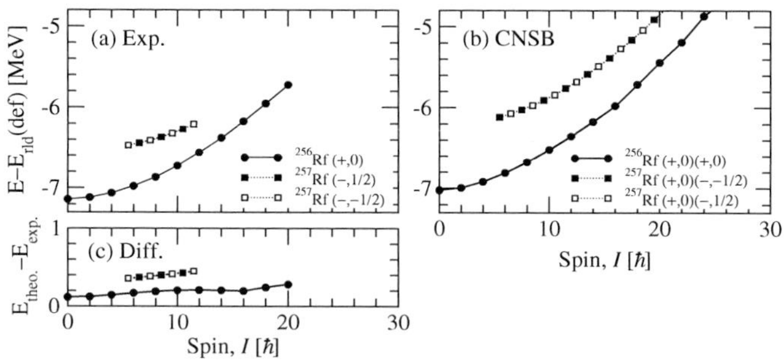

Fig. 1. The experimental (a) and theoretical (b) energies respect to the rotational rigid body reference and their differences (c) of the yrast bands in ^{256,257}Rf.

^{257}Rf are chosen in the following CNSB calculations. However, it should be noted that although the quadrupole deformations in this paper are very close to those in Refs. 8,15, the hexadecapole deformation of ^{256}Rf in the CNS calculations are substantially smaller. In Ref.16 the instabilities of the folded Yukawa potential which is used by the finite range liquid drop model was discussed. It was pointed out that sometimes the high multipolarities could cause unphysical shapes. Thus the authors in Ref. 12 reach the conclusion that usually the LSD model is generally safer to use.

The theoretical energies are compared with the experimental data in Fig. 1. The CNSB model has well reproduced the experimental energies within 0.5 MeV. However inconsistent energy differences $\sim$ 0.2 MeV between ^{256}Rf and ^{257}Rf shown in Fig. 1(c) indicates the misplacement of Nilsson orbitals around $N = 152$ in the present parametrization.

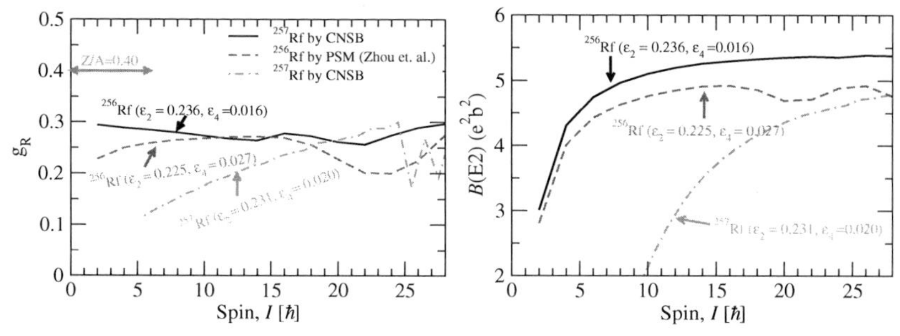

Fig. 2. The g-factors and $B(E2, I \rightarrow I-2)$ values of the yrast bands in ^{256,257}Rf.

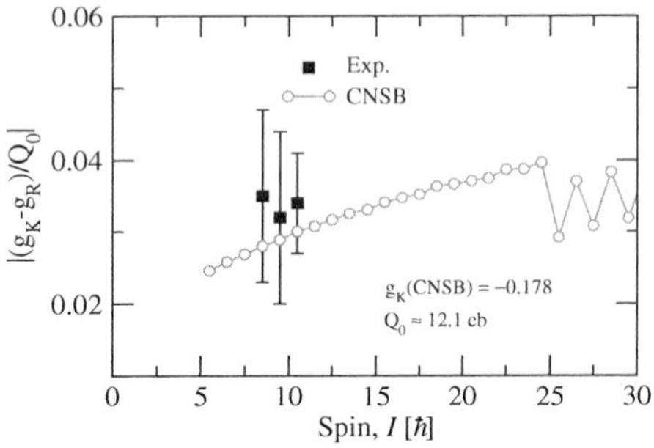

Fig. 3. The experimental and theoretical $|(g_K - g_R)/Q_0|$ values of the yrast bands in ^{257}Rf.

The gyromagnetic factors calculated by the following formula,

$$g(I) = \frac{\langle \hat{\mu}_x \rangle}{\langle \hat{I}_x \rangle} = \left[g_{l,\pi} \langle \hat{j}_{\pi,x} \rangle + (g_{s,\pi} - g_{l,\pi}) \langle \hat{S}_{\pi,x} \rangle + g_{s,\nu} \langle \hat{S}_{\nu,x} \rangle \right] / I, \quad (1)$$

where $\langle \hat{j}_{\pi,x} \rangle$, $\langle \hat{S}_{\pi,x} \rangle$ and $\langle \hat{S}_{\nu,x} \rangle$ are expectation values of the total spin and intrinsic spin projections on the cranking axis of protons and neutrons. Throughout this work we have used $g_s = 0.7 g_s^{\text{free}}$. The electric quadrupole transition probabilities $B(E2, I \to I-2)$ are calculated from the transition quadrupole moment Q_t.

The calculated g-factors and $B(E2)$ values are shown in Fig. 2. We also compared our results of ^{256}Rf with the projected shell model calculations.[8] It is seen from Fig. 2(a) that the g-factors of ^{256}Rf calculated by both the CNSB and PSM do not change too much with the increasing spin. They generally agree each other and they deviate significantly from the ideal rotor value Z/A. For ^{257}Rf, the unpaired neutron contributes significantly to the total spin at low spins, hence the g-factors are much smaller in the low-spin region. As spin increases, the g-factor increases gradually and approaches the values in ^{256}Rf. For the $B(E2)$ values in Fig. 2(b), they generally increases with spin due to the Clebsch-Gordan coefficients and approaches $\frac{5}{16\pi} Q_t^2 \times \frac{3}{8}$. Since we have used somewhat larger quadrupole deformation than that in Ref. 8 for ^{256}Rf, our $B(E2)$ values are also larger than the PSM results.

Recently J. Rissanen, *et. al.* reported the first electromagnetic information on the rotational bands of ^{257}Rf.[2] The ratios of $|(g_K - g_R)/Q_0|$, where the g_K and g_R are the intrinsic and rotational gyromagnetic factors and Q_0 is the intrinsic quadrupole moment, are obtained from the experimental γ energies and branching ratios. We have compared the CNSB results with the experimental data in Fig. 3. The CNSB results are seen within the experimental error bars. Because Q_0 is almost constant if the nuclear deformation is fixed and g_K is negative, the trend of $|(g_K - g_R)/Q_0|$ is the same as the g_R values shown in Fig. 2(b). However the experimental errors are so large that it is impossible to make any conclusion.

4. Summary

In summary, the rotational structures of ^{256,257}Rf are investigated with the particle-number-projected CNSB model. The experimental energies of the yrast bands are reproduced within 0.5 MeV. However inconsistent energy differences between ^{256}Rf and ^{257}Rf indicate that the further refinement of the Nilsson parameters for the superheavy nuclei is needed. We also calculated the g-factors and $B(E2)$ values with the CNSB model. The calculated electromagnetic properties are generally in agreement with the projected shell model results for ^{256}Rf. The calculated g-factor for ^{257}Rf increases with the increasing spin and approaches the value of ^{256}Rf for $I \gtrsim 20$. The resulted theoretical $|(g_K - g_R)/Q_0|$ values in ^{257}Rf are in agreement of experimental data extracted from the γ energies and branching ratios.

Acknowledgments

The authors acknowledge the support from the NSF of China under Contract Nos. 11205245, 11375266, 10975189 and 11405273.

References

1. P. T. Greenlees, *Physica Scripta* **2013**, 014016 (2013).
2. J. Rissanen, *et al.*, *Phys. Rev. C* **88**, 044313 (2013).
3. P. T. Greenlees, *et al.*, *Phys. Rev. Lett.* **109**, 012501 (2012).
4. J. S. Berryman, *et al.*, *Phys. Rev. C* **81**, 064325 (2010).
5. Z.-H. Zhang, *et al.*, *Phys. Rev. C* **87**, 054308 (2013).
6. W Hua-Lei, *et al.*, *Chinese Physics C* **38**, 074101 (2014).
7. H. Liu and F. Xu, *Chinese Science Bulletin* **59**, 11 (2014).
8. Z. Zhou, *et al.*, *Chinese Science Bulletin* **59**, 3853 (2014).
9. B. G. Carlsson,*et al.*, *Phys. Rev. C* **78**, 034316 (2008).
10. H. L. Ma, *et al.*, *Phys. Rev. C* **90**, 014316 (2014).
11. T. Bengtsson, *Nucl. Phys. A* **496**, 56 (1989).
12. B. G. Carlsson and I. Ragnarsson, *Phys. Rev. C* **74**, 011302 (2006).
13. S. G. Nilsson, *et al.*, *Nucl. Phys. A* **131**, 1 (1969).
14. T. Bengtsson and I. Ragnarsson, *Nucl. Phys. A* **436**, 14 (1985).
15. P. Möller, *et al.*, *Atomic Data and Nuclear Data Tables* **59**, 185 (1995).
16. L. O. Jönsson, *Nucl. Phys. A* **608**, 1 (1996).

OBSERVATION OF DOUBLET BANDS IN ^{130}Pr

K. Y. Ma, J. B. Lu*, D. Yang, Y. Z. Liu, Y. J. Ma
Department of Physics, Jilin University,
Changchun, 130012, China

X. G. Wu, G. S. Li, C. Y. He
China Institute of Atomic Energy,
Beijing, 102413, China

J. B. Lu
Department of Physics, Jilin University,
Changchun, 130012, China
**E-mail: ljb@jlu.edu.cn*

High-spin states of ^{130}Pr have been populated using the ^{99}Ru(^{35}Cl, 2n2p)^{130}Pr reaction at a beam energy of 151 MeV. A positive-parity side band with the same $\pi h_{11/2} \otimes \nu h_{11/2}$ configuration as that of the yrast band has been identified. The separation energy $\Delta E(I)$ between the states in the side band and the yrast band at the same spin has been studied systematically in odd-odd La and Pr isotopes. The variation trend of $\Delta E(I)$ indicates that it is reasonable to interpret the second $\pi h_{11/2} \otimes \nu h_{11/2}$ band in ^{130}Pr as the excited $\pi h_{11/2} \otimes \nu h_{11/2}$ band as did in ^{124}La and ^{126}La, rather than to interpret it as the partner band of near degenerate chiral doublet band as did in $^{128-134}$La and $^{132-134}$Pr.

Keywords: High-spin states, separation energy, chiral doublet band.

1. Introduction

In the $A \sim 130$ mass region, the nearly degenerate $\Delta I = 1$ doublet bands built on the $\pi h_{11/2} \otimes \nu h_{11/2}$ configuration have attracted significant attention and intensive discussion in odd-odd $N = 77$ (^{132}Cs, ^{134}La),[1,2] $N = 75$ (^{130}Cs, ^{132}La, ^{134}Pr, ^{136}Pm, ^{138}Eu),[3-5] $N = 73$ (^{128}Cs, ^{130}La, ^{132}Pr),[6,7] $N = 71$ (^{126}Cs, ^{128}La),[8,9] $N = 69$ (^{124}Cs, ^{126}La, ^{128}Pr)[10-12] and $N = 67$ (^{122}Cs, ^{124}La)[13,14] isotones. These $\Delta I = 1$ doublet bands with the same positive parity are mostly interpreted as a manifestation of "chirality" in the sense of the angular-momentum coupling.[15] In addition, several competitive mechanisms have been also proposed, including γ band, shape coexistence, quasiparticle excitation and pseudospin partner bands.[16]

Recently, we have reported the positive-parity doublet bands with the configuration of $\pi h_{11/2} \otimes \nu h_{11/2}$ in ^{126}La[11] and ^{128}La[9] nuclei. So far, the doublet bands have been systematically observed in odd-odd La isotopes from ^{124}La to ^{134}La. However, for Pr isotope chain, the doublet bands have been observed from ^{128}Pr to ^{134}Pr except for ^{130}Pr. Meanwhile, the doublet bands have also been observed in the ^{126}Cs[10] and ^{128}La[12] isotones of ^{130}Pr. Hence, we infer that the similar bands may appear in ^{130}Pr. In addition, the validity of theories is often tested by how well the theory reproduces the systematic behavior of isotopes or isotones, and thus it is important to establish the doublet bands systematically appeared in the Pr isotopic chain. Based on the above considerations, a new high-spin state experiment of ^{130}Pr was completed in the HI-13 tandem accelerator at CIAE in Beijing, and the main purpose of the experiment was trying to search for the positive-parity doublet bands.

2. Experiment and Results

High-spin states of ^{130}Pr were populated through the ^{99}Ru(^{35}Cl, 2n2p)^{130}Pr reaction at a beam energy of 151 MeV. The ^{99}Ru target, with an enrichment of 92.8% and a thickness of 2.24 mg/cm^2, was rolled onto a 15 mg/cm^2 lead backing. The beam was provided by the HI-13 tandem accelerator at CIAE in Beijing. The γ–γ coincidence data were recorded by the use of the detecting system consisting of nine compton-suppressed HPGe detectors, two HPGe planar detectors and one clover-type detector. These Ge detectors in the array were placed at 90°, 37°, 30°, and 60° relative to the beam direction. A total of 4.5×10^8 γ–γ coincidence events were recorded. The data were sorted into a symmetrized γ–γ coincidence matrix and a DCO (directional correlation from oriented states) matrix, and the DCO matrix was created by sorting the detectors at 30° and 37° on one axis and the detectors at $\sim$ 90° on the other. DCO ratios were obtained from spectra gated either on quadrupole or dipole transitions. For our detector array, when gating on a quadrupole transition, the DCO ratio of the measured transition is around 1.0 for quadrupole transition and around 0.6 for dipole transition, and when gating on a dipole transition, the DCO ratio of measured transition becomes around 1.0 for dipole transition and around 1.7 for quadrupole transition.

The positive-parity yrast band 1 is the most intensely populated in the present experiment, and its configuration has been assigned to $\pi h_{11/2} \otimes$

$\nu h_{11/2}$ in the previous work.[17,18] In present work, two new 199.5 and 53.5 keV transitions have been observed at the bottom of this band, and the spin of the lowest observed state is extended to $6\hbar$. Parities and spins of this band are assigned based on the proposed configuration and the systematics of Liu et al.[19]

Band 2 (side band) has been reported for the first time, and several new linking transitions between bands 1 and 2 are observed. Multipolarity analysis indicates that 1080.5 and 1030.0 keV linking transitions are of $\Delta I = 1$ character with DCO ratios of 0.95 and 1.09, and 1294.0 keV linking transitions is of $\Delta I = 2$ character with DCO ratio of 1.62 (DCO ratios from gating on the dipole transition). The observation of both $\Delta I = 1$ and $\Delta I = 2$ linking transitions between bands 1 and 2 implies that band 2 has positive parity as that of band 1. Meanwhile, the energies and spins of band 2 are fixed relative to the levels of band 1 as shown in Fig. 1. A sample γ–γ coincidence spectrum supporting the level scheme of Fig. 1 is shown in Fig. 2.

3. Results and Discussion

All these experimental observations and theoretical calculations suggest that band 2 has the same $\pi h_{11/2} \otimes \nu h_{11/2}$ configuration as that of band 1. A possible interpretation of the band 2 is that it may result from chiral symmetry breaking. For chiral interpretation, the separation energy between the states in the side band and the yrast band at the same spin I, $\Delta E(I) = E(I)_{side} - E(I)_{yrast}$, is an important quantity for the interpretation of the side band. Meanwhile, for practical near degenerate chiral doublet bands, their corresponding properties should be somewhat different from each other depending on the degree of the deviation from the ideal chiral doublet bands. Therefore, in order to explain the doublet bands, the separation energy between the states in the side band and the yrast band at the same spin, $\Delta E(I) = E(I)_{side} - E(I)_{yrast}$, in $^{124-134}$La are compared in Fig. 3. Within the observed spin region, the magnitude and variation trend of $\Delta E(I)$ of ^{130}Pr are very similar to those of ^{124}La[14] and ^{126}La,[11] and quite different from those of $^{128-134}$La and $^{132-134}$Pr.[2–4,7,9] In the later cases the twin $\pi h_{11/2} \otimes \nu h_{11/2}$ bands have been cited as nearly degenerate chiral doublet bands or their candidates. In the cases of ^{130}Pr, ^{124}La and ^{126}La, $\Delta E(I)$ is too large for the twin bands to interpret as nearly degenerate doublet bands. This fact suggests that it is reasonable to interpret the second $\pi h_{11/2} \otimes \nu h_{11/2}$ band in ^{130}Pr as the excited $\pi h_{11/2} \otimes \nu h_{11/2}$ band, as did in ^{124}La[14] and ^{126}La.[11]

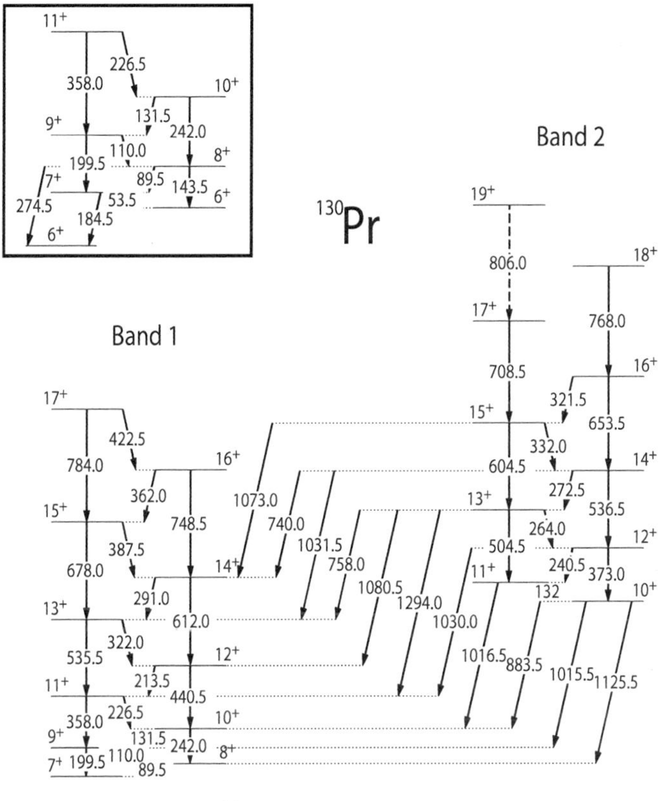

Fig. 1. Partial level scheme of ^{130}Pr. The insertion shows the bottom portion of band 1.

In ^{124}La and ^{126}La, the two signature components of the side $\pi h_{11/2} \otimes \nu h_{11/2}$ band (the excited $\pi h_{11/2} \otimes \nu h_{11/2}$ band) are interpreted from the coupling of the unfavored signature component of the $\pi h_{11/2}$ proton orbital with the two signature components of the $\nu h_{11/2}$ neutron orbital, while the two signature components of the yrast $\pi h_{11/2} \otimes \nu h_{11/2}$ band result from the coupling of the favored signature component of the $\pi h_{11/2}$ proton orbital with the two signature components of the $\nu h_{11/2}$ neutron orbital. Thus, the similar variation trends of $\Delta E(I)$ suggest that it is reasonable to interpret the second $\pi h_{11/2} \otimes \nu h_{11/2}$ band of ^{130}Pr as proposed in ^{124}La and ^{126}La

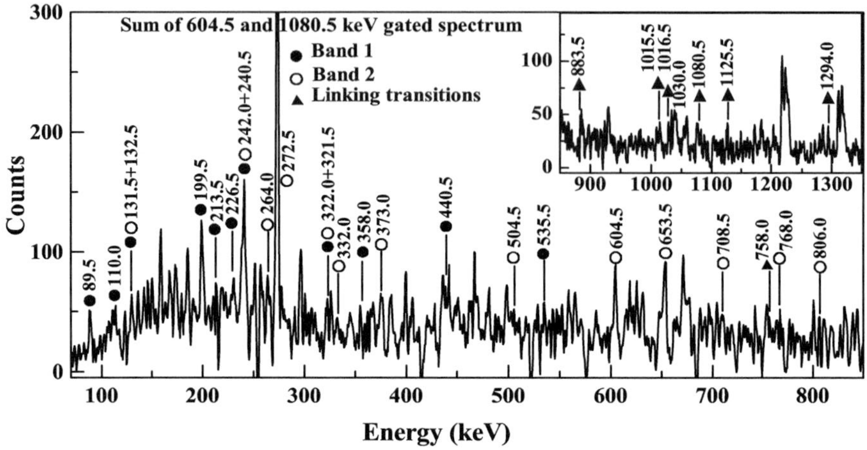

Fig. 2. Sample $\gamma - \gamma$ coincidence spectrum supporting the partial level scheme of ^{130}Pr as shown in Fig. 1.

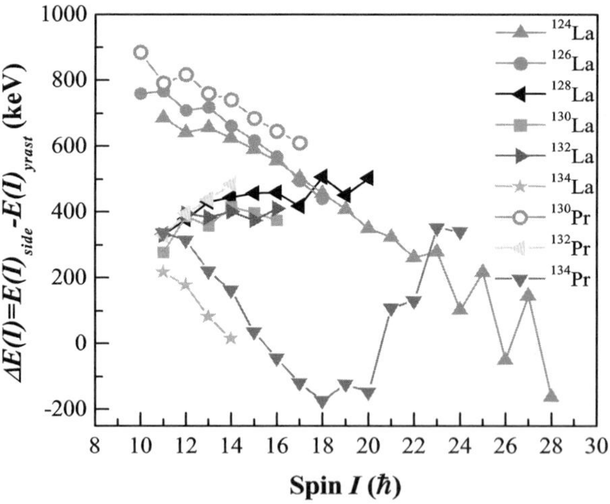

Fig. 3. (Color online) $\Delta E(I) = E(I)_{side} - E(I)_{yrast}$ of the two bands in ^{130}Pr compared with those in ^{124}La,[14] ^{126}La,[11] ^{128}La,[9] ^{130}La,[7] ^{132}La,[3] ^{134}La,[2] ^{132}Pr[7] and ^{134}Pr.[4]

rather than to interpret it as the partner band of near degenerate chiral doublet bands as proposed in $^{128-134}$La and $^{132-134}$Pr.

4. Summary

A positive-parity side band built on the $\pi h_{11/2} \otimes \nu h_{11/2}$ configuration has been observed, and linking transitions between the yrast band and the side band are identified. The separation energy for the side band and the yrast band has been studied systematically in odd-odd La isotopes. It is noted that, within the observed spin region, the magnitude and variation trend of the separation energy $\Delta E(I)$ between the second $\pi h_{11/2} \otimes \nu h_{11/2}$ band and the yrast $\pi h_{11/2} \otimes \nu h_{11/2}$ band at the same spin in ^{130}Pr are very similar to those of ^{124}La and ^{126}La and quite different from those of $^{128-134}$La and $^{132-134}$Pr. The variation trend of $\Delta E(I)$ suggests that it is reasonable to interpret the second $\pi h_{11/2} \otimes \nu h_{11/2}$ band in ^{130}Pr as the excited $\pi h_{11/2} \otimes \nu h_{11/2}$ band as did in ^{124}La and ^{126}La, rather than to interpret it as the partner band of near degenerate chiral doublet bands as proposed in $^{128-134}$La and $^{132-134}$Pr.

Acknowledgments

This work is supported by the National Basic Research Programme of China under Grant No.2007CB815005, the National Natural Science Foundation of China under Grant No.11075064 and No.11405072, and China Postdoctoral Science Foundation No.2013M541285.

References

1. G. Rainovski *et al.*, Phys. Rev. C **68**, 024318 (2003).
2. R. A. Bark, A. M. Baxter, A. P. Byrne, G. D. Dracoulis, T. Kibédi, T. R. McGoram, and S. M. Mullins, Nucl. Phys. A **691**, 577 (2001).
3. K. Starosta *et al.*, Phys. Rev. Lett. **86**, 971 (2001).
4. C. M. Petrache *et al.*, Nucl. Phys. A **597**, 106 (1996).
5. A. A. Hecht *et al.*, Phys. Rev. C **63**, 051302 (2001).
6. T. Koike, K. Starosta, C. J. Chiara, D. B. Fossan, and D. R. LaFosse, Phys. Rev. C **67**, 044319 (2003).
7. T. Koike, K. Starosta, C. J. Chiara, D. B. Fossan, and D. R. LaFosse, Phys. Rev. C **63**, 061304 (2001).
8. S. Y. Wang, Y. Z. Liu, T. Komatsubara, Y. J. Ma, and Y. H. Zhang, Phys. Rev. C **74**, 017302 (2006).
9. K. Y. Ma, J. B. Lu, D. Yang, H. D. Wang, Y. Z. Liu, X. G. Wu, Y. Zheng, and C. Y. He, Phys. Rev. C **85**, 037301 (2012).
10. A. Gizon *et al.*, Nucl. Phys. A **694**, 63 (2001).
11. K. Y. Ma, J. B. Lu, S. P. Ruan, D. Yang, J. Li, Y. Z. Liu, Y. J. Ma, X. G. Wu, Y. Zheng, and C.Y.He, Phys. Rev. C **88**, 057302 (2013).
12. D. J. Hartley *et al.*, Phys. Rev. C **65**, 044329 (2002).
13. Yong-Nam U *et al.*, J. Phys. G **31**, B1 (2005).

14. H. J. Chantler *et al.*, Phys. Rev. C **66**, 014311 (2002).
15. S. Frauendorf and J. Meng, Nucl. Phys. A **617**, 131 (1997).
16. J. Meng and S. Q. Zhang, J. Phys. G **37**, 064025 (2010).
17. R. Ma *et al.*, Phys. Rev. C **37**, 1926 (1988).
18. C. M. Petrache *et al.*, Nucl. Phys. A **635**, 361 (1998).
19. Y. Z. Liu, J. B. Lu, Y. J. Ma, S. G. Zhou, and H. Zheng, Phys. Rev. C **54**, 719 (1996).

ALPHA-DECAY HALF-LIFE WITH A GENERALIZED LIQUID DROP MODEL BY USING A PRECISE DECAY ENERGY

Na Na Ma, Hai Fei Zhang, Jian Min Dong and Hong Fei Zhang*

School of Nuclear Science and Technology, Lanzhou University, Lanzhou 730000, China

**E-mail: zhanghongfei@lzu.edu.cn*

Incorporating with the radial basis function (RBF) correction, the rms deviation of a new Weizsäcker-Skyrme-type nuclear mass formula [Haifei Zhang, *et al.*, Nucl. Phys. A 929, 38 (2014)] has been significantly reduced from 493 keV to 323 keV by up to 34% to 2267 calculated atomic masses comparing with experimental data abstracted from AME12. With the Q_α values calculated by this hybrid formula as inputs, the systematically evaluated α decay half-lives within the GLDM for the isotopic nuclei agree with the experimental ones.

Keywords: Nuclear mass; radial basis function (RBF); superheavy nuclei; GLDM; α-decay.

1. Introduction

The accuracy of nuclear mass have a very high demand in many aspects, specifically for the *rp*-process path[1] in the astrophysics.[2] Recently we proposed a mass formula which[3] belongs to a macroscopic-microscopic WS-type model. The radial basis function (RBF) approach[4] have been advocated to improve the accuracy of mass models. Nowadays, α decay as a tool can be used to the prediction of "island of stability" of superheavy nuclei (SHNs) and the synthesis of SHNs. Here the generalized liquid drop model (GLDM)[5] which is one of the most successful macroscopic models be used to study α decay.

2. The Improved Nuclear Mass Formula

The total binding energy of a nucleus consists of two parts:

$$E(Z, A, \beta_k) = E_{\text{LDM}} \prod_{k\geq 2} \left(1 + b_k^2 \beta_k^2\right) + E_{\text{mic}}(Z, A, \beta_k), \qquad (1)$$

where $E_{\rm LDM}$ is the liquid drop energy, Strutinsky[12] and BCS approach are used to deal with the microscopic part $E_{\rm mic}(Z, A, \beta_k)$.

The RBF approach is written as:[4]

$$S(x) = \sum_{i=1}^{m} \omega_i \phi(||x - x_i||), \tag{2}$$

The revised binding energies for nucleus is given by

$$E_{\rm Cal.}^{\rm RBF}(Z, N) = E_{\rm Cal.}(Z, N) + S(Z, N). \tag{3}$$

3. Results and Discussion

3.1. *Deviations of binding energy*

The rms deviation for 3255 nuclei[13] reduces to 0.339 MeV and to 2267 experimental data has reduced from 0.493 MeV to 0.323 MeV by up to 34%. In order to facilitate the interpretation of the article, our formula is labeled as the WS-LZ model and the one after RBF correction is called WS-LZ+RBF model.

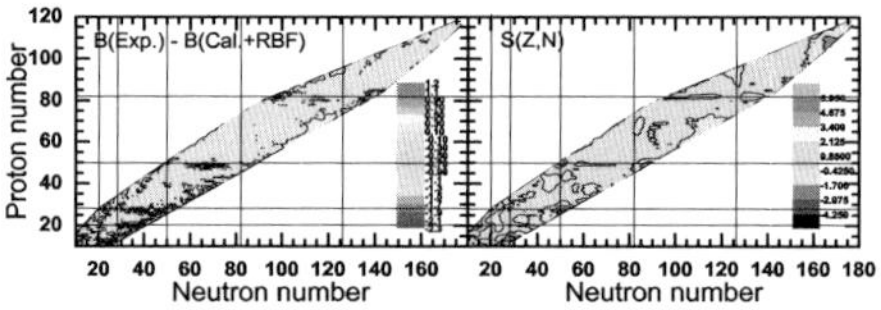

Fig. 1. (Color online) (a) Differences between the calculated binding energies be improved by the RBF approach and the experimental ones of 3255 nuclei. (b) The reconstructed function $S(N, Z)$.

In Fig. 1, almost all differences are located in the region of ± 0.5 MeV. The results have verified that RBF correction is an effective tool to reconstruct the deviations. However, there still exists a relatively larger difference in the region close to the magic numbers and the neutron /or proton drip line may be because the pairing correction in WS-LZ model is treated by the BCS method which becomes gradually invalid as approaching the drip lines.

3.2. α *decay half-life with* Q_α

The Q_α values as an input to calculate the α decay half-lives of nuclei chains $Z = 74 - 118$ in the framework of GLDM. The Wentzel-Kramers-Brillouin

(WKB) method is used to here

$$T_{1/2} = \frac{\ln 2}{P_0 \upsilon_0 P}, \tag{4}$$

$$Q_\alpha = B(\alpha) + B(Z-2, A-4) - B(Z, A). \tag{5}$$

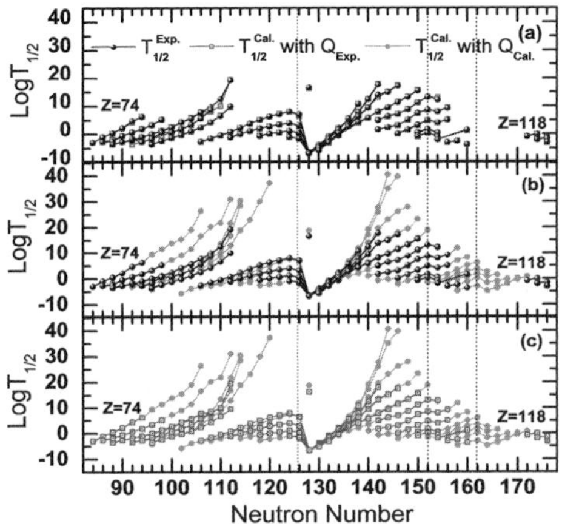

Fig. 2. (Color online)(a) The measured decimal logarithm of alpha decay half-lives and the calculated ones using the $Q_\alpha^{\mathrm{Expt.}}$. (b) Similar as (a) but with the calculated results using the $Q_\alpha^{\mathrm{Cal.}}$. (c) The results with the $Q_\alpha^{\mathrm{Expt.}}$ and the $Q_\alpha^{\mathrm{Cal.}}$ values, respectively.

One can find that the experimental data are found to be in surprising agreement with the corresponding calculated half-lives using the experimental Q_α and $Q_\alpha^{\mathrm{Cal.+RBF}}$ values with WS-LZ+RBF formula, indicating that GLDM model is powerful to reproduce the α-decay half-life and further pointing that the as inputs are very effective. The α decay half-lives generally increase with increasing neutron number rapidly up to the spherical shell closure $N = 126$. At the neutron number $N = 128$ appears the minimum half-lives. Similar phenomenon reappears at the neutron submagic numbers but not so obvious, the half-lives also have a drop at $N = 152$, $N = 162$ and the minimum half-lives appear at $N = 154$, $N = 164$.

4. Summary

In summary, WS-LZ mass formula as a useful model has been certificated. That is using a unified approach called Strutinsky-like BCS method is used to deal with pairing and shell corrections spontaneously. Consequently, the rms deviation with respect to 2267 measured nuclear masses is 0.493 MeV. The deviation successfully falls to 0.323 MeV by up to 34% with the WS-LZ-RBF model. In the framework of GLDM within the above calculated Q_α values as inputs, the calculated results are in excellent agreement with the experimental data. The strong shell effect appears at neutron number $N = 126, 152$ and $N = 162$. The extrapolations of WS-LZ+RBF are tested and verified. There is no doubt that GLDM is an indubitably successful macroscopic model calculating the α decay half-life.

References

1. H. Schatz, K. E. Rehm, Nucl. Phys. A **777** 601 (2006).
2. A. Arcones and G. F. Bertsch, Phys. Rev. Lett. **108**, 151101 (2012).
3. Haifei Zhang, Jianmin Dong, Nana Ma, G. Royer, Junqing Li, Hongfei Zhang, Nucl. Phys. A **929**, 38 (2014).
4. Ning Wang and Min Liu, Phys. Rev. C **84**, 051303(R) (2011).
5. G. Royer, J. Phys. G: Nucl. Part. Phys. **26**, 1149 (2000).
6. G. Royer, Nucl. Phys. A **807** 105 (2008).
7. N. H. Allal, M. Fellah, Phys. Rev. C **48** 1656 (1993).
8. N. Tajima, Y. R. Shimizu, S. Takahara, Phys. Rev. C **82** 034316 (2010).
9. J. M. Pearson, Hyperfine Interact. **132** 59 (2001).
10. Z. M. Niu, Z. L. Zhu, Y. F. Niu, B. H. Sun, T. H. Heng, and J. Y. Guo, Phys. Rev. C **88**, 024325 (2013) .
11. J. S. Zheng, N. Y. Wang, Z. Y. Wang, Z. M. Niu, Y. F. Niu, and B. Sun, Phys. Rev. C **90**, 014303 (2014).
12. M. Brack, H. C. Pauli, Nucl. Phys. A **207** 401 (1973).
13. M. Wang, G. Audi, A.H. Wapstra, F. G. Kondev, M. MacCormick, X. Xu, B. Pfeiffer, Chin. Phys. C **36** (2012) 1603.

INFLUENCE OF SYMMETRY POTENTIAL ON THE SHELL-CORRECTIONS OF SUPER-HEAVY NUCLEI

Qiuhong Mo, Min Liu and Ning Wang*

Department of Physics, Guangxi Normal University, Guilin 541004, P. R. China
**E-mail: wangning@gxnu.edu.cn*

Based on the Woods-Saxon single-particle potential, we study the single-particle levels and the shell-corrections of super-heavy nuclei. Through comparing the results from the parameter set "universal" and "WS", we find that the isospin dependence of symmetry potential κ had a great influence on the shell-corrections of super-heavy nuclei.

Keywords: Woods-Saxon potential; symmetry potential; shell correction; spin-orbit potential.

1. Introduction

The Woods-Saxon[1] single-particle potential is widely used in the study of nuclear structures. The Woods-Saxon potential with parameter set "universal"[2] is successful for reproducing the single particle levels of stable nuclei, especially the high-spin spectra in the lead region which is attracted much attention in recent years.[3] In addition, it can interpret the structure of yrast states identically compared to the calculated single-particle configurations[4] with the shell model[5] results. In "universal", the strength of spin-orbit potential and the symmetry potential are taken as constants. On the other hand, Wang et al proposed a macroscopic-microscopic mass model in which the Woods-Saxon potential is adopted for calculating the microscopic shell corrections.[6] Through adjusting the potential parameters and fitting the masses of 2149 nuclei,[7] the optimal parameter set "WS"[6] is obtained. In "WS", the isospin dependence of the spin-orbit potential and symmetry potential is taken into account, which is different from those in "universal". It is therefore interesting to investigate how the Woods-Saxon potential parameters influence the shell-corrections of super-heavy nuclei.

2. The Woods-Saxon Potential Parameters

The axially deformed Woods-Saxon potential is expressed as

$$V(\vec{r}) = \frac{V_q}{1+\exp[\frac{r-\mathcal{R}(\theta)}{a}]}, \tag{1}$$

where a denotes the surface diffuseness of single-particle potential. V_q denotes the depth of the central potential ($q = p$ for protons and $q = n$ for neutrons):

$$V_q = V_0(1 \pm \kappa I) = V_0 \pm V_s I, \tag{2}$$

the plus sign for neutrons and minus for protons. $I = \frac{N-Z}{N+Z}$ and $V_s = V_0\kappa$ denote the isospin asymmetry and the symmetry potential, respectively.$\mathcal{R}$ defines the distance from the origin of the coordinate system to the point on the nuclear surface

$$\mathcal{R}(\theta) = c_0 R\,[1 + \beta_2 Y_{20}(\theta) + \beta_4 Y_{40}(\theta) + ...], \tag{3}$$

with the scale factor c_0 which represents the effect of incompressibility of nuclear matter in the nucleus and is determined by the so-called constant volume condition.[8] $Y_{lm}(\theta, \phi)$ are the spherical harmonics. $R = r_0 A^{1/3}$ denotes the radius of the single particle potential. In "WS", $r_0 = 1.384$ fm is for both protons and neutrons. The spin-orbit potential is written as

$$V_{s.o.} = -\lambda(\frac{\hbar}{2Mc})^2 \nabla V \cdot (\vec{\sigma} \times \vec{p}), \tag{4}$$

where M is the free nucleonic mass. $\vec{\sigma}$ and $\vec{p}$ are the Pauli spin matrix and the nucleon momentum. λ denotes the strength of the spin-orbit potential. As mentioned previously, λ is a constant in "WS", whereas in "WS", $\lambda = \frac{3}{2}\lambda_0\left(1 \pm \frac{1}{3}I\right)$ with $\lambda_0 = 26.3$, the plus sign for neutrons and the minus sign for protons, in which the isospin dependence of the strength of the spin-orbit potential is considered according to the Skyrme energy-density functional.[9,10] In addition, V_0 and κ are set as constants in "universal", so that V_s is a constant. Whereas in "WS", the mass and isospin dependence of V_s is considered, $V_s = 29.3\left[1 - \frac{1.45}{A^{1/3}} + \frac{2-|I|}{2+|I|A}\right]$. The depth V_0 and surface diffuseness a adopted in "universal" and "WS" are comparable. $V_0 = -49.6$ MeV and $a = 0.7$ fm in "universal", while $V_0 = -47.5$ MeV and $a = 0.78$ fm in "WS".

3. Results and Discussions

In this section, we study the single-particle levels of super-heavy nuclei based on the Woods-Saxon potential parameters. The single-particle levels will be used to calculate the Strutinsky shell-corrections of super-heavy nuclei.

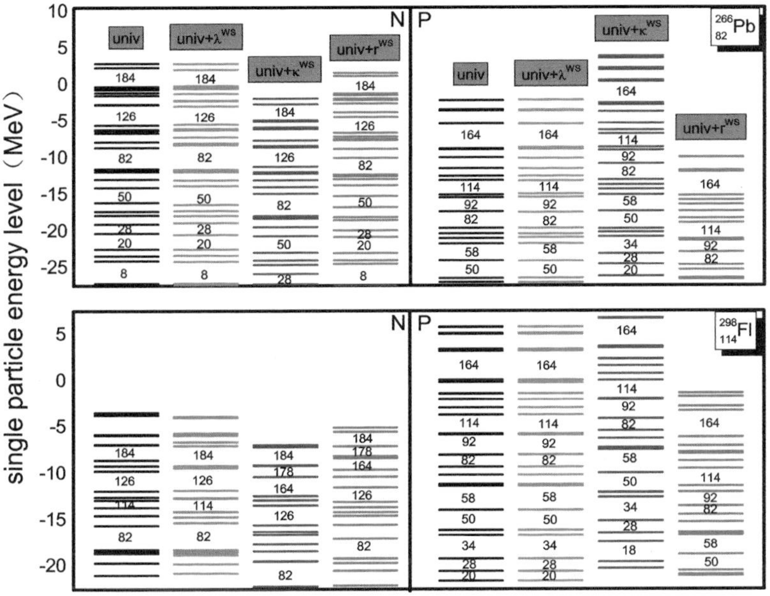

Fig. 1. (Color online) Single particle energy levels of nuclei ^{266}Pb ($Z = 82, N = 184$) and ^{298}Fl ($Z = 114, N = 184$).

In order to see the influence of potential parameters on the single-particle levels, especially the gaps at magic numbers, we replace the parameters adopted in "universal" one by one with the corresponding parameters adopted in "WS". Figure 1 shows the calculated single-particle levels of the nuclei ^{266}Pb ($Z = 82, N = 184$) and ^{298}Fl ($Z = 114, N = 184$). The first column (black) in each sub-figures presents results from the parameter set "universal". The second column (red) means the results with the same parameters in "universal", but the strength of spin-orbit potential λ is replaced with those in "WS". The third and the fourth columns denotes the results with the symmetry potential coefficient $\kappa = V_S/V_0$ and the radius coefficient r_0 are replaced with those in "WS", respectively. One sees that the change of potential parameters will lead to the changing of the single-particle energy levels. In particular, the gaps at $N = 164, 178$ for neutrons in ^{266}Pb evidently appear after adopting the symmetry potential parameter κ in "WS".

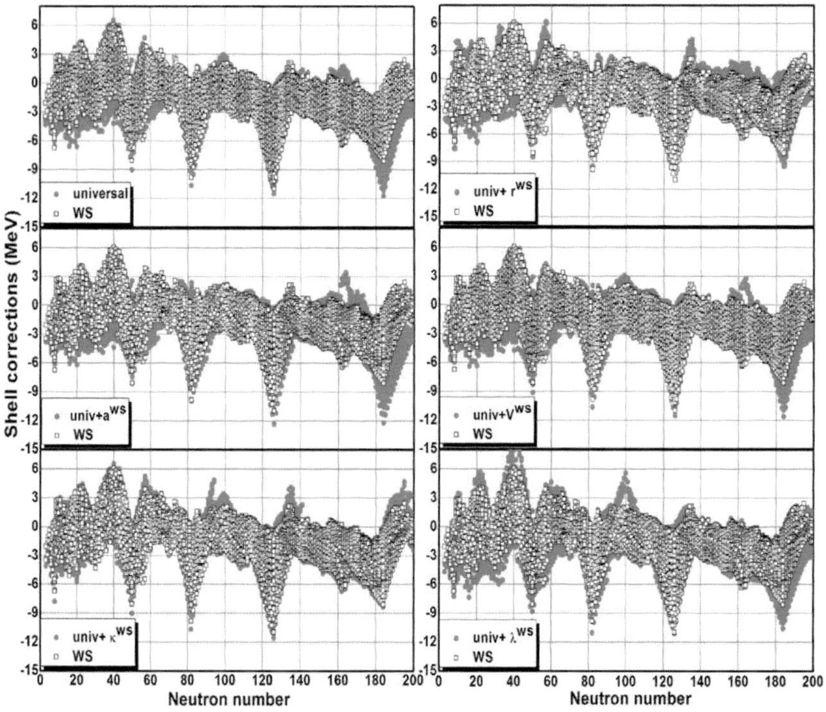

Fig. 2. (Color online) Shell-correction of nuclei as a function of neutron number. The black squares denote the results with the potential parameters in "WS". The red circles denote the results with the same parameters in "WS", but one parameter replaced by the corresponding value adopted in "WS".

Fig. 2 shows that the calculated shell-corrections of nuclei as the function of neutron number with different potential parameters adopted. One sees that the symmetry potential parameter κ significantly influences shell-corrections of super-heavy nuclei around $N = 184$. The isospin dependence of symmetry potential seems influence the shell evolution.

4. Summary

In this brief report, we investigate the influence of parameters of Woods-Saxon single-particle potential on the energy levels of heavy and superheavy nuclei. By replacing the Woods-Saxon potential parameters adopted in "WS" with the corresponding values adopted in "WS" one by one, we find that the isospin dependence of symmetry potential significantly influences the single particle levels of heavy nuclei, especially the levels around $N = 164, 178$. More information about the symmetry potential should be further investigated, since it is of great importance not only for the study of superheavy nuclei, but also for the study of nuclear symmetry energy.

References

1. M. Cerkaski, J.Dudek, Z. Szymalski, P. Rozmej, and 8. G. Nilsson, Nucl. Phys. A315, 269 (1979).
2. S. Kahane, S. Zaman, and J. Dudek, Phys. Rev. C 40, 2282 (1989).
3. J. Dudek, Z. Szymanski, and T. Werner, Phys. Rev. C 23, 920 (1981).
4. G. Andersson, S. E. Larsson, G. Leander, P. Moiler, 8. G. Nilsson, I. Ragnarsson, 8. Aberg, R. Bengtsson, J. Dudek, B.Nerlo-Pomorska, K. Pomorski, and Z. Szyma6ski, Nucl. Phys. A268, 205 (1976).
5. J. Blomqvist, Proceedings of the International Conference on High Spin Phenomena, Argonne, 1979.
6. Ning Wang, Min Liu, and Xizhen Wu, Phys. Rev. C 81, 04432 (2010).
7. G. Audi, A.H. Wapstra and C. Thibault, Nucl. Phys. A. 2003, 729: 337.
8. S. Cwoik, J. Dudek, et al., Comp. Phys. Commun. 1987, 46: 379.
9. J. Bartel and K. Bencheikh, Eur. Phys. J, 2002, 14: 179.
10. Min Liu, Ning Wang, et al., Nucl.Phys. A. 2006, 768: 80.

COUPLED-CHANNELS STUDY OF α-DECAY RATES FOR DEFORMED NUCLEI

Dongdong Ni[1,2,*] and Zhongzhou Ren[1,2,3]

[1] *Department of Physics and Key Laboratory of Modern Acoustics, Institute of Acoustics, Nanjing University, Nanjing 210093, China*

[2] *Joint Center of Nuclear Science and Technology, Nanjing University, Nanjing 210093, China*

[3] *Center of Theoretical Nuclear Physics, National Laboratory of Heavy-Ion Accelerator, Lanzhou 730000, China*

* *E-mail: dongdongnick@gmail.com*

The generalized density-dependent cluster model is devoted to calculate α-decay half-lives of spherical and deformed nuclei. The multi-channel cluster model is developed to describe the α-decay fine structure in heavy deformed nuclei, including half-lives and branching ratios. After a brief review of these two models, special cases of the α-decay fine structure are presented. Calculations are separately performed using the coupled-channels and WKB approaches.

Keywords: α-decay fine structure; rotational bands; coupled-channels.

1. Introduction

The study on α decay dates back to the early days of nuclear physics. Nowadays it is still an interesting problem for current researches of unstable nuclei. For example, combining α and γ spectra is a useful tool to probe the structure properties of nuclei, such as shape coexistence, new isomers, and energy spectrum of heavy and even superheavy nuclei.[1–5] Since the work of Gamow, there have been many approaches for α decay. They fall into into three categories: empirical study,[6,7] semiclassical approximation,[8–11] and solving Schrödinger equation for quasibound states.[12–15] Considering α decay is a pure quantum tunneling effect, the last group is more consistent from the theoretical standpoint. The generalized density-dependent cluster model (GDDCM) and multi-channel cluster model (MCCM) belong to the last group. In this work, we briefly introduce these two models and give some numerical results for deformed nuclei. Then special cases of the α-decay fine structure are presented and tentative calculations are separately performed within the MCCM and using the WKB barrier penetration approach.

2. Theoretical Framework for Deformed Nuclei

Within the GDDCM, the nuclear and Coulomb potentials between α particles and daughter nuclei are numerically constructed using the double-folding method with the popular Michigan three-range Yukawa (M3Y) nucleon-nucleon (NN) interaction. Then the single or coupled-channel Schrödinger equations are solved with outgoing wave boundary conditions to calculate α-decay half-lives.[15] Two points should be emphasized: (1) The number of internal nodes in the cluster radial wave function n is chosen in terms of the Wildermuth rule,[16]

$$G = 2n + L = \sum_{i=1}^{4} g_i. \tag{1}$$

In this expression, L is the orbital angular momentum carried by the emitted α particles, and g_i are the corresponding oscillator quantum numbers for the constituent nucleons of the α cluster in the parent nucleus, whose values are chosen to guarantee the α cluster entirely outside of the shell occupied by the core nucleus. In this way, the main requirements of the Pauli exclusion principle between a nucleon forming the α cluster and the one composing the residual core nucleus are satisfied. (2) The M3Y-Reid-type NN interaction used in the double-folding model includes a zero-range potential for the single-nucleon exchange. This term guarantees the anti-symmetrization of identical nucleons in the α cluster and in the core nucleus to some extent. In this way, the remaining effects of the Pauli principle are largely absorbed into such effective interactions.

For deformed emitters, including two or three important decay channels is sufficient for proper convergence in calculating α-decay half-lives, as shown in the GDDCM.[15] However, this procedure is clearly insufficient for convergence in describing the α-decay fine structure. So the MCCM was established, where much more decay channels are considered for convergence in calculating α-decay rates concerning half-lives and branching ratios.

The α decay of deformed even-even nuclei generally proceeds from ground 0^+ states to various members of the ground-state rotational band (i.e., 0^+, 2^+, 4^+, 6^+,...). Each member of the rotational band exhibit one single decay channel. Such α transitions have been widely investigated for even-even nuclei.[17–22] The situation for the case of odd-A and odd-odd emitters is much more complicated owing to last unpaired nucleons. There are various rotational bands built on single-particle states or two-quasiparticle states in daughter nuclei. Also, the structure of the parent state is generally different from that of the ground-state rotational band. There exits another rotational band whose structure is similar to that of the parent state, called the favored rotational band. In this case, the α transitions to the ground-state band is preferred in energy, while the transitions to the favored band is preferred in structure. So there is competition between them. In general, the transitions to the favored band play a major role in the α decay of odd-A and odd-odd nuclei.[23] They are of great interest in theoretical studies.[24,25] It should be particularly noted that each member of the rotational band exhibit several decay channels with the even ($\pi\pi' = 1$) or odd ($\pi\pi' = -1$) L values in the range of $|J - I| \leq L \leq (J + I)$, where J^{π} ($I^{\pi'}$) is the spin-parity of the parent (daughter) state.

3. Numerical Results and Discussion

Both the GDDCM and MCCM can well reproduce the experimental α-decay half-lives of deformed nuclei. Table 1 displays the goodness of the GDDCM and MCCM calculations for deformed even-even (e-e) and odd-A nuclei. Here we do not list the results for deformed odd-odd nuclei due to their rare experimental information.[26] The average and standard deviations are used to estimate the goodness. One can see that the calculated half-lives in these two models agree with the experimental data within a factor of less than two.

Table 1. Agreement between the calculated α-decay half-lives and the experimental data for deformed even-even (e-e) and odd-A nuclei. Calculations are separately performed within the GDDCM and MCCM.

	GDDCM		MCCM	
	e-e	odd-A	e-e	odd-A
Number	65	35	35	32
$\langle\sigma\rangle$	0.209	0.230	0.153	0.236
$\sqrt{\langle\sigma^2\rangle}$	0.250	0.266	0.192	0.279

Besides α-decay half-lives, the MCCM also gives the branching ratios to various daughter states. In Ref. 22, five-channel calculations were performed for 35 well-deformed even-even nuclei with $Z = 92 - 106$. The calculation well reproduces the experimental branching ratios to 0^+, 2^+, 4^+, 6^+, and 8^+ states with mean factors of 1.1, 1.3, 2.7, 3.3, and 3.2, respectively. It should be pointed out that the WKB calculations have a tendency to overestimate the branching ratios to excited 4^+ stats especially for the α decay of Pu and Cm isotopes, by more than one order of magnitude. In Ref. 24, multi-channel calculations were performed for 32 deformed odd-mass nuclei with $Z = 93$-102. The branching ratios for various daughter states are well reproduced within a factor of about 2.66. The detailed results can be found in Tables (II-III) of Ref. 24. In particular, the coupled-channel calculations successfully interpret the abnormal behavior that the branching ratio for excited $(I+1)$ states is larger than that for excited I states in the α decay of ^{251}Cf and ^{253}Fm. In Ref. 25, the MCCM was also extended for the α-decay fine structure in heavy odd-odd nuclei.

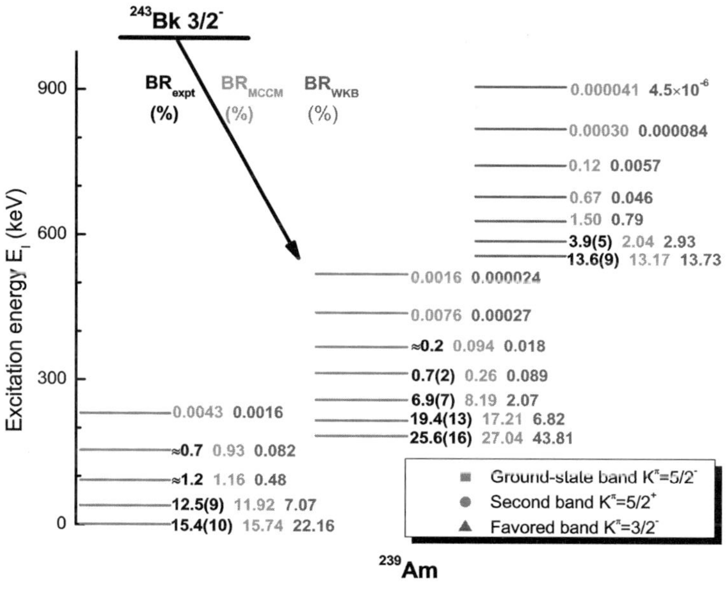

Fig. 1. (Color online) Comparison of the calculated branching ratios with the experimental data for the α decay of ^{243}Bk from its $3/2^-$ ground state to the three rotational bands in ^{239}Am. The experimental data, MCCM results, and WKB results are, respectively, shown in black, red, and blue colors.

In some cases of odd-mass nuclei, the favored rotational band in daughter nuclei is located at rather high excitation energies so that the α transitions to the favored band are not preferential. As a result, the component to the low-lying rotational bands is dominant or comparable with respect to that to the favored band.[5] The representative example is the α decay of ^{243}Bk from its ground state $3/2^-$ to three rotational bands in ^{239}Am (the ground-state rotational band $K^\pi = 5/2^-$, the second rotational band $K^\pi = 5/2^+$, and the favored rotational band $K^\pi = 3/2^-$). Fig. 1 shows the MCCM results for ^{243}Bk, compared with the available experimental data. For comparison, the WKB barrier penetration approach is also used to calculate the branching ratios, where the coupling effects among decay channels are ignored. As one can see, the MCCM results show better agreement with the available data than the WKB results. For the high-lying members of the rotational band, there are rare experimental data on the branching ratios and the MCCM seems to yield the larger values than the WKB approach. New measurements of these branching ratios in present facilities would be most welcome to discern the validity and reliability of the two different approaches.

In these calculations, the mixing of the rotational bands is out of this investigation and the α-preformation factor P_α is assumed to vary with the rotational band. By dividing the experimental α-decay widths by the computed results, one obtains the α-preformation factors for the ground-state and second bands with respect to the favored band, $P_\alpha^{\text{g.s.}}/P_\alpha^{\text{favored}}$ and $P_\alpha^{\text{second}}/P_\alpha^{\text{favored}}$. The WKB calculations yield the relative values of 2.08×10^{-7} and 5.77×10^{-5}, while the MCCM calculations yield the values of 0.0151 and 0.102. As one would expect, the α-preformation factor for the favored band is the largest one among them. There is a clear discrepancy between the WKB and MCCM results. The WKB calculations exhibit a rapid variation of the P_α factor. To be specific, the P_α factor is decreased by about five orders of magnitude from the favored to second band, and it is then decreased by about two orders of magnitude from the second to ground-state band. By contrast, the MCCM results for the P_α factor change smoothly with the rotational bands, by roughly one order of magnitude. This is worth further investigations, since microscopic evaluations of α-preformation factors are still an open problem for nuclear structure theory.[27]

Acknowledgments

This work is supported by the National Natural Science Foundation of China (Grant Nos. 10735010, 10975072, 11035001, 11120101005, and 11375086), by 973 National Major State Basic Research and Development of China (Grants Nos. 2013CB834400 and 2010CB327803), by a Project Funded by the Priority Academic Programme Development of JiangSu Higher Education Institutions (PAPD), by the Research Fund of Doctoral Point (RFDP; Grant No. 20100091110028), by the Science and Technology Development Fund of Macau (Grant No. 068/2011/A), and by a China Postdoctoral Science Foundation funded project (Grant Nos. 2013M530246 and 2014T70499).

References

1. A. N. Andreyev *et al.*, *Nature (London)* **405**, 430 (2000).
2. R. D. Page *et al.*, *Phys. Rev. C* **53**, 660 (1996).
3. A. Lopez-Martens *et al.*, *Nucl. Phys. A* **852**, 15 (2011).
4. F. P. Heßberger *et al.*, *Eur. Phys. J. A* **48**, 75 (2012).
5. I. Ahmad, J. P. Greene, F. G. Kondev *et al.*, *Phys. Rev. C* **87**, 054328 (2013).
6. Dongdong Ni, Zhongzhou Ren, Tiekuang Dong, and Chang Xu, *Phys. Rev. C* **78**, 044310 (2008).
7. Yuejiao Ren and Zhongzhou Ren, *Phys. Rev. C* **85**, 044608 (2012).
8. Chang Xu and Zhongzhou Ren, *Phys. Rev. C* **73**, 041301(R) (2006); *Phys. Rev. C* **74**, 014304 (2006).
9. N. G. Kelkar and H. M. Castañeda, *Phys. Rev. C* **76**, 064605 (2007).
10. J. C. Pei, F. R. Xu, Z. J. Lin, and E. G. Zhao, *Phys. Rev. C* **76**, 044326 (2007).
11. L. L. Li, S. G. Zhou, E. G. Zhao *et al.*, *Int. J. Mod. Phys. E* **19**, 359 (2010).
12. V. I. Furman, S. Holan, S. G. Kadmensky, and G. Stratan, *Nucl. Phys. A* **226**, (1976) 131.
13. T. Fliessbach, H. J. Mang, and J. O. Rasmussen, *Phys. Rev. C* **13**, 1318 (1976).
14. B. Sahu, *Phys. Rev. C* **78**, 044608 (2008).
15. Dongdong Ni and Zhongzhou Ren, *Phys. Rev. C* **80**, 051303(R) (2009); *Phys. Rev. C* **81**, 024315 (2010).
16. K. Wildermuth and Y. C. Tang, *A Unified Theory of the Nucleus*, (Academic Press, New York, 1997).
17. V. Yu. Denisov and A. A. Khudenko, *Phys. Rev. C* **80**, 034603 (2009).
18. Y. Z. Wang, *et al.*, *Phys. Rev. C* **79**, 014316 (2009).
19. K. P. Santhosh, S. Sahadevan, and J. G. Joseph, *Nucl. Phys. A* **850**, 34 (2011).
20. S. Peltonen, D. S. Delion, and J. Suhonen, *Phys. Rev. C* **78**, 034608 (2008).
21. Dongdong Ni and Zhongzhou Ren, *Phys. Rev. C* **81**, 064318 (2010).
22. Dongdong Ni and Zhongzhou Ren, *Phys. Rev. C* **83**, 067302 (2011).
23. NNDC of the Brookhaven National Laboratory, http://www.nndc.bnl.gov
24. Dongdong Ni and Zhongzhou Ren, *Phys. Rev. C* **86**, 054608 (2012).
25. Dongdong Ni and Zhongzhou Ren, *Phys. Rev. C* **87**, 027602 (2013).
26. G. Audi, F. G. Kondev, M. Wang *et al.*, *Chinese Phys. C* **36**, 1157 (2012).
27. R. G. Lovas, *et al.*, *Phys. Rep.* **294**, 265 (1998).

SYSTEMATIC CALCULATIONS ON HALF-LIVES OF VARIOUS CLUSTER EMISSIONS IN TRANS-LEAD NUCLEI

Yibin Qian[1,2,a] and Zhongzhou Ren[2,3,b]

[1] *Department of Applied Physics, Nanjing University of Science and Technology, Nanjing 210094, China*
[2] *Department of Physics, Nanjing University, Nanjing 210093, China*
[3] *Center of Theoretical Nuclear Physics, National Laboratory of Heavy-Ion Accelerator, Lanzhou 730000, China*
[a] *E-mail: qyibin@njust.edu.cn*
[b] *E-mail: zren@nju.edu.cn*

We systematically investigate cluster radioactivity in trans-lead nuclei within the density-dependent cluster model. The cluster-daughter potential is constructed by the double-folding integral of density distributions of daughter nuclei and clusters and the effective nucleon-nucleon interactions. The half-life of cluster emission is then obtained by the modified two-potential approach plus a well defined cluster-preformation factor based on the available experimental facts. It is found that the calculated results are in good agreement with the experimental data, including α decay half-lives in this region. This indicates that α decay and cluster radioactivity are described in unified framework.

Keywords: Cluster radioactivity; half-lives; cluster model.

1. Introduction

Cluster radioactivity is the spontaneous fission of charged cluster heavier than α particle but lighter than fission fragments, which was first predicted by Săndulescu, Poenaru and Greiner in 1980.[1] Subsequently, Rose and Jones in 1984 experimentally observed ^{14}C emission from ^{223}Ra,[2] confirming this new kind of radioactivity. Since then, other emissions with light to heavy (^{14}C to ^{34}Si) clusters, decaying from various heavy nuclei to ^{208}Pb and its neighboring nuclei, have been observed. The cluster radioactivity of ^{34}Si emission from ^{238}U and ^{14}C and ^{15}N emissions from ^{223}Ac were reported in the recent experiments.[3,4] Very strikingly, it is of great interest to denote that the concept of cluster radioactivity has been changed by allowing the emissions with $Z_c > 28$ from superheavy parents in very recent studies.[5] Considering that the cluster emission is located between α decay and nuclear fission, there are two directions in explaining the exotic process, i.e., the fissionlike treatment[1,5–10] and the traditional α decay approach.[11–19] In the preformed cluster model (PCM),[12,13] the stationary Schrödinger equation for the dynamical flow of mass and charge is solved to evaluate the cluster-preformation probability. The generalized density-dependent cluster model, along with a simple formula of cluster-preformation factor, has been recently extended to give a good description of cluster radioactivity in heavy emitters.[16] As well, there are several phenomenological investigations on unified analytical expressions for half-lives of α decay and cluster radioactivity.[20–22]

In previous studies, we have combined the modified two-potential approach with the cluster model to perform systematic calculations on α decay half-lives for a wide range of nuclei.[23,24] Encouraged by the good agreement between theory and experiment for these α emitters, we hope to achieve a unified description of α decay and cluster radioactivity, which are physically similar processes triggered by the quantum tunneling effect. In this paper, a brief description of the theoretical framework is presented in Section 2. In Section 3, numerical results of cluster emitters in the trans-lead region are presented, including several α decaying nuclei. A summary is given in Section 4.

2. Theoretical Framework

Given that the emitted cluster interacts with the residual daughter nucleus, the nuclear and Coulomb potentials are constructed in the double folding model:[25,26]

$$V_{N or C}(\mathbf{r}) = \int\int d\mathbf{r_1}\, d\mathbf{r_2}\, \rho_1(\mathbf{r_1})\rho_2(\mathbf{r_2})v(\mathbf{s} = |\mathbf{r_2} + \mathbf{r} - \mathbf{r_1}|), \quad (1)$$

where $v(\mathbf{s})$ is the effective M3Y-Reid-type nucleon-nucleon interaction and the standard Coulomb proton-proton interaction, respectively for the nuclear potential and the Coulomb term. The density distributions $\rho_{1,2}$ of the cluster and the residual daughter nucleus are curved in the Fermi form.[15,16,26] Sequentially, the total interaction potential consisting of the nuclear and Coulomb components is given by

$$V(r) = \lambda V_N(r) + V_C(r), \quad (2)$$

where λ is the renormalization factor of the nuclear potential. Within the two-potential approach, the total potential is then divided into the "inner" part and the "outer" part by a separation radius R, which is taken reasonably inside the potential barrier. Subsequently, the radial Schrödinger equation for the relative motion between the cluster and the daughter nucleus, is numerically solved under the inner potential for the bound-state wavefunction $\phi_{n\ell j(r)}$ vanishing sharply exponentially from the separation radius R.[23,24] During the computing process, the λ factor is adjusted to reproduce the decay energy Q, and the special quantum number n (namely the number of internal nodes in the bound solution) is restricted by the Wildermuth condition[27] to reflect the main effect of the Pauli principle. Once the bound-state wave function $\phi_{n\ell j}$ is obtained, one can achieve the decay width (see Refs. 23,24 and references therein)

$$\Gamma = \frac{\hbar^2 k}{\mu}\left[\frac{\phi_{n\ell j}(\bar{r})}{G_\ell(k\bar{r})}\right]^2, \tag{3}$$

where $k = \sqrt{2\mu Q}/\hbar$, ℓ is the angular momentum carried by the cluster, and G_ℓ is the irregular Coulomb wave function. The value of $\bar{r}$, smaller than R_0 is chosen well in such a way that the total potential $V(r)$ can be well approximated by the repulsive component (i.e., the attractive part vanishes) for $r \geq \bar{r}$. Then the half-life for cluster emissions is related to the decay width as

$$T_{1/2} = \frac{\hbar \ln 2}{P_c \Gamma}. \tag{4}$$

The cluster-preformation factor P_c, measuring the probability for the formation of the cluster at nuclear surface, is indispensable to calculate the absolute half-life of cluster radioactivity. In view of the fact that the preformation factor of a heavier cluster should be associated with the size of the cluster and the daughter nucleus, P_c can be depicted in the following way:

$$\log_{10} P_c = a\sqrt{\mu}(Z_c Z_d)^{1/2} + b, \tag{5}$$

where Z_c and Z_d, respectively, denote the atomic numbers of the cluster and the residual daughter nucleus. In this study, the parameters a and b are taken the same as in Ref. 16. Within this choice, the α-preformation factors of ^{212}Po and ^{211}Po are, respectively, 0.23 and 0.012, which agree well with the microscopically calculated results[28] and previous studies on exotic α decays around the shell closure $N = 126$.[23]

3. Numerical Results and Discussions

Following the above procedure, we have systematically calculated half-lives of cluster radioactivity in trans-lead nuclei, including α emitters decaying to the double magic nucleus ^{208}Pb or its neighboring nuclei. For comparison, we also employ the unified formula for half-lives of α decay and cluster radioactivity previously proposed by our group,[21]

$$\log_{10} T_{1/2} = c\sqrt{\mu} Z_c Z_d Q^{-1/2} + d\sqrt{\mu}(Z_c Z_d)^{1/2} + e. \quad (6)$$

In the present calculations, the decay energies Q are obtained as follows:

$$Q = M_p - M_d - M_c, \quad (7)$$

Table 1. Comparison of the present results with the experimental data and those in the analytical formula (6) for various cluster emissions in the trans-lead region (in seconds).

Transition	Q (MeV)	$\log_{10} P_c$	$\log_{10} T_{1/2}^{\rm expt}$	$\log_{10} T_{1/2}^{\rm calc}$	$\log_{10} T_{1/2}^{\rm form}$
^{210}Pb→^{206}Hg+^{4}He	3.792	-0.62	16.57	16.24	15.28
^{211}Bi→^{207}Tl+^{4}He	6.751	-1.91	2.18	1.84	1.10
^{211}Po→^{207}Pb+^{4}He	7.594	-1.92	-0.28	-0.10	-1.30
^{212}Po→^{208}Pb+^{4}He	8.954	-0.63	-6.52	-6.58	-6.49
^{214}Po→^{210}Pb+^{4}He	7.833	-0.63	-3.78	-3.57	-3.52
^{221}Fr→^{207}Tl+^{14}C	31.29	-4.75	14.56	15.02	15.27
^{221}Ra→^{207}Pb+^{14}C	32.40	-4.78	13.39	13.89	14.09
^{222}Ra→^{208}Pb+^{14}C	33.05	-3.49	11.22	10.89	11.35
^{223}Ra→^{209}Pb+^{14}C	31.83	-4.78	15.05	14.90	15.19
^{224}Ra→^{210}Pb+^{14}C	30.54	-3.49	15.87	15.70	16.31
^{226}Ra→^{212}Pb+^{14}C	28.20	-3.49	21.20	20.82	21.50
^{228}Th→^{208}Pb+^{20}O	44.72	-5.00	20.73	20.30	21.80
^{230}Th→^{206}Hg+^{24}Ne	57.76	-6.13	24.63	24.60	24.66
^{231}Pa→^{208}Pb+^{23}F	51.87	-7.03	26.02	25.29	26.09
^{231}Pa→^{207}Tl+^{24}Ne	60.41	-7.46	22.89	23.04	23.34
^{230}U →^{208}Pb+^{22}Ne	61.39	-5.95	19.57	19.67	20.09
^{232}U →^{208}Pb+^{24}Ne	62.31	-6.22	20.39	20.30	20.32
^{233}U →^{209}Pb+^{24}Ne	60.49	-7.51	24.84	24.36	24.69
^{233}U →^{208}Pb+^{25}Ne	60.73	-7.63	24.84	24.70	25.02
^{234}U →^{210}Pb+^{24}Ne	58.83	-6.22	25.93	25.47	25.90
^{234}U →^{208}Pb+^{26}Ne	59.42	-6.47	25.93	25.86	26.49
^{234}U →^{206}Hg+^{28}Mg	74.11	-7.31	25.53	25.31	25.10
^{236}Pu→^{208}Pb+^{28}Mg	79.66	-7.41	21.52	20.91	20.48
^{238}Pu→^{210}Pb+^{28}Mg	75.91	-7.42	25.70	25.78	25.80
^{238}Pu→^{208}Pb+^{30}Mg	76.81	-7.66	25.70	25.24	25.85
^{238}Pu→^{206}Hg+^{32}Si	91.19	-8.47	25.28	25.72	25.19
^{242}Cm→^{208}Pb+^{34}Si	96.51	-8.83	23.15	23.30	23.00

where the masses of parent nuclei, daughter nuclei and clusters, namely M_p, M_d, and M_c, are all taken from the recent mass table.[29] Moreover, decay products of even-even parent nuclei are usually even-even, resulting in the zero angular momentum carried by the emitted cluster (i.e., favored transitions). For odd-A nuclei, the situation is a little more complicated, and the ℓ value of the cluster is chosen as the smallest one obeying the spin-parity selection rule.

Table 1 presents the detailed results of various cluster emissions from even-even and odd-A trans-lead nuclei. The first column denotes the transitions, the decay energy Q is in the second column, and the third column presents the calculated preformation factors in Eq. (5). The experimental half-lives, calculated results in this model and the formula (6) are respectively listed in the last three columns. As can be seen, although there is a large measuring error in the present data for many cluster emitters due to only a few events observed in the experiment,[3] our calculated results are found to well agree with the experimental data within a mean factor of about 2. Meanwhile, the two theoretical results obtained in the present framework and the unified formula (6) are consistent with each other. It is interesting to denote that the P_c value of odd-A emitters are in general smaller by one order of magnitude than that of even-even ones for one given emitter. This may be caused by the structural hindrance from the different states of odd-A nucleus and its daughter nucleus, and the block effect of the odd nucleon.

4. Summary

In summary, α and heavier cluster radioactivities of trans-lead nuclei are studied in the framework of the modified two-potential approach combining with the density-dependent cluster model. The bound-state solution of the Schrödinger equation is performed to get the decay width, leading to the final half-lives along with the reasonable assumption of the cluster-preformation factor. The present calculated results agree well with the available measured values, and are consistent with other theoretical evaluations.

Acknowledgments

This work is supported by the National Natural Science Foundation of China (Grant Nos. 11035001, 10975072, 10735010, 11375086, and 11120101005), by the 973 National Major State Basic Research and Development of China (Grant Nos. 2010CB327803 and 2013CB834400), by CAS Knowledge Innovation Project No. KJCX2-SW-N02, by Research Fund of Doctoral Point (RFDP), Grant No. 20100091110028, by a Project Funded by the Priority Academic Programme Development of JiangSu Higher Education Institutions (PAPD).

References

1. A. Săndulescu, D. N. Poenaru and W. Greiner, *Sov. J. Part. Nucl.* **11**, 528 (1980).
2. H. J. Rose and G. A. Jones, *Nature* (London) **307**, 245 (1984).
3. R. Bonetti and A. Guglielmetti, *Romanian Rep. Phys.* **59**, 301 (2007).
4. A. Guglielmetti *et al.*, *J. Phys.: Conf. Ser.* **111**, 012050 (2008).
5. D. N. Poenaru, R. A. Gherghescu and W. Greiner, *Phys. Rev. Lett.* **107**, 062503 (2011); *Phys. Rev. C* **85**, 034615 (2012).
6. W. Greiner, M. Ivascu, D.N. Poenaru and A. Sandulescu, *Z. Phys. A* **320**, 347 (1985).
7. Yi-Jin Shi and W. J. Swiatecki, *Phys. Rev. Lett.* **54**, 300 (1985).
8. G. Royer and R. Moustabchir, *Nucl. Phys. A* **683**, 182 (2001).
9. K. P. Santhosh, R. K. Biju and S. Sahadevan, *Nucl. Phys. A* **838**, 38 (2010).
10. Z. Sheng, D. Ni and Z. Ren, J. Phys. G: Nucl. Part. Phys. **38**, 055103 (2011).
11. R. Blendowske and H. Walliser, *Phys. Rev. Lett.* **61**, 1930 (1988).
12. S. S. Malik and R. K. Gupta, *Phys. Rev. C* **39**, 1992 (1989).
13. B. Singh, S. K. Patra and R. K. Gupta, *Phys. Rev. C* **82**, 014607 (2010).
14. B. Buck and A. C. Merchant, *Phys. Rev. C* **39**, 2097 (1989).
15. Zhongzhou Ren, Chang Xu and Zaijun Wang, *Phys. Rev. C* **70**, 034304 (2004).
16. Dongdong Ni and Zhongzhou Ren, *Phys. Rev. C* **82**, 024311 (2010).
17. Yuejiao Ren and Zhongzhou Ren, *Phys. Rev. C* **85**, 044608 (2012).
18. H. F. Zhang, J. M. Dong, G. Royer, W. Zuo and J. Q. Li, *Phys. Rev. C* **80**, 037307 (2009).
19. S. N. Kuklin, G. G. Adamian and N. V. Antonenko, *Phys. Rev. C* **71**, 014301 (2005).
20. D. S. Delion, *Phys. Rev. C* **80**, 024310 (2009).
21. Dongdon Ni, Zhongzhou Ren, Tiekuang Dong and Chang Xu, *Phys. Rev. C* **78**, 044310 (2008).
22. D. N. Poenaru, R. A. Gherghescu and W. Greiner, *Phys. Rev. C* **83**, 014601 (2011).
23. Yibin Qian, Zhongzhou Ren and Dongdong Ni, *J. Phys. G: Nucl. Part. Phys.* **38**, 015102 (2011).
24. Yibin Qian, Zhongzhou Ren and Dongdong Ni, *Phys. Rev. C* **83**, 044317 (2011).
25. A. M. Kobos, B. A. Brown, P. E. Hodgson, G. R. Satchler and A. Budzanowski, Nucl. Phys. A 384, 65 (1982).
26. Chang Xu and Zhongzhou Ren, Phys. Rev. C 73, 041301(R) (2006); 74, 014304 (2006).
27. K. Wildermuth and Y. C. Tang, *A Unified Theory of the nucleus* (Academic Press, New York, 1997).
28. K. Varga, R. G. Lovas and R. J. Liotta, *Phys. Rev. Lett.* **69**, 37 (1992).
29. G. Audi, F. G. Kondev, M. Wang, B. Pfeiffer, X. Sun, J. Blachot and M. MacCormick, *Chin. Phys. C* **36**, 1157 (2012); M. Wang, G. Audi. A. H. Wapstra, F. G. Kondev, M. MacCormick, X. Xu and B. Pfeiffer, *Chin. Phys. C* **36**, 1603 (2012).

NEW GAMMAS AND LEVELS OF ^{76}Se OBSERVED IN THE ^{76}Br(β^++EC)^{76}Se DECAY

S. F. Shen*

Key Laboratory of Neutronics and Radiation Safety, Institute of Nuclear Energy Safety Technology, Chinese Academy of Sciences,
Hefei, Anhui 230031, China,
** E-mail: shuifa.shen@fds.org.cn*
www.inest.cas.cn

Y. P. Yan

School of Physics, Suranaree University of Technology,
Nakhon Ratchasima 30000, Thailand
E-mail: yupeng@sut.ac.th

The decay of bromine-76 is studied by means of γ-ray spectroscopy. Bromine-76 is produced via two reactions: ^{75}As(α, 3n)^{76}Br and ^{76}Se(p, n)^{76}Br. The results of these experiments verify the previously reported levels of ^{76}Se. In addition to the previously described transitions and levels, 37 new γ-rays and 15 new energy levels are found for the first time. In addition, in order to study the deformation of ^{76}Se which lies in transitional region between strongly deformed and spherical nuclei, we calculate the total Routhian surfaces (TRS) of positive-parity states by cranking shell model formalism.

Keywords: γ-ray; level; coincidence; total Routhian surface.

1. Introduction

A strongly deformed structure in ground state was not expected in the $A = 70$ region until nuclear-shape coexistence was first discovered in ^{72}Se by Hamilton *et al.*[1–3] With the beginning of ^{72}Se work, a new very strong deformation region in $Z = 38$ was discovered. The experimental data show that the near-spherical ground states, the low-lying well-deformed 0^+ state and deformed bands coexist in ^{72,74}Se. Up to now, the level structure from ^{76}Br decay have been studied with Ge(Li) detectors.[4–6] Detailed properties of the level structure of doubly even nuclei in this mass region are currently needed to provide systematic knowledge for the further development of nuclear models. The present work is to study the decay of ^{76}Br with a high-purity Ge Compton-suppression spectrometer, and a HPGe-HPGe-T coincidence spectrometer.

2. Experimental Details

2.1. *Source preparation*

Two methods of source preparation are used to make ^{76}Br. First, the sources of ^{76}Br are produced at the cyclotron at Shanghai Institute of Applied Physics via the reaction ^{75}As(α, 3n)^{76}Br. The α beam energy used is 32 MeV. The target monocrystal GaAs is welded onto the copper backing. The radioactive target material is put into a high temperature distiller. The radioactive Br distilled from the target is dissolved in 0.005 mol NaOH solution. The solution is dropped onto a lucite source holder. After drying, the source is covered with another lucite film. In the second method, the ^{76}Br source is produced via the ^{76}Se(p, n)^{76}Br reaction using a 16 MeV proton beam from the same cyclotron and with selenium as the target with an enrichment of 99.99%. After irradiation, approximately 23 h is allowed for short-lived isotopes to decay out before measurements are performed.

2.2. *γ-ray singles spectra measurements*

The Compton-suppressed spectrometer consisted of a 130 cm^3 HPGe detector and 6 NaI detectors(made by Canberra Company) is employed to study the singles γ-ray spectra of ^{76}Br decay. The resolution of system is 1.9 keV at 1332 keV. No Compton-suppressed measurements are carried out at the same time. The total background counts are 0.7 cps (Compton-suppressed) and 1.3 cps (none Compton-suppressed)in the energy region of 30 to 2500 keV. Spectra are taken over a period of several days to ensure that the observed photopeaks are decaying with the half-life of ^{76}Br. The spectra are obtained with and without a lead absorber between the source and HPGe detector. This lead plate is used as an absorber to suppress β rays. Thus, both energies and relative intensities of the gamma rays can be determined more accurately. Energies and intensities are determined by energy calibration curve and efficiency curve made with the energy standard ^{152}Eu, ^{56}Co sources.

2.3. *Coincidence spectra*

A series of coincidence measurements are carried out in order to establish the relation among the γ-rays observed. The experiments are performed with two HPGe detectors coupled to three-parameter γ-γ-T data acquisition system. The above two HPGe detectors are employed at 90° to each other. The resolving time of the coincidence circuit used is 20 ns when gated by the ^{60}Co full energy peaks. For low-energy transitions it can be much larger. One of the detectors is a coaxial detector of 115 cm^3 HPGe with 0.5 mm beryllium absorbing layers, and FWHM=1.9 keV at 1332 keV of ^{60}Co. The other is a coaxial detector of 232 cm^3 HPGe. Its FHWM is 2.0 keV at 1332 keV. Total 2.1×10^7 coincidence events are recorded in event by event.

3. Decay Scheme

The decay scheme constructed from the present experiments is shown in Fig. 3 of our preceding publication.[7] The total γ-ray transitions measured are 138, of them 37 are observed for the first time. 120 γ-rays are placed in the level scheme with 46 excited states, including 15 new levels. The energies of the new excited states are 1791.31, 2812.0, 2817.02, 3220.3, 3269.29, 3312.8, 3527.6, 3637.4, 3915.0, 3965.7, 4205.0, 4411.7, 4453.5, 4489.3 and 4778.5keV. The new γ-rays in the level scheme are seen in coincidence spectra. A new level is confirmed when at least one γ-ray depopulating or populating the level is found in coincidence spectra. The level energies are obtained by the energy sum relations of the γ-ray transitions. From the γ-ray and internal-conversion electron intensities (only the K-conversion electron is considered),[5] the β^++EC branching to excited levels are calculated from the $I_{\gamma+ce}$ intensity balance. Finally, based on the (6±1)% of β^++EC-decay to the ground state of ^{76}Se,[8] and the 4963±9 keV Q-value for β^++EC-decay,[9] the log*ft* values and the intensities of the β^+ and electron-capture branches to the ^{76}Se levels are calculated by using the ENSDF(the Evaluated Nuclear Structure Data File) analysis and checking program offered by NNDC(the National Nuclear Data Center) at Brookhaven National Laboratory. The ground state spin parity of ^{76}Br is 1^-.[5] The result is presented in the decay scheme (Fig. 3 of Ref. [7]). The β^+(EC) branching ratio and $I_{\gamma+ce}$ are absolute in Fig. 3 of Ref. [7]. Dashed arrows denote the placements of the γ-ray transitions are uncertain.

4. Deformation of Nucleus ^{76}Se

The even-even isotopes of selenium ($Z = 34$) lie in the mass region between the semi-closed shell at $N = 40$ and the magic $N = 50$ and their low-lying states have resisted description within the framework of either the phenomenological or the microscopic models. We may expect, since these isotopes are nearly spherical, that their level structure should show vibrational features. Lecomte *et al.*.[10] suggested, as a possible explanation for the particular features of the selenium isotopes, that these nuclei show an extreme softness in the γ direction with significant triaxial deformation.

In the experiment carried out by Wells *et al.*,[11] at an energy slightly higher than the ground state, the γ-band is observed (the 2^+ and 3^+ states of γ-band correspond to the member of two-phonon and three-phonon multiplets, respectively, whereas the 0^+, 2^+ and 4^+ states of ground-state band correspond to zero-phonon, one-phonon and the member of two-phonon multiplets, respectively). This proves ^{76}Se is deformed in the low-energy states.

Total Routhian surface (TRS) calculations by means of the pairing-deformation-frequency self-consistent cranked shell model have been carried out for positive-parity states in ^{76}Se in order to analyze the deformation of this nucleus in detail. Samples of total Routhian surfaces are presented in figure 1 at specific rotational frequencies, $\hbar\omega$=0.00, 0.20, 0.40, and 0.60 MeV, corresponding to $I\sim$(0-10.5)$\hbar$, and the energy difference between neighboring contours is 200 keV. It is noted in the TRS calculations, at rotational frequency $\hbar\omega$=0, which corresponds to ground state in ^{76}Se, the quadrupole deformation parameter β_2 is 0.245 (see the upper left part of figure 1), which indicates it is a deformed nucleus and the shape is oblate. It should be pointed out here that when the nucleus is static, the TRS is reflection symmetric about γ=0° and γ=-60° for nonrotational states. When $\hbar\omega$=0.341 MeV, the absolute minimum moves to β_2=0.320, $\gamma=-29.582°$ where the maximum triaxial deformation happens. Then this minimum persists up to rotational frequency $\hbar\omega$ $\sim$0.8MeV. However, its asymmetry continues to decrease, and it becomes steeper (stiffer) with increasing rotational frequency. Moreover, while ω is zero or smaller (namely the nucleus does not rotate or rotates slowly), it is rather soft around the absolute minimum point in the γ direction, which indicates it is a typical γ-soft vibrational nucleus (at least at the low-spin region of I$\leq$10). This is consistent with the experimental result carried out by Wells et al.[11] as mentioned above, who observed only the γ-band. It turns out that the result of our calculation is very close to those evaluated in the work of Möller *et al.*,[12] in which the deformation of ground state in ^{76}Se has been calculated to be $\beta_2=-0.241$ using a finite-range droplet macroscopic model and a folded-Yukawa single-particle microscopic model. In fact, these two values are not significantly different from one another, but are different from those

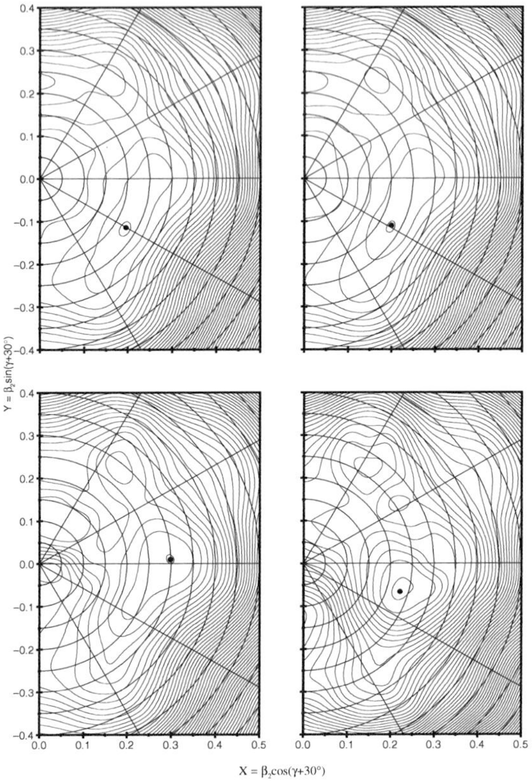

Fig. 1. Calculated TRS's for ^{76}Se positive-parity states, at $\hbar\omega$=0.00 MeV (upper left) and 0.20 MeV (upper right), 0.40 (lower left) and 0.60 MeV (lower right) corresponding to $I\sim$(0-10.5)$\hbar$. The black dot indicates the lowest minimum, and the energy difference between neighboring contours is 200 keV.

calculated by Subber *et al.*[13] who calculated the potential energy surface by the dynamic deformation model (DDM). It is only a saddle point (not a minimum point) at the oblate shape position calculated in the framework of DDM in the work of Subber *et al.* The DDM can not reproduce the sign of quadrupole moment $Q(2_1^+)$ in any of the Se isotopes. In our previous work,[7] we constructed the shell model basis at the deformation ε_2=0.30 proposed by Subber *et al.*,[13] and calculated the positive-parity yrast band in the framework of projected shell model (PSM). Now we adopt quadrupole deformation parameter β_2=−0.245 obtained in the present work to calculate them once again, the conclusion is well consistent with Ref. [7]. Obviously the energies are overpredicted in ^{76}Se. We emphasize here that all these states have been obtained by a single diagonalization, without any adjustment for individual states. It seems we should include the triaxial deformation parameter γ as the result of TRS calculations has already shown its importance. It is expected that this will lower the level energies.[14]

References

1. D. A. Bromley, *Treatise on Heavy-ion Physics*, Vol. 8 (Plenum Press, London, 1989).
2. J. H. Hamilton, A. V. Ramayyga, W. T. Pinkston *et al.*, *Phys. Rev. Lett.* **32** (1974) 239.
3. J. H. Hamilton, H. L. Crowell, R. L. Robinson *et al.*, *Phys. Rev. Lett.* **36** (1976) 340.
4. B. Singh and D. A. Viggars, *Nucl. Data Sheet* **42** (1984) 233.
5. B. Singh, *Nucl. Data Sheet* **74** (1995) 63.
6. I. M. Ladenbauer-Bellis, H. Bakhru and R. Bakhru, *Can. J. Phys.* **49** (1971) 54.
7. S. Shen, Y. Li, W. Huang *et al.*, *J. Phys. Soc. Jpn.* **73** (2004) 1180.
8. B. S. Dzhelepov, A. G. Dmitriev and N. N. Zhukovskii, *Izv. Akad. Nauk SSSR, Ser. Fiz.* **35** (1971) 2249; *Bull. Acad. Sci. USSR, Phys. Ser.* **35** (1972) 2043.
9. G. Audi and A. H. Wapstra, *Nucl. Phys.* **A565** (1993) 1.
10. R. Lecomte, P. Paradis, J. Barrette, M. Barrette, G. Lamoureux and S. Monaro, *Nucl. Phys.* **A284** (1977) 123.
11. J. C. Wells, Jr., R. L. Robinson, H. J. Kim, R. O. Sayer, R. B. Piercey, A. V. Ramayya, J. H. Hamilton and C. F. Maguire, *Phys. Rev.* **C22** (1980) 1126.
12. P. Möller, J. R. Nix, W. D. Myers and W. J. Swiatecki, *At. Data Nucl. Data Tables* **59** (1995) 185.
13. A. R. H. Subber, S. J. Robinson, P. Hungerford, W. D. Hamilton, P. Van Isacker, K. Kumar, P. Park, K. Schreckenbach and G. Colvin, *J. Phys.* **G13** (1987) 807.
14. X. Kang, S. Shen, J. Gu, Y. Yan, X. Wu, L. Zhu, S. Yan and T. Wen, *J. Phys. Soc. Jpn.* **80** (2011) 044201.

BANDHEAD ENERGIES IN ^{125}Cs

Ji Sun[1], Xue-yuan Hu[1], Ying-jun Ma[1], Yun-zuo Liu[1], Komatsubara Tetsuro[2], Furuno Kohei[2], Yu-hu Zhang[3], Wen-ping Zhou[4], and Shou-yu Wang[5]

[1] *College of Physics, Jilin University, Changchun 130012, China*

[2] *Institute of Physics and Tandem Accelerator Center, University of Tsukuba, Ibaraki 305-8577, Japan*

[3] *Institute of Modern Physics, Chinese Academy of Sciences, Lanzhou 730000, China*

[4] *Shenyang Institute of Engineering, Shenyang 110136, China*

[5] *Shandong Provincial Key Laboratory of Optical Astronomy and Solar-Terrestrial Environment,School of Space Science and Physics,Shandong University, Weihai 264209, China*

Excited states in ^{125}Cs have been studied with the fusion-evaporation-reaction ^{116}Cd(^{14}N,5n)^{125}Cs at 65 MeV beam energy, using the Nordball-multidetector-system at the Niels-Bohr-Institute in Denmark. The level scheme of ^{125}Cs was extended with the addition of more than 40 new γ-transitions. Moreover, the bandhead excitation energies of the previously known $g_{9/2}$ and $h_{11/2}$ bands were unambiguously corrected with plenty of hard evidence.

The nucleus ^{125}Cs, with three valence protons and eight valence neutrons outside the spherical ^{114}Sn core, is known to have sizeable quadrupole deformation ($\beta \approx 0.2$) at low spins and, as a consequence, exhibit predominantly collective excitations. In vicinity of the proton Fermi surface, there are the extruder orbitals originating from the $g_{7/2}$, $d_{5/2}$ and $g_{9/2}$ subshells and the unique-parity intruder orbitals from the $h_{11/2}$ subshell. From atomic-beam magnetic resonance[1] and from ^{125}Ba ε decay,[2] the ground state of ^{125}Cs is known[3] to have spin/parity of $1/2^+$ and a half-life of 46.7 min. On the other hand, spectroscopic studies using in-beam techniques revealed collectively rotating bands built on several different orbitals available near the proton Fermi surface.

As early as in 1979, U Garg et al,[4] utilizing the ^{118}Sn(^{10}B,3n) reaction, identified the $\pi h_{11/2}$, $\pi g_{9/2}$, $\pi g_{7/2}$ bands and established their decay paths down to the ground state. Later, J R Hughes et al,[5] utilizing the ^{110}Pd(^{19}F, 4n) reaction, confirmed the previous work by U Garg et al and extended the level scheme significantly. Then, the bandhead excitation energy of the $\pi h_{11/2}$ band determined by U Garg et al was confirmed again by M. Sugawara et al,[6] who made a lifetime measurement specially for the $\pi h_{11/2}$ bandhead and found that the $11/2^-$ state is an isomer with $T_{1/2} \approx 0.9$ms. Recently, K Singh et al[7] made a re-investigation following the ^{110}Pd(^{19}F, 4n) reaction. In addition to significant extension of the level scheme, they confirmed again the decay pattern established previously for the $\pi h_{11/2}$ bandhead. At the same time, based on comparison with adjacent odd-A Cs isotopes, they proposed a new possibility for the decay of the $\pi g_{9/2}$ bandhead, namely, it should decay down to states of the $\pi g_{7/2}$ band rather than to the states of the $\pi h_{11/2}$ band as established in earlier studies. However, the authors failed in finding direct evidence in support of their tentative idea. Instead, they attributed the failure to possible long-lifetime isomerism of the $\pi g_{9/2}$ bandhead. In the present work, we have re-investigated the ^{125}Cs nucleus and made significant extension of the level scheme. Moreover, there are serious discrepancies between this work and all previous studies about the bandhead energies in ^{125}Cs. Since reliable fundamental experimental results can be very crucial for discussing further topics, it is felt necessary to provide relevant arguments in great detail. Preliminary results of the present work were previously reported in Ref.[8]

In the current work, the nucleus ^{125}Cs was a by-product of a standard in-beam experiment aiming at ^{126}Cs.[9] An isotopically-enriched ^{116}Cd target was bombarded with 65 MeV ^{14}N beams, which were provided by the tandem-linac accelerator complex at Niels-Bohr-Institute in Denmark. The Nordball detector system, which consisted of 19 Compton-suppressed detectors and one LEPS (low-energy photon spectrometer) detector, was employed for recording the γ-ray spectra. Further experimental details can be found in Refs.[9,10]

The level scheme of ^{125}Cs deduced from this study is shown in Fig. 1. It preserves the major features of the level scheme established from previous studies[4,5,7] and contains more than 40 transitions newly observed from this work. In particular, newly-observed transitions in sequences 10 and 11 form two new bands of ^{125}Cs. Furthermore, as mentioned previously, there are substantial disagreements between this study and previous studies about the bandhead energies of the $\pi g_{9/2}$ and $\pi h_{11/2}$ bands (bands 1 and 5 in Fig. 1).

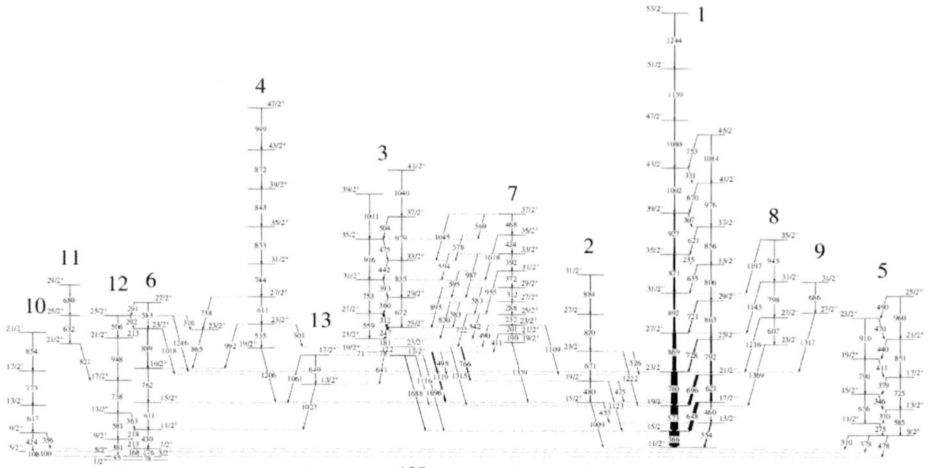

Fig. 1. The level scheme of ^{125}Cs established from present work.

Based on an observation of a 189 keV $E3$ transition from the $\pi h_{11/2}$ bandhead to a known state at 77 keV, together with an assumption that a 13 keV $M2$ transition exists between the $\pi h_{11/2}$ bandhead and a known state at 253 keV, the $\pi h_{11/2}$ bandhead was located at excitation energy of 266 keV by U Garg et al.[4] This assignment was confirmed or adopted by the later studies[5–7] and compiled into Nuclear Data Sheets.[3] In the present work, evading the difficulty to find those high-degree internal-conversion transitions, we have established several high-lying selection-rule-allowed γ-transitions, such as the 1018, 1246, 992, 1206, 1023, 649, 641, and 71 keV transitions, which link the high-lying bands 3 and 4 to the low-lying $\pi g_{7/2}$ and $\pi h_{11/2}$ bands. With the observation of so many cross-talking transitions, the $\pi h_{11/2}$ bandhead is found lying at excitation energy of 293 keV instead of the previously reported 267 keV. As proofs in support of our assignment, Fig. 2 presents two examples of coincidence spectra gating on the 1023 and 1206 keV linking transitions.

The $\pi g_{9/2}$ band was connected directly to the $\pi h_{11/2}$ bandhead through two linking transitions with energies of 275 and 584 keV in earlier studies by U. Garg et al[4] and J. R. Hughes et al.[5] Such an assignment is apparently questionable as no cross evidence is available. The later study by K Singh et al[7] re-assigned the 275 and 584 keV transitions as in-band transitions of the $\pi g_{9/2}$ band. At the same time, they presumed that the $\pi g_{9/2}$ band should de-excite into states of the $\pi g_{7/2}$ or $\pi d_{5/2}$ bands. Unfortunately,

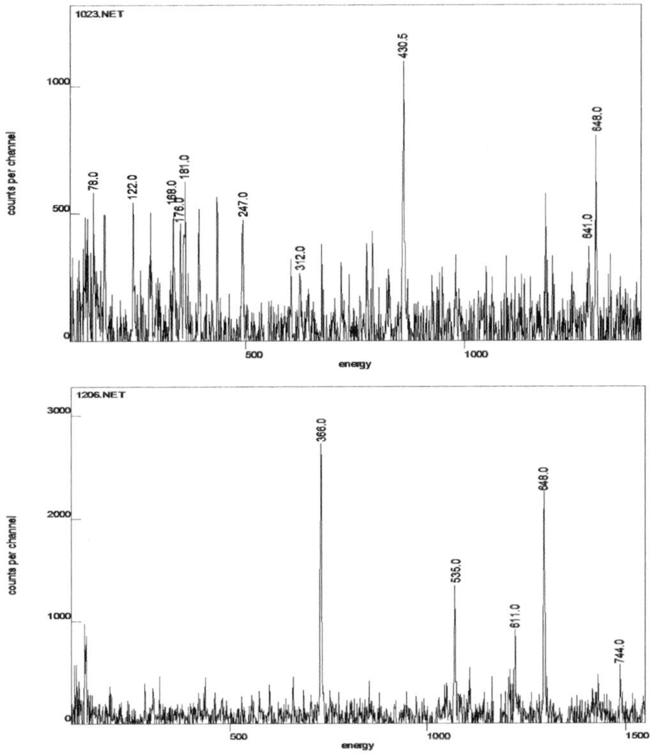

Fig. 2. Examples of coincidence spectra with gate set at (a) 1023 keV and (b) 1206 keV. Transition energies are labeled in keV. Unmarked peaks are contaminations.

their presumption was made solely according to systematics of odd-A Cs isotopes rather than direct experimental evidence and were not adopted in the recent compilation of nuclear data of ^{125}Cs.[3] In the present work, after careful analysis of the coincidence data, we have been able to establish the decay of the $\pi g_{9/2}$ bandhead (cf. Fig. 1). It decays into the lower-lying $7/2^+$, $5/2^+$ and $5/2^+$ states via three γ-transitions with energies of 310, 378 and 478 keV. Among the three linking transitions, the 310 and 378 keV transitions are doublet; transitions with nearly equal energies occur also within the $\pi g_{9/2}$ band. In spite of such a trouble, the observation of several linking transitions lends firm evidence to the decay pattern as established in Fig. 1. Fig. 3 presents two examples of coincidence spectra supporting our results. With the establishment of decay of the $\pi g_{9/2}$ bandhead, its excitation energy is found to be 563 keV. The $7/2^+$ and $5/2^+$ states fed by the 310 and 478 keV transitions are know from previous in-beam studies[4,5,7]

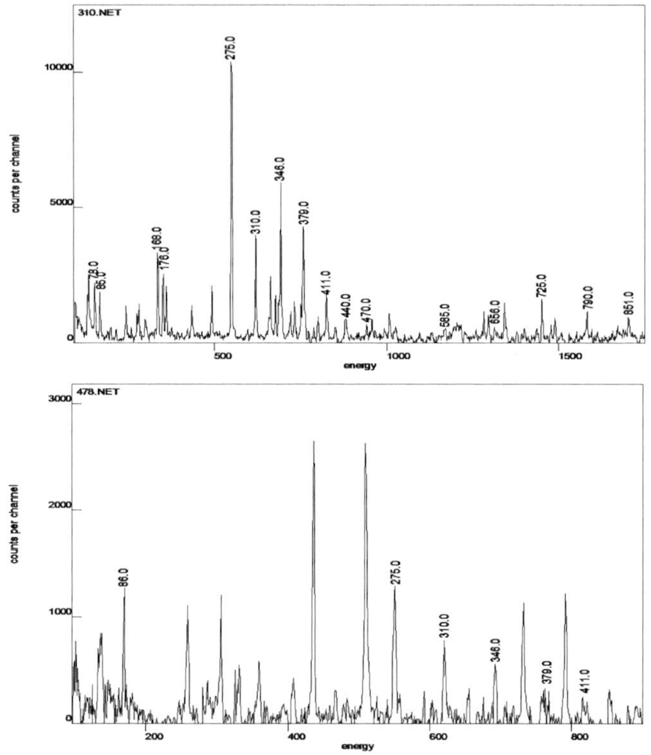

Fig. 3. Examples of coincidence spectra with gate set at (a) 310 keV and (b) 478 keV. Transition energies are labeled in keV. Unmarked peaks are contaminations.

while the state fed by the 378 keV transition has an excitation energy (185 keV) which is exactly equal to that of a state observed in the decay of ^{125}Ba.[2,3] Three possible spin values, i.e. 1/2, 3/2 and 5/2, were suggested[2,3] for this state. In view of the easiness in observing the 378 keV γ-transition from the $\pi g_{9/2}$ bandhead to this state, it should have spin/parity $5/2^+$. The band growing from this state is believed to be the signature partner of the previously known $\pi g_{7/2}$ band (band 6), as will be discussed in detail in a forthcoming paper.

In brief summary, excited states in ^{125}Cs have been studied using the $^{116}Cd(^{14}N, 5n)$ reaction. More than 40 new transitions have been added into its level scheme and some new transitions form two new collective bands of ^{125}Cs. The bandhead energies determined in previous studies have been revised based on direct and firm experimental evidence.

Acknowledgment

This work is supported in part by Research Fund for the Doctoral Program of Higher Education of China (20120061110008).

References

1. O. B. Dabbousi, M. H. Prior, H. A. Shugart, Phys. Rev. C. **3**, 1326 (1971)
2. R. Arlt, A. Jasinski, W. Neubert *et al.*, Acta Phys.Pol. B. **6**, 433 (1975).
3. J. Katakura, Nucl. Data Sheets **112**, 495 (2011).
4. U. Garg *et al.*, Phys. Rev. C **19**, 217 (1979).
5. J. R. Hughes *et al.*, Phys. Rev. C **44**, 2390 (1991).
6. M. Sugawara *et al.*, Eur. Phys. J. A **2**, 237 (1998).
7. K. Singh *et al.*, Eur. Phys. J. A 27 **27**, 321 (2006).
8. Y-J Ma *et al.*, Nucl. Phys. Rev. **21**, 331 (2004), (in Chinese).
9. T. Komatsubara *et al.*, Nucl. Phys. A **557**, 419c (1993).
10. S. Y. Wang *et al.*, Phys. Rev. C **75**, 037302 (2007).

NEW GLOBAL FORMULA OF THE NUCLEAR ELECTRIC QUADRUPOLE MOMENTS

Xiao-jun Sun*, Chun-xing Chen, Ning Wang

Department of Physics, Guangxi Normal University,
Guilin 541004, P. R. China
State Key Laboratory of Theoretical Physics, Institute of Theoretical Physics,
Chinese Academy of Sciences, Beijing 100190, P. R. China
**E-mail: sxj0212@gxnu.edu.cn*

In this work, a new global formula of nuclear electric quadrupole moments is explicitly derived with respect to the spin, charge and mass number, charge radius and deformation parameters. The deformation parameters extracted from the mass model WS4 and HFB27* are used to calculate the nuclear electric quadrupole moments at the ground states. The calculated results reproduce the measured data well for many isomers. The comparisons between the theoretical results and 527 measured data ($Z \geq 8, A \geq 16$) show that the root mean square deviations are 0.552 barn for WS4 and 0.657 barn for HFB27*, respectively. The results of this work are also compared with that of the microscopic shell-model calculations and the least-squares fitting to the experimental data.

Keywords: Nuclear electric quadrupole moment; charge radius; nuclear deformation.

1. Introduction

Nuclear electric quadrupole moment (NQM), which describes the effective non-spherical shape of nuclear charge distribution, is one of the important bulk properties of nuclei.[1] It plays an important role to understand matter structures,[2,3] nuclear models[4,5] and nucleon-nucleon interaction.[6–8]

Recently, the study of the NQM is one of the hot topics in nuclear physics, including exotic nuclei,[9] shape coexistence,[10] triaxiality deformation,[11] and shell evolution.[12] Furthermore, NQM can be considered as the local probes to obtain microscopic information on the tunneling motion in amorphous materials,[13] and it may induce the Cotton-Mouton effect in noble gas atoms.[14]

A large number of the available NQMs was tabulated by P. Pyykkö[15–17] and by N. J. Stone,[18–22] respectively. Furthermore, there are some updated

NQMs published in recent papers.[23–26] But there are only 540 measured NQMs of the ground state, and are greatly less than the number of the measured mass isotopes in AME2012[27] as shown in Fig. 1. So the theoretical predicts and as well as more experimental measurements, are necessary.

Most theoretical calculations of NQMs focus on the medium weight nuclei, using the shell model (such as 2p-1f shell,[28] f_5pg_9-shell[29] and 2p, $1f_{5/2}$, $1g_{9/2}$ shell,[30] et al.), the single-quasiparticle approximation of odd nuclei,[31] the approximation of non-interacting odd-odd near-magic nuclei[32] and the self-consistent theory of finite Fermi systems of spherical semi-magic nuclei.[33] In this work, a new global NQM formula is explicitly derived with respect to the spin, charge and mass number, charge radius and deformation parameters.

This paper is organized as follows. In the next section, we will explicitly give the new global NQM formula. In the 3rd section, the calculated results using the deformation parameters from the mass model WS4[34] and HFB27*[35] are compared with the measured data. The comparisons are also given between this work and other theoretical predictions for some medium weight nuclei. Finally, a summary is given in the last section.

2. Global Formula

Assuming an axially symmetric static deformation with the nuclear spin I along the symmetry axis, the measured NQM is expressed as

$$Q = \frac{I(2I-1)}{(I+1)(2I+3)} Q_0, \tag{1}$$

where Q_0 is the intrinsic electric quadrupole moment. It is obvious that Q=0 if I=0 or 1/2, and this case is verified by experiments. One can see that Q is almost linear with respect to Q_0 when $I \geq 1$. Furthermore, Q is close to Q_0 if spin I is enough large.

We assume that nuclear charges uniformly distribute as an axially-deformed liquid drop in this work. $R(\theta)$ is defined the distance from the origin of the coordinate system to the point on the nuclear surface as

$$R(\theta) = c_0 R_c[1 + \beta_2 Y_{20}(\theta) + \beta_4 Y_{40}(\theta)]. \tag{2}$$

Where, the scale factor c_0 represents the effect of incompressibility of nuclear matter in the nucleus and is determined by the so-called constant volume condition.[36] $Y_{20}(\theta)$ and $Y_{40}(\theta)$ are the spherical harmonic functions. β_2 and β_4 are the deformation parameters. Here, the higher orders of the

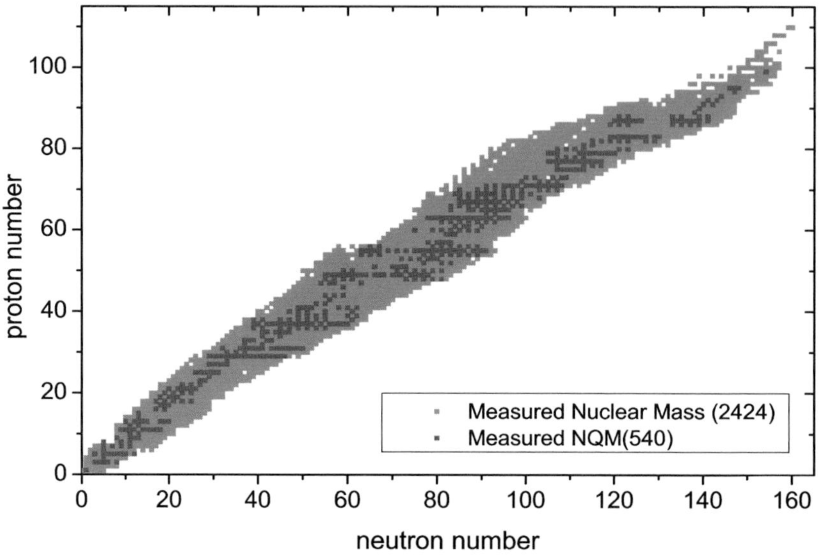

Fig. 1. (Color online) The isotope scheme of the measured nuclear mass and NQM of the ground states. The gray squares denote the measured nuclear mass,[27] and blue squares denote the measured NQMs,[22–26] respectively.

deformation are not considered the NQMs. The nuclear charge radius R_c is expressed as[37]

$$R_c = r_0 A^{1/3} + r_1 A^{-2/3} + r_s \frac{N-Z}{A}(1 - \frac{N-Z}{A}) + r_d \frac{\Delta E}{A}. \tag{3}$$

Where ΔE denotes the shell corrections of the nuclei, and A, Z, N denote the mass, proton and neutron number, respectively. This charge radius formula, considering the shell effect and isospin effect, is proposed by W. Ning, et al.[37] Its rms deviation with respect to the 885 measured charge radii ($Z \geq 8, A \geq 16$) attains to 0.022 fm.

After multiple and complex integral, the intrinsic electric quadrupole moment Q_0 can be expressed as

$$Q_0 = \frac{3}{5} c_0^2 R_c^2 Z \frac{f(\beta_2, \beta_4)}{k(\beta_2, \beta_4)}, \tag{4}$$

$$
\begin{aligned}
f(\beta_2,\beta_4) = {} & 2\sqrt{\frac{5}{\pi}}\beta_2 + \frac{20\beta_2^2}{7\pi} + \frac{15\sqrt{5}\beta_2^3}{7\pi^{3/2}} + \frac{125\beta_2^4}{7\pi^2} + \frac{24\sqrt{5}\beta_2\beta_4}{7\pi} \\
& + \frac{540\beta_2^2\beta_4}{77\pi^{3/2}} + \frac{3060\sqrt{5}\beta_2^3\beta_4}{1001\pi^2} + \frac{200\beta_4^2}{77\pi} + \frac{5367\sqrt{5}\beta_2\beta_4^2}{1001\pi^{3/2}} \\
& + \frac{9255\beta_2^2\beta_4^2}{1001\pi^2} + \frac{110185\sqrt{5}\beta_2^3\beta_4^2}{68068\pi^{5/2}},
\end{aligned} \tag{5}
$$

and

$$
\begin{aligned}
k(\beta_2,\beta_4) = {} & 2 + \frac{3\beta_2^2}{2\pi} + \frac{\sqrt{5}\beta_2^3}{14\pi^{3/2}} + \frac{9\beta_2^2\beta_4}{14\pi^{3/2}} + \frac{3\beta_4^2}{2\pi} \\
& + \frac{15\sqrt{5}\beta_2\beta_4^2}{77\pi^{3/2}} + \frac{243\beta_4^3}{2002\pi^{3/2}}.
\end{aligned} \tag{6}
$$

Thus, one can see from Eqs. (1) - (6) that the NQM intensively depends on the spin, charge number, mass number, charge radius and deformation parameters.

3. Results and Discussion

The root mean square deviation (rms) σ and mean deviation $\bar{\epsilon}$ of the ground state NQMs ($Z \geq 8, A \geq 16$) between the measured data and predictions in this work are tabulated in Table 1. Here, the deformation parameters are taken from the WS4 model[34] and HFB27* model,[35] respectively. The experimental data are compiled by N. J. Stone in different years (see the 1st column).[18–22] The 2nd column is the number of the ground state NQMs. The 3rd-6th columns are the root mean square σ and mean $\bar{\epsilon}$ deviations corresponding to WS4 and HFB27* models, respectively. From table 1, one can see that the predictive precision is higher in recent years both for WS4 and HFB27* models. Taken as an example, Fig. 2 shows the difference between the experimental data[22–26] and the theoretical predictions. The black squares and red circles denote the results of the HFB27* and WS4, respectively. From Fig. 2 one can see that the rms deviations are bigger for heavy nuclei, as verified in Refs.[38,39] From Table 1 and Fig 2, one can see that the obtained σ and $\bar{\epsilon}$ by using WS4 is significantly smaller than those of HFB27*. It is partly due to the fact that the deformations of the ground nuclei are more reasonably described by WS4 mass model[34] with which the rms deviations to 2353 measured masses ($Z \geq 8, A \geq 16$) in AME2012[27] is smaller than 0.3 MeV.

Figure 3 shows the comparisons between the theoretical results of M. Honma[29] and this work. The theoretical values Q_{WS4} (red circles) and $Q_{\mathrm{HFB27^*}}$ (black squares) are calculated using the deformation parameters

Table 1. Root mean square σ and mean $\bar{\epsilon}$ deviations of the ground state NQMs ($Z \geq 8, A \geq 16$) between the experimental data measured in different years and the theoretical predictions, of which the deformation parameters are taken from the WS4 model[34] and HFB27* model,[35] respectively. The units are barn both for σ and mean $\bar{\epsilon}$.

Year	Number	σ_{WS4}	$\sigma_{\mathrm{HFB27*}}$	$\bar{\epsilon}_{\mathrm{WS4}}$	$\bar{\epsilon}_{\mathrm{HFB27*}}$
1989[18]	389	0.618	0.723	0.382	0.522
2005[19]	463	0.610	0.706	0.372	0.499
2011[20]	509	0.568	0.669	0.323	0.448
2013[21]	525	0.558	0.651	0.311	0.424
2014[22]	527	0.552	0.657	0.305	0.431

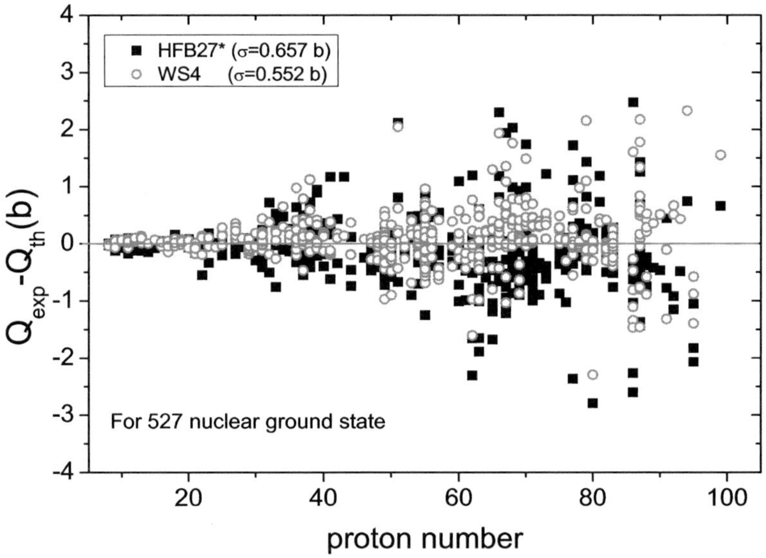

Fig. 2. (Color online) Difference between the experimental data and the theoretical predictions for 527 NQMs ($Z \geq 8, A \geq 16$) of the ground states. The black squares and red circles denote the results of the HFB27* and WS4, respectively.

of WS4 and HFB27*, respectively. The values Q_{th1} (blue circles) and Q_{th2} (green circles) denote the results of the shell-model calculation and the least-squares fitting to the experimental data.[29] From Fig. 3, one can see that the results of WS4 is very close to both the shell-model calculation and the least-square fitting to the experimental data.

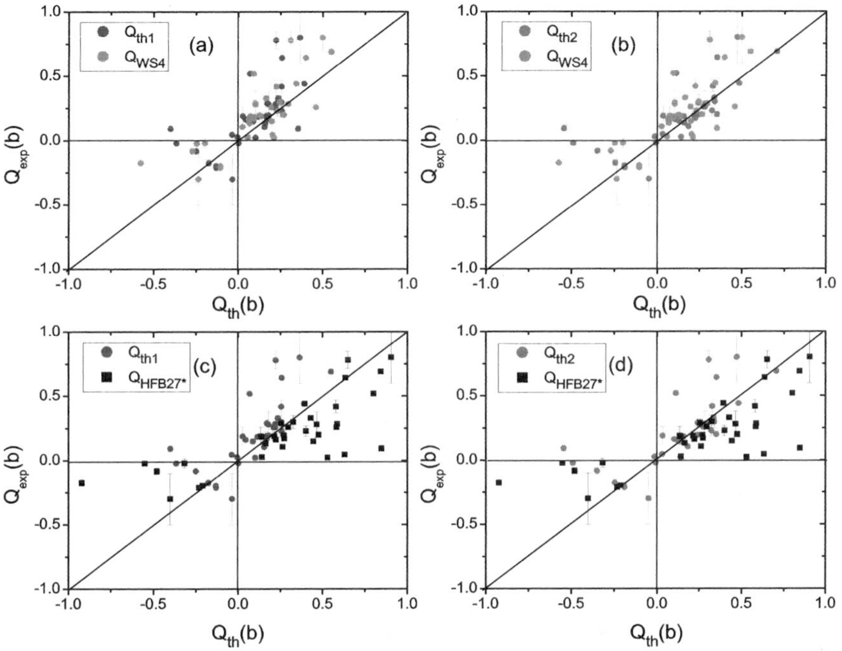

Fig. 3. (Color online) Comparisons between the theoretical results of M. Honma[20] and this work. The theoretical values Q_{WS4} (red circles) and $Q_{\mathrm{HFB27^*}}$ (black squares) are calculated using the deformation parameters of WS4 and HFB27*, respectively. The values Q_{th1} (blue circles) and Q_{th2} (green circles) denote the results of the shell-model calculation and the least-squares fitting to the experimental data.[29]

4. Summary and Conclusions

In this work, a new global formula of NQM is explicitly given as a function of spin, charge and mass number, charge radius and deformation parameters. The deformation parameters extracted from the mass model WS4 and HFB27* are used to calculate NQMs at the ground states. The comparisons between the theoretical results and 527 recently measured data ($Z \geq 8, A \geq 16$) show that the root mean square deviations are 0.552 barn for WS4 and 0.657 barn for HFB27*, respectively. The results of this work are also compared with that of the microscopic shell-model calculations and the least-squares fitting to the experimental data. These comparisons show that the rms and mean deviations are bigger for heavy nuclei. The rms deviation obtained by using WS4 is significantly smaller than those of HFB27*, which implies that the deformations of the ground nuclei are more reasonably described by WS4 mass model.

Acknowledgements

This work has been supported by the National Natural Science Foundation of China (No. 11465005, 11422548, 11475004); the Open Project Program of State Key Laboratory of Theoretical Physics, Institute of Theoretical Physics, Chinese Academy of Sciences, China (No. Y4KF041CJ1), the Natural Science Foundation of Guangxi (No. 2014GXNSFDA118003, 2012GXNSFAA053008), and the project of outstanding young teachers' training in higher education institutions of Guangxi.

References

1. E.A.C.Lucken, Nuclear Quadrupole Coupling Constants, Academic, London, (1969).
2. Takaharu Otsuka,Toshio Suzuki, *at al.*, *Phys. Rev. Lett.*, **95**, 232502 (2005).
3. P.W. Zhao, S.Q. Zhang and J.Meng, *Phys. Rev. C*, **89**, 011301(R) (2014).
4. A.Bohr, *Rev. Mod. Phys.*, **48**, 365 (1976).
5. K.Heyde, *et al.*, *Hyperfine Interactions*, **75**, 69 (1992).
6. J.Papuga, *et al.*, *Phys. Rev. Lett.*, **110**, 172503 (2013).
7. T.Ohtsubo, *et al.*, *Phys. Rev. Lett.*, **109**, 032504 (2012).
8. D.T.Yordanov, *et al.*, *Phys. Rev. Lett.*, **110**, 192501 (2013).
9. G.Neyens, *Rep. Prog. Phys.*, **66**, 633 (2003).
10. K.Heyde and J.L.Wood, *Rev. Mod. Phys.*, **83**, 1467 (2011).
11. X.Wang, *et al.*, *Phys. Lett. B*, **702**, 127 (2011).
12. R.Chevrier, *et al.*, *Phys. Rev. Lett.*, **108**, 162501 (2012).
13. M.Bartkowiak, M.Bazrafshan, C.Fischer, *et al.*, *Phys. Rev. Lett.*, **110**, 205502 (2013).
14. Li-juan Fu and Juha Vaara, *J. Chem. Phys.*, **140**, 024103 (2014).
15. P.Pyykkö. The Nuclear-Quadrupole Moments of the 20 1st Elements-High-Precision Calculations on Atoms and Small Molecules, *Zeitschrift Fur Naturforschung Secyion a Journal of Physical Sciences*, **47**, 189 (1992).
16. P.Pyykkö. Spectroscopic nuclear quadrupole moments. *Molecular Physics*, **99**, 1617 (2001).
17. P.Pyykkö, Year-2008 nuclear quadrupole moments, *Molecular Physics*, **106**, 1965 (2008).
18. Pramila Raghavan, Table of nuclear moments, *Atomic Data and Nuclear Data Tables*, **42**, 189 (1989).

19. N.J.Stone. Table of nuclear magnetic dipole and electric quadrupole moments, *Atomic Data and Nuclear Data Tables*, **90**, 75 (2005).
20. N.J.Stone, Table of nuclear magnetic dipole and electric quadrupole moments, *INDC(NDS)*, 0594 (2011).
21. N.J.Stone, Table of nuclear magnetic dipole and electric quadrupole moments, *INDC(NDS)*, 0650 (2013).
22. N.J.Stone, Table of nuclear magnetic dipole and electric quadrupole moments, *INDC(NDS)*, 0658 (2014).
23. R.T.Santiago, T.Q.Teodoroa, R.L.Haiduke, The nuclear electric quadrupole moment of copper, *Phys. Chem. Chem. Phys.*, **16**, 11590 (2014).
24. M.S.Safronova, U.I.Safronova, A.G.Radnaev, *et al.*, Magnetic dipole and electric quadrupole moments of the ^{229}Th nucleus, *Phys. Rev. A*, **88**, 060501(R) (2013).
25. S.Stopkowicz, L.Cheng, M.E.Harding, *et al.*, The bromine nuclear quadrupole moment revisited, *Molecular Physics*, **111**, 1382 (2013).
26. D.T.Yordanov, D.L.Balabanski, J.Bieroń, *et al.*, Spins, Electromagnetic Moments, and Isomers of $^{107-129}$Cd, *Phys. Rev. Lett.*, **110**, 192501 (2013).
27. G.Audi, *et al.*, *Chin. Phys. C*, **36**, 1287 (2012).
28. S.J.Q.Robinson, T.Hoang, L.Zamick, A.Escuderos and Y.Y.Sharon, *Phys. Rev. C*, **89**, 014316 (2014).
29. Honma M, Otsuka T, Mizusaki T, *et al.*, *Phys. Rev. C*, **80**, 064323 (2009).
30. P.Vingerhoets, K.T.Flanagan, M.Avgoulea, *et al.*, *Phys. Rev. C*, **82**, 064311 (2010).
31. S.V.Tolokonnikov, *Phys. Rev. C*, **84**, 064324 (2011).
32. D.Voitenkov, O.Achakovskiy, S.Kamerdzhiev and S.Tolokonnikov, *EPJWeb of Conferences*, **38**, 17012 (2012).
33. S.V.Tolokonnikov, S.Kamerdzhiev, S.Krewald, *et al.*, *Eur. Phys. J. A*, **48**, 70 (2012).
34. Ning Wang, Min Liu, Xizhen Wu and Jie Meng, *Phys. Lett. B*, **734**, 215 (2014).
35. S. Goriely, *et al.*, *Phys. Rev. C*, **88**, 061302(R) (2013)
36. S.Cwiok, J.Dudek, W.Nazarewicz, J.Skalski, T.Werner, *Comput. Phys. Commun.*, **46**, 379 (1987).
37. Wang N, Li T. Shell and isospin effects in nuclear charge radii, *Phys. Rev. C*, **88**, 011301(R) (2013).
38. M.Pernpointner, *J. Phys. B*, **35**, 383 (2002).
39. Belpassi L, Tarantelli F, Sgamellotti A, *et al.*, Nuclear electric quadrupole moment of gold, *Journal of chemical physics*, **126**, 064314 (2007).

SYSTEMATICAL BEHAVIOR OF BREAKUP EFFECTS ON COMPLETE FUSION AT ENERGIES ABOVE THE COULOMB BARRIER

Bing Wang[a], Wei-Juan Zhao[a], P. R. S. Gomes[b],
En-Guang Zhao[c,d], and Shan-Gui Zhou[c,d,*]

[a] *Department of Physics, Zhengzhou University, Zhengzhou 450001, China*
[b] *Instituto de Física, Universidade Federal Fluminense, Avenida Litoranea s/n, Gragoatá, Niterói, Rio de Janeiro, 24210-340, Brazil*
[c] *State Key Laboratory of Theoretical Physics, Institute of Theoretical Physics, Chinese Academy of Sciences, Beijing 100190, China*
[d] *Center of Theoretical Nuclear Physics, National Laboratory of Heavy Ion Accelerator, Lanzhou 730000, China*
[*] *E-mail: sgzhou@itp.ac.cn*

We investigate the systematical behavior of the breakup effects on the complete fusion (CF) cross sections at energies above the Coulomb barrier. The CF cross sections are suppressed by the prompt breakup of the projectiles. This suppression effect, expressed as the ratio of the reduced fusion function and the universal fusion function (UFF), for reactions induced by the same projectile, is independent of the target and mainly determined by the lowest energy breakup channel of the projectile. There holds a good exponential relation between the suppression factor and the energy corresponding to the lowest breakup threshold.

Keywords: Breakup; complete fusion; breakup threshold; suppression.

1. Introduction

In recent years, the study of breakup effects on fusion reactions around the Coulomb barrier has been a subject of intense experimental and theoretical interests.[1] When a reaction involves weakly bound nuclei, various processes can occur. One is called the incomplete fusion (ICF) in which part of the breakup fragments is absorbed by the target. When all the fragments fuse with the target, the process is called sequential complete fusion (SCF). Experimentally, the SCF cannot be distinguished from the direct complete fusion (DCF) in which the projectile fuses with the target without breakup. Therefore, only the complete fusion (CF) cross section, which includes both DCF and SCF cross sections, can be measured.

To investigate the influence of the breakup on CF, the most widely employed approach is to compare the data with either predictions of coupled channel calculations without the breakup channels or predictions of a single barrier penetration model. It was confirmed that the CF cross sections are suppressed at energies above the Coulomb barrier.[1] Many efforts have been made to investigate this suppression effect. Conclusions obtained by different methods are different. Recently, we have made a systematic study of the breakup effects on the CF cross sections at energies above the Coulomb barrier.[2] In this contribution, we present briefly these results with some updated experimental information.

2. Methods

It is necessary to reduce the data to eliminate the geometrical factors in different reaction systems. Then the CF data for different reaction systems can be compared directly and the systematics can be investigated. After the reduction, the data should be compared with the theoretical predictions without the breakup channels. Deviations of CF data, if exist, from the predictions mainly arise from the effects of the breakup on CF cross sections. In order to eliminate completely the geometrical factors and static effects of the potential, the method proposed in Ref. 3 is adopt to reduce the data. According to this prescription, the fusion cross section $\sigma_{\rm F}$ and the collision energy $E_{\rm c.m.}$ are reduced to the dimensionless variables,

$$F(x) = \frac{2E_{\rm c.m.}}{R_{\rm B}^2 \hbar\omega}\sigma_{\rm F}, \quad x = \frac{E_{\rm c.m.} - V_{\rm B}}{\hbar\omega}, \tag{1}$$

where $R_{\rm B}$, $V_{\rm B}$, and $\hbar\omega$ denote the radius, height, and curvature of the barrier which is approximated by a parabola. These three parameters are obtained from São Paulo potential (SPP).[4] If the fusion cross section can be accurately described by the Wong's formula,[5]

$$\sigma_{\rm F}^{\rm W}(E_{\rm c.m.}) = \frac{R_{\rm B}^2 \hbar\omega}{2E_{\rm c.m.}} \ln\left[1 + \exp\left(\frac{2\pi(E_{\rm c.m.} - V_{\rm B})}{\hbar\omega}\right)\right]. \tag{2}$$

The $F(x)$ can be reduced to $F_0(x) = \ln[1 + \exp(2\pi x)]$. Note that $F_0(x)$ is a universal fusion function (UFF), which is independent of reaction systems.[3] So, the $F_0(x)$ is used as a uniform standard reference to explore the breakup effects on CF cross sections. Note that we focus on the behavior of the CF excitation function at energies above the Coulomb barrier, an energy region where the Wong's formula can be applied.

3. Results and Discussions

We collected 41 sets of CF excitation functions of the reactions with ^{6,7}Li, ^{9}Be, ^{10,11}B, ^{12,13}C, and ^{16}O as projectiles. The suppression for the reactions involving the same projectile is roughly independent of the target charge. For more details, please refer to Ref. 2. Here we select the reactions induced by ^{9}Be as an example. Moreover, three sets of new measured CF excitation functions are included.[6,7] Figure 1(a) shows the CF functions for the reactions with ^{9}Be as a projectile. The lowest breakup threshold of ^{9}Be is 1.573 MeV for its breakup into $n + 2\alpha$. It can be seen that the CF functions are very close to each other, except for the reactions of ^{9}Be on ^{144}Sm and ^{186}W, and below the UFF. This means that the suppression is roughly independent of the target charge. The suppression factor [defined as $F_{\text{B.U.}} = F(x)/F_0(x)$] is 0.67. This is consistent with the results obtained in our preceding study.[2] The reason why the reactions of ^{9}Be on ^{144}Sm and ^{186}W are out of the systematic is discussed in Ref. 2. Hinde *et al.*[8] performed coincidence experiments to determine time scales in the breakup of ^{9}Be. They were able to disentangle prompt ^{9}Be breakup from the delayed breakup of ^{8}Be, triggered by a one-neutron stripping process. In the latter case, the transfer reaction produces the unstable ^{8}Be nucleus, with a $T_{1/2} = 10^{-16}$ s, several orders of magnitude longer than the collision time, and so a process of this kind cannot influence fusion.

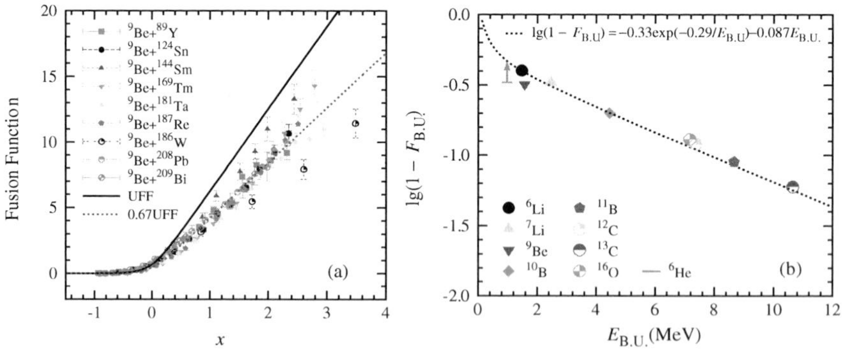

Fig. 1. (Color online) (a) Comparison of the UFF with the CF function $F(x)$ for reactions with ^{9}Be as a projectile. The dotted line is the UFF multiplied by the suppression factor 0.67. (b) The CF suppression factors for projectiles ^{6,7}Li, ^{9}Be, ^{10,11}B, ^{12,13}C, and ^{16}O as a function of the lowest projectile breakup threshold ($E_{\text{B.U.}}$). The solid line represents the suppression factor of total fusion for reactions with ^{6}He as a projectile. The dotted line denotes the results of the empirical formula shown in the figure.

The CF suppression factors for projectiles ^{6,7}Li, ^{9}Be, ^{10,11}B, ^{12,13}C, and ^{16}O are shown as a function of the lowest projectile breakup threshold in Fig. 1(b). It can be seen that there exists an exponential relation between the suppression factor and the breakup threshold energy. This suggests that the influence of the breakup channel on the CF is a threshold effect. One can find that the $F_{\mathrm{B.U.}}$ for ^{9}Be is a little larger than that suggested from the empirical formula shown in Fig. 1(b). A possible explanation might be that the events corresponding to ICF of 2α or ^{8}Be fusing with target were considered as CF in the measurement of the cross sections.

4. Summary

The CF cross sections are suppressed owing to the prompt breakup of projectiles. For the reactions involving the same projectile, the suppression is independent of target charge. Furthermore, the suppression for different projectiles are mainly determined by the lowest breakup threshold. There holds a good exponential relation between the suppression factor and the energy corresponding to the lowest breakup threshold.

Acknowledgments

P.R.S.G. acknowledges partial support from CNPq, FAPERJ, and the PRONEX. This work has been supported by MOST of China (973 Program: 2013CB834400), NSF of China (11121403, 11175252, 11120101005, 11211120152, and 11275248), and the KIP of CAS (KJCX2-EW-N01).

References

1. L. F. Canto, P. R. S. Gomes, R. Donangelo and M. S. Hussein, *Phys. Rep.* **424**, 1 (2006); L. F. Canto, P. R. S. Gomes, R. Donangelo, J. Lubian, M. S. Hussein, Phys. Rep. (in press)
2. Bing Wang, Wei-Juan Zhao, P. R. S. Gomes, En-Guang Zhao and Shan-Gui Zhou, *Phys. Rev. C* **90**, 034612 (2014) and references therein.
3. L. F. Canto, P. R. S. Gomes, J. Lubian, L. C. Chamon and E. Crema, *J. Phys. G* **36**, 015109 (2009); *Nucl. Phys. A* **821**, 51 (2009).
4. L. C. Chamon, B. V. Carlson, L. R. Gasques *et al.*, *Phys. Rev. C* **66**, 014610 (2002).
5. C. Y. Wong, *Phys. Rev. Lett.* **31**, 766 (1973).
6. Y. D. Fang, P. R. S. Gomes, J. Lubian *et al.*, *Phys. Rev. C* **91**, 014608 (2015).
7. N. T. Zhang, Y. D. Fang, P. R. S. Gomes *et al.*, *Phys. Rev. C* **90**, 024621 (2014).
8. D. J. Hinde, M. Dasgupta, B. R. Fulton, C. R. Morton, R. J. Wooliscroft, A. C. Berriman and K. Hagino, *Phys. Rev. Lett.* **89**, 272701 (2002).

DECAY STUDIES OF PROTON EMITTER: ^{151}Lu

F. Wang, B. Sun* and L. Zhu

School of Physics and Nuclear Energy Engineering, Beihang University,
Beijing 100191, China
**E-mail: bhsun@buaa.edu.cn*
www.buaa.edu.cn

Z. Liu

Institute of Modern Physics, Chinese Academy of Sciences,
Lanzhou 730000, China
liuzhong@impcas.ac.cn

An experiment aiming to search for new isomers in the region of proton emitter ^{151}Lu was done in the Accelerator Laboratory of the University of Jyväskylä (JYFL). Rich information on ^{151}Lu and ^{151m}Lu has been obtained from our data analysis. In this work, we revisit the level scheme of ^{151}Lu by using the proton-tagging technique and measure the half-lives of ^{151}Lu and ^{151m}Lu are 82.8±0.7 ms and 15.4±0.8 μs, respectively.

Keywords: Proton emitter; ^{151}Lu; level scheme; half-life.

1. Introduction

The study of proton emitting nuclei has provided very important information on the structure of nuclei around the proton drip-line.[1] ^{151}Lu was the first established case with proton decay from a nuclear ground state,[2] and had attracted lots of attentions in both theories and experiments. Only in the last decade was it possible to measure and identify γ rays decaying from the excited states of ^{151}Lu at ORNL.[3] A short-lived low-spin isomer with the proton energy of 1310(10) keV and half-life of 16(1) ms[4] was reported as well in the same experiment. A later experiment[5] conducted at ORNL with the same setup, but no gamma detectors were placed at the focal plane of the mass separator, resulted in a tentative level scheme. In 2013 an updated level scheme was reported [6] by using the proton-tagging and γ-ray coincidence techniques, and furthermore determined the lifetime of the first excite state above the proton-emitting ground state using the recoil-distance Doppler shift method. The latest experiments[5,6] argued rather different excited states for the higher lying states.

2. Experimental Details

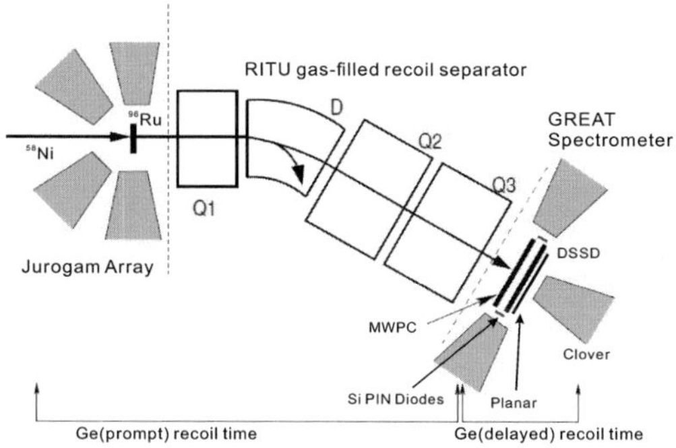

Fig. 1. Schematic view of the recoil-decay tagging setup in the experiment.

The experiments were performed in the Accelerator Laboratory of the University of Jyväskylä (JYFL). A 2×500 μg/cm^2 thick target of isotopically enriched ^{96}Ru deposited on a 2 mg/cm^2 C foil was bombarded with 266 MeV and 274 MeV ^{58}Ni ions extracted from the K130 cyclotron. The interesting recoil nuclei were separated from the beam by the gas-filled separator RITU(Recoil Ion Transport Unit).[7] The recoils are implanted in the DSSDs of the GREAT(Gamma Recoil Electron Alpha Tagging) spectrometer.[8] Prompt γ rays were recorded with the JUROGAM array[9] utilising the recoil decay tagging(RDT) method. A schemetic view of the experimental stepup is shown in Fig. 1.

3. Results and Discussion

3.1. *Level scheme*

The statistics obtained in this work is higher than that of the work reported in Ref. 6 by a factor of two. Our preliminary analysis produced a level scheme of ^{151}Lu consistent with Ref. 6, and in addition three newly observed gamma-ray transitions with energies of 242 keV, 625 keV and 704 keV are assigned and higher in the level scheme. Fig. 2 summarizes the updated level scheme for ^{151}Lu. Dedicated model calculations are in progress, which hopefully could help to interpret the level scheme.

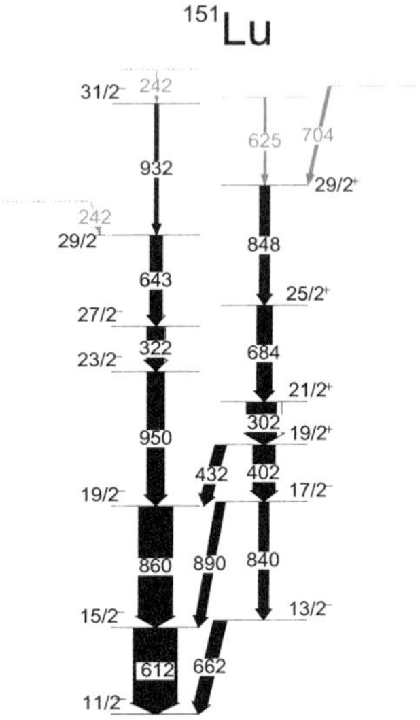

Fig. 2. Level scheme of ^{151}Lu. The red ones are the newly identified transitions from this work. It is not sure where to put the 242keV for the low statisctis.

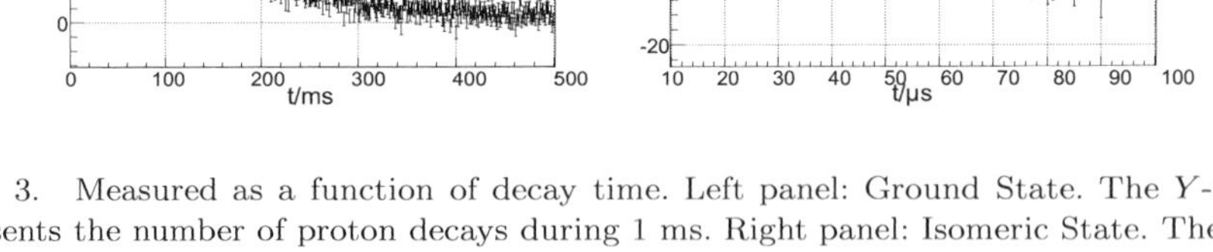

Fig. 3. Measured as a function of decay time. Left panel: Ground State. The Y-axis presents the number of proton decays during 1 ms. Right panel: Isomeric State. The Y-axis presents the number of proton decays during 1 μs .Based on the decay time spectrum of proton emitter, the half-lives of the previously known ground state and isomeric state were observed to be 82.8±0.7 ms and 15.4±0.8 μs, respectively.

3.2. *Half-life measurments*

For each implant on the DSSDs, the energy of the implant, the pixel in which the implant occurred and the time were recorded. Decays within a given pixel were correlated with the previous implants in the same pixel, so we can determine the decay time of the radioactivity, which made it possible for us to measured the half-lives of interesting decays. By using the recoil-decay tagging technique, it is found that the half-lives of two proton emitters can be reproduced. We fit a exponential function to Fig. 3 to get the half-lives of ground state and isomeric state, which are 82.8±0.7 ms and 15.4±0.8 μs, respectively.

4. Summary

In conclusion, we report the level scheme of ^{151}Lu agrees well with the latest experimental results. In additions, the half-lives of the previously known ground state and isomeric state were also observed.

References

1. P. J. Woods and C. N. Davids, *Annual Review of Nuclear and Particle Science* **47**, 541 (1997).
2. S. Hofmann, G. Munzenberg, F. Hebberger, W. Reisdorf, P. Armbruster and B. Thuma, *Zeitschrift fur Physik A Atoms and Nuclei* **299**, 281 (1981).
3. C.-H. Yu, J. C. Batchelder, C. R. Bingham, R. Grzywacz, K. Rykaczewski, K. S. Toth, Y. Akovali, C. Baktash, A. Galindo-Uribarri, T. N. Ginter, C. J. Gross, M. Karny, S. H. Kim, B. D. MacDonald, S. D. Paul, D. C. Radford, J. Szerypo and W. Weintraub, *Phys. Rev. C* **58**, R3042(Dec 1998).
4. C. R. Bingham, J. C. Batchelder, K. Rykaczewski, K. S. Toth, C.-H. Yu, T. N. Ginter, C. J. Gross, R. Grzywacz, M. Karny, S. H. Kim, B. D. MacDonald, J. Mas, J. W. McConnell, P. B. Semmes, J. Szerypo, W. Weintraub and E. F. Zganjar, *Phys. Rev. C* **59**, R2984(Jun 1999).
5. L. Zhong, , Tech. Rep. 961, AIP Conf. Proc (2007).
6. M. Procter, D. Cullen, M. Taylor, G. Alharshan, L. Ferreira, E. Maglione, K. Auranen, T. Grahn, P. Greenlees, U. Jakobsson, R. Julin, A. Herzan, J. Konki, M. Leino, J. Pakarinen, J. Partanen, P. Peura, P. Rahkila, P. Ruotsalainen, M. Sandzelius, J. Sarn, S. Stolze, C. Scholey, J. Sorri, J. Uusitalo, T. Braunroth, E. Ellinger, A. Dewald, D. Joss, C. McPeake and B. Saygi, *Physics Letters B* **725**, 79 (2013).
7. M. Leino, J. Äystö, T. Enqvist, P. Heikkinen, A. Jokinen, M. Nurmia, A. Ostrowski, W. Trzaska, J. Uusitalo, K. Eskola *et al.*, *Nuclear Instruments and Methods in Physics Research Section B: Beam Interactions with Materials and Atoms* **99**, 653 (1995).
8. R. Page, A. Andreyev, D. Appelbe, P. Butler, S. Freeman, P. Greenlees, R.-D. Herzberg, D. Jenkins, G. Jones, P. Jones *et al.*, *Nuclear Instruments and Methods in Physics Research Section B: Beam Interactions with Materials and Atoms* **204**, 634 (2003).
9. C. Beausang, S. Forbes, P. Fallon, P. Nolan, P. Twin, J. Mo, J. Lisle, M. Bentley, J. Simpson, F. Beck *et al.*, *Nucl. Instr. Meth. A* **313**, p. 37 (1992).

PRODUCTION CROSS SECTIONS FOR SYNTHESIS OF NUCLIDES WITH $Z = 118$ IN LARGE MASS TRANSFER REACTIONS

Nan Wang, Xin Jin, Shuang Yang

College of Physics, Shenzhen University, Shenzhen 518060, China

En-Guang Zhao

State Key Laboratory of Theoretical Physics, Institute of Theoretical Physics, Chinese Academy of Sciences, Beijing 100190, China

Center of Theoretical Nuclear Physics, National Laboratory of Heavy Ion Accelerator, Lanzhou 730000, China

The production cross sections for the synthesis of superheavy nuclei with charge number 118 are studied with the di-nuclear system model with dynamical potential surface (DNS-DyPES model). The dynamical potential energy surface (PES) and the fusion probabilities for ^{48}Ca bombarding Cf nuclei reactions are studied. By multiplying the capture cross section, fusion probability and survival probability, the evaporation residue(ER) cross sections for superheavy nuclei with $Z = 118$ are obtained. And the excitation functions for the reactions with a mixture target of Cf isotopes are also shown.

Keywords: Superheavy nuclei; production cross section; fusion mechanism.

1. Introduction

In recent years, the explorations in the production of super-heavy nuclei (SHN) is an exciting field in nuclear physics. In experimental aspect, people have already synthesized the SHN (Z = 107-112) in cold fusion reactions.[1,2] And also the hot fusion reactions were adopted to synthesize the SHN (Z = 112-118) where some actinide nuclei are bombarded by ^{48}Ca.[3–8] In theoretical aspect, in order to understand the mechanism of the fusion reaction and predict the ER cross sections of SHN, people have proposed several theoretical models, such as the fusion-by-diffusion model,[9–15] nuclear collectivization model,[16,17] di-nuclear system (DNS) model[18–26] and other model.[27] In the newly developed DNS-DyPES model,[26] the dynamical deformation is taken into consideration in the DNS model. Recently, an experiment which is aimed to produce the isotopes of element $Z = 118$ is under way at the Flerov Laboratory of Joint Institute for Nuclear Research in Dubna, Russia. In order to predict the evaporation residue cross sections and the optimal excitation energies for the reaction, some theoretical calculations are performed.[28–30] However some differences for the results[28,29] can be found and more investigations for ER cross sections should be carried out.

In this paper, the effect of the dynamical deformation on PES is investigated. Then the fusion probabilities for the ^{48}Ca bombarding Cf isotopes are calculated. Finally, the theoretical results of the excitation functions of ^{48}Ca bombarding the Cf isotopes with the DNS-DyPES model[26] are shown and the results for the experiment with a mixture of Cf isotopes target which is under way at the Flerov Laboratory are also discussed.

2. Description of the DNS-DyPES Model

In DNS model, the fusion is considered as a diffusion of the di-nuclear configuration in the mass asymmetry coordinate. The production cross section in a heavy-ion fusion reaction can be written as the summation over all partial wave J,

$$\sigma_{\mathrm{ER}}(E_{\mathrm{c.m.}}) = \frac{\pi\hbar^2}{2\mu E_{\mathrm{c.m.}}} \sum_J (2J+1) T(E_{\mathrm{c.m.}}, J) P_{\mathrm{CN}}(E_{\mathrm{c.m.}}, J) W_{\mathrm{sur}}(E_{\mathrm{c.m.}}, J), \tag{1}$$

where $E_{\mathrm{c.m.}}$ is the incident energy in the c.m.frame. $T(E_{\mathrm{c.m.}}, J)$ is the transmission probability of the colliding nuclei overcoming the potential barrier to form the DNS. P_{CN} and W_{sur} represent the fusion probability for the di-nuclear system to form compound nucleus and the survival probability for the compound nucleus in de-excitation process, respectively.

The fusion probability can be described with the master equation,[26]

$$\begin{aligned} \frac{dP(A_1,t)}{dt} &= \sum_{A_1'} W_{A_1 A_1'}(t) \left[d_{A_1}(t) P(A_1',t) - d_{A_1'}(t) P(A_1,t) \right] \\ &\quad - \Lambda_{A_1}^{\mathrm{qf}}(t) P(A_1,t), \end{aligned} \tag{2}$$

where $P(A_1, t)$ denotes the distribution function to find fragment 1 with A_1 nucleons at time t. $W_{A_1,A_1'}$ is the mean transition probability from A_1 to A_1' and d_{A_1} denotes the microscopic dimension. The sum should be taken over all mass numbers (from 0 to $A_1 + A_2$). The potential energy in the mass asymmetry degree of freedom is defined as

$$U_{\mathrm{PES}}(A_1, t) \equiv V(A_1, t) + (M_1 + M_2 - M_{\mathrm{T}} - M_{\mathrm{P}})c^2. \qquad (3)$$

$V(A_1, t) = V_{\mathrm{N}}(A_1, t) + V_{\mathrm{C}}(A_1, t)$ and $V_{\mathrm{N}}(A_1, t)$ and $V_{\mathrm{C}}(A_1, t)$ are the nuclear and the Coulomb interactions between the two nuclei. M_{T}, M_{P} are the masses for the projectile and target, respectively. M_1 and M_2 are the masses for the two fragments in di-nuclear system, respectively. The nuclei in a DNS are distorted and dynamical deformations develop during the process of nuclear transfers.[31–33] This results in the time-dependence of the potential energy surface. More details about the dynamical deformation in this work can be found in Ref. 26.

3. Results and Discussions

The potential energy surface, defined as Eq. (3), is shown as a function of the mass asymmetry degree of freedom for the reactions ^{48}Ca+^{250}Cf in Fig. 1. The dynamical PESs at $t = \infty$ and at $t = 0$ are shown with the

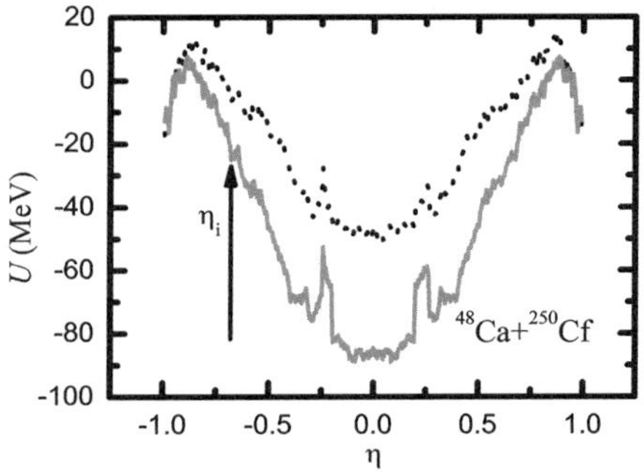

Fig. 1. (Color online) PES for reaction ^{48}Ca + ^{250}Cf as a function of the mass asymmetry. The arrow denotes the entrance channel.

solid line and dotted line, respectively, in Fig. 1. The mass asymmetry for the entrance channel $\eta_i = -0.678$ is indicated with the arrow. The highest point for the PES decreases significantly due to the development of the dynamical deformation. And the inner fusion barrier at $t = \infty$ and at $t = 0$ are 32.66 MeV and 20.91 MeV, respectively. Thus, the fusion probability decreases because of the higher inner fusion barrier for the inclusion of the dynamical deformation.

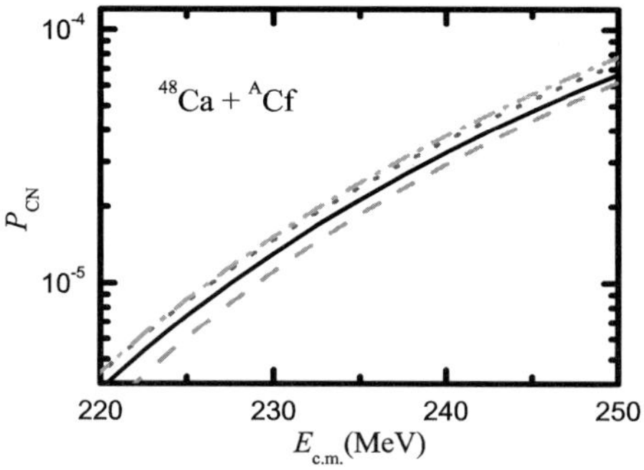

Fig. 2. (Color online) Fusion probability as a function of the incident energy for the reactions ^{48}Ca $+^{249,250,251,252}$Cf, respectively.

The fusion probabilities as a function of incident energy for ^{48}Ca $+^{249,250,251,252}$Cf at central collision are depicted in Fig. 2 with the solid, dashed, dotted and dash dotted lines, respectively. It can be seen that the fusion probabilities increase with the incident energy, e.g. $P_{\rm CN}$ for ^{48}Ca $+^{250}$Cf increases from 3×10^{-6} at 220 MeV to 6×10^{-5} at 250 MeV. The fusion probabilities for the four channels are generally close to one another. The differences among them can be attributed to the differences of the structures of PES.

The excitation functions for the reactions ^{48}Ca + ^{249,250,251,252}Cf can be found in Ref. 30. It is found that the theoretical results of the ER cross section and the optimal incident energy are in good agreement with the experimental results for ^{48}Ca + ^{249}Cf. The maximum ER cross section for $3n$ channel is about 0.5 pb. The calculated maximum ER cross section for ^{48}Ca + ^{250}Cf is about 0.4 pb. For ^{251}Cf and ^{252}Cf induced reactions, the maximum ER cross sections in $3n$ channels are about 1 pb and the maximum ER cross section in $4n$ channel for ^{48}Ca + ^{252}Cf is also close to 1 pb.

As shown in Fig. 2 of Ref. 30 and Fig.3 in this paper, the weighted excitation functions for the production of isotopes ^{293,294,295,296,297}118 for the reactions ^{48}Ca bombarding the mixture target of ^{249,250,251,252}Cf. The dash dot dot, dash dot, solid, short dashed and dashed lines represent the excitation functions for producing the element 118 with the mass numbers 293, 294, 295, 296 and 297, respectively. The numbers in Fig. 3 are the mass numbers of the SHN. The isotopic composition of the Cf target (42.31% of ^{249}Cf, 21.76% of ^{250}Cf, 35.64% of ^{251}Cf, and 0.29% of ^{252}Cf) are taken from Ref. 28,29. The results show that the maximum ER cross section for ^{296}118

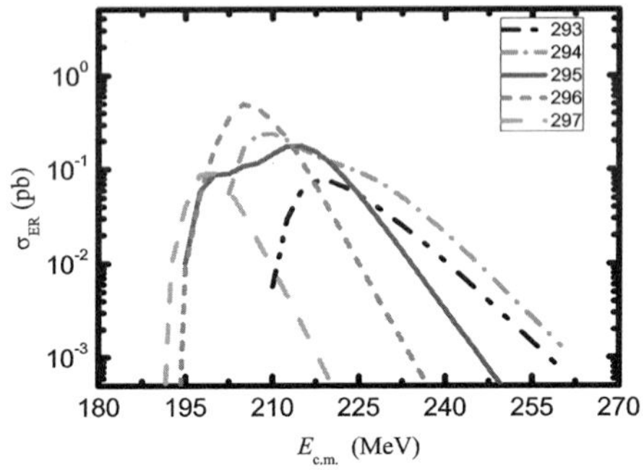

Fig. 3. (Color online) As shown in Fig. 2 of Ref. 30, weighted evaporation residue cross sections for the production of isotopes $^{293,294,295,296,297}118$ in the ^{249,250,251,252}Cf(^{48}Ca, xn) reactions, predicted for the experiment at FLNR in Dubna with isotopic composition of the Cf target. See text for details.

is about 0.5 pb at 205 MeV. The largest ER cross sections for $^{294}118$ and $^{295}118$ which are about 200 fb appear at $E_{\mathrm{c.m.}}$= 210 MeV and 215 MeV, respectively. For $^{293}118$ and $^{297}118$, the largest ER cross sections are only about 100 fm which is far below the current experimental limit.

4. Summary

In this paper, the production cross sections for the synthesis of superheavy nuclei with charge number 118 are studied with the DNS-DyPES model. It can be seen that the dynamical deformation affects the potential energy surface significantly. The fusion probabilities for ^{48}Ca bombarding Cf isotopes are calculated. The excitation functions for the production of $Z = 118$ isotopes in the reactions with a mixture target of Cf isotopes are shown. It is seen that the ER cross section for $^{296}118$ is the largest among all the isotopes while those for $^{293}118$ and $^{297}118$ are relatively low (smaller than 0.1 pb).

5. Acknowledgement

The work was supported by NSF China (11475115, 11275098, 10975100, 10979066, 11175252), MOST of China (2007CB815000), the Knowledge Innovation Project of CAS (KJCX2-EW-N01 and KJCX2-YW-N32). Part of the numerical results is obtained on the ScGrid of Supercomputing Center, CNIC of CAS.

References

1. S. Hofmann and G. Münzenberg, *Rev. Mod. Phys.* **72**, 733(July 2000).
2. K. Morita, K. Morimoto, D. Kaji, T. Akiyama, S.-i. Goto, H. Haba, E. Ideguchi, R. Kanungo, K. Katori, H. Koura, H. Kudo, T. Ohnishi, A. Ozawa, T. Suda, K. Sueki, H.-S. Xu, T. Yamaguchi, A. Yoneda, A. Yoshida and Y.-L. Zhao, *J. Phys. Soc. Jpn.* **73**, 2593 (2004).
3. Y. T. Oganessian, V. K. Utyonkov, Y. V. Lobanov, F. S. Abdullin, A. N. Polyakov, R. N. Sagaidak, I. V. Shirokovsky, Y. S. Tsyganov, A. A. Voinov, G. G. Gulbekian, S. L. Bogomolov, B. N. Gikal, A. N. Mezentsev, V. G. Subbotin, A. M. Sukhov, K. Subotic, V. I. Zagrebaev, G. K. Vostokin, M. G. Itkis, R. A. Henderson, J. M. Kenneally, J. H. Landrum, K. J. Moody, D. A. Shaughnessy, M. A. Stoyer, N. J. Stoyer and P. A. Wilk, *Phys. Rev. C* **76**, 011601(R)(July 2007).
4. Y. T. Oganessian, V. K. Utyonkov, Y. V. Lobanov, F. S. Abdullin, A. N. Polyakov, R. N. Sagaidak, I. V. Shirokovsky, Y. S. Tsyganov, A. A. Voinov, G. G. Gulbekian, S. L. Bogomolov, B. N. Gikal, A. N. Mezentsev, S. Iliev, V. G. Subbotin, A. M. Sukhov, K. Subotic, V. I. Zagrebaev, G. K. Vostokin, M. G. Itkis, K. J. Moody, J. B. Patin, D. A. Shaughnessy, M. A. Stoyer, N. J. Stoyer, P. A. Wilk, J. M. Kenneally, J. H. Landrum, J. F. Wild and R. W. Lougheed, *Phys. Rev. C* **74**, 044602(Octtober 2006).
5. Y. Oganessian, *J. Phys. G: Nucl. Phys.* **34**, R165(April 2007).
6. Y. T. Oganessian, F. S. Abdullin, P. D. Bailey, D. E. Benker, M. E. Bennett, S. N. Dmitriev, J. G. Ezold, J. H. Hamilton, R. A. Henderson, M. G. Itkis, Y. V. Lobanov, A. N. Mezentsev, K. J. Moody, S. L. Nelson, A. N. Polyakov, C. E. Porter, A. V. Ramayya, F. D. Riley, J. B. Roberto, M. A. Ryabinin, K. P. Rykaczewski, R. N. Sagaidak, D. A. Shaughnessy, I. V. Shirokovsky, M. A. Stoyer, V. G. Subbotin, R. Sudowe, A. M. Sukhov, Y. S. Tsyganov, V. K. Utyonkov, A. A. Voinov, G. K. Vostokin and P. A. Wilk, *Phys. Rev. Lett.* **104**, 142502(April 2010).
7. P. A. Ellison, K. E. Gregorich, J. S. Berryman, D. L. Bleuel, R. M. Clark, I. Dragojević, J. Dvorak, P. Fallon, C. Fineman-Sotomayor, J. M. Gates, O. R. Gothe, I. Y. Lee, W. D. Loveland, J. P. McLaughlin, S. Paschalis, M. Petri, J. Qian, L. Stavsetra, M. Wiedeking and H. Nitsche, *Phys. Rev. Lett.* **105**, p. 182701(Oct 2010).
8. S. Hofmann, S. Heinz, R. Mann, J. Maurer, J. Khuyagbaatar, D. Ackermann, S. Antalic, W. Barth, M. Block, H. Burkhard *et al.*, *The European Physical Journal A-Hadrons and Nuclei* **48**, 1 (2012).
9. K. Siwek-Wilczyńska, T. Cap, M. Kowal, A. Sobiczewski and J. Wilczyński, *Phys. Rev. C* **86**, p. 014611(Jul 2012).
10. Z. H. Liu and J.-D. Bao, *Phys. Rev. C* **80**, 054608(November 2009).
11. K. Siwek-Wilczynska, T. Cap and J. Wilczynski, *Int. J. Mod. Phys. E* **19**, 500 (2010).
12. Z.-H. Liu and J.-D. Bao, *Phys. Rev. C* **83**, 044613(April 2011).
13. Z.-H. Liu and J.-D. Bao, *Phys. Rev. C* **84**, 031602(R)(September 2011).
14. C. Wang, J. Zhang, Z. Z. Ren and C. W. Shen, *Phys. Rev. C* **82**, p. 054605(Nov 2010).

15. C. Shen, G. Kosenko and Y. Abe, *Phys. Rev. C* **66**, 061602(R) (2002).
16. V. I. Zagrebaev, *Phys. Rev. C* **64**, 034606(August 2001).
17. V. Zagrebaev and W. Greiner, *Phys. Rev. C* **78**, 034610(September 2008).
18. G. Adamian, N. Antonenko and W. Scheid, *Nuclear Physics A* **618**, 176 (1997).
19. A. Zubov, G. Adamian and N. Antonenko, *Phys. Part. Nucl.* **40**, 847(November 2009).
20. A. Nasirov, K. Kim, G. Mandaglio, G. Giardina, A. Muminov and Y. Kim, *The European Physical Journal A* **49**, p. 147 (2013).
21. W. Li, N. Wang, J. F. Li, H. Xu, W. Zuo, E. Zhao, J. Q. Li and W. Scheid, *Europhys. Lett.* **64**, 750(December 2003).
22. W. Li, N. Wang, F. Jia, H. Xu, W. Zuo, Q. Li, E. Zhao, J. Li and W. Scheid, *J. Phys. G: Nucl. Phys.* **32**, 1143(June 2006).
23. M. Huang, Z. Gan, X. Zhou, J. Li and W. Scheid, *Phys. Rev. C* **82**, p. 044614(Oct 2010).
24. Z.-G. Gan, X.-H. Zhou, M.-H. Huang, Z.-Q. Feng and J.-Q. Li, *Sci. China-Phys. Mech. Astron.* **54 (Supp. 1)**, s61 (2011).
25. Z.-Q. Feng, G.-M. Jin, F. Fu and J.-Q. Li, *Nucl. Phys. A* **771**, 50(May 2006).
26. N. Wang, E.-G. Zhao, W. Scheid and S.-G. Zhou, *Phys. Rev. C* **85**, 041601(R)(Apr 2012).
27. N. Wang, J. Tian and W. Scheid, *Phys. Rev. C* **84**, 061601(R)(December 2011).
28. G. Mandaglio, G. Giardina, A. K. Nasirov and A. Sobiczewski, *Phys. Rev. C* **86**, p. 064607(Dec 2012).
29. T. Cap, K. Siwek-Wilczyńska, M. Kowal and J. Wilczyński, *Phys. Rev. C* **88**, p. 037603(Sep 2013).
30. N. Wang, E.-G. Zhao and W. Scheid, *Phys. Rev. C* **89**, p. 037601(Mar 2014).
31. C. Riedel, G. Wolschin and W. Noerenberg, *Z. Phys. A* **290**, 47(March 1979).
32. C. Riedel and W. Nörenberg, *Z. Phys. A* **290**, 385(December 1979).
33. G. Wolschin, *Phys. Lett. B* **88**, 35(December 1979).

EFFECTS OF TENSOR FORCE AND PAIRING CORRELATION ON BUBBLE STRUCTURE OF EXOTIC NUCLEUS

Yanzhao Wang

Department of Mathematics and Physics, Shijiazhuang TieDao University, Shijiazhuang 050043, China

Institute of Applied Physics, Shijiazhuang TieDao University, Shijiazhuang 050043, China

Jianzhong Gu*

China Institute of Atomic Energy, P. O. Box 275 (10), Beijing 102413, China

**E-mail: gujianzhong2000@aliyun.com.cn*

The proton density distribution of ^{46}Ar is calculated in the framework of the Skyrme-Hartree-Fock (SHF) approach with the SLy5, SLy5+T and SLy5+T$_w$ interactions. It is shown that the tensor force in the SLy5+T$_w$ interaction favors its bubble structure due to the inversion between the single-proton states $2s_{1/2}$ and $1d_{3/2}$. However, the SLy5+T interaction can not lead to any $2s_{1/2}$-$1d_{3/2}$ inversion, consequently no bubble structure can be found. In addition, a detailed discussion on the competition between tensor force and pairing correlation on the bubble formation is performed within the Skyrme-Hartree-Fock-Bogoliubov (SHFB) approach.

Keywords: Tensor force, pairing correlation, bubble structure, SHF(B) approach.

1. Introduction

In the nucleon-nucleon interaction the tensor force is an important ingredient.[1–4] However, in the last century it was usually neglected in the studies of nuclear structure because of its complexity. In recent years, people have studied the shell structure and excitation properties of exotic nuclei within the shell model and various mean-field approaches by taking into account the tensor force.[5–9] In the Skyrme interaction, the tensor force was usually added perturbatively to the existing standard parametrizations,[3,4,6–9] such as SLy5+T[6] and SLy5+T$_w$.[3,10] The calculations indicate that the properties of exotic nuclei are described successfully when the tensor force is included.[6–9]

Recent years, the study on bubble nuclei has attracted much attention.[11–15] A bubble nucleus is an exotic one, which is characterized by a depleted central density. The bubble structure originates from a very low occupation probability of the $s_{1/2}$ state. Studies indicate that the bubble structure requires weak pairing and small deformation,[15] to ensure a low occupation probability of the s state. The bubbles have been predicted in various nuclei.[11–15] Study on bubble nuclear structure began with the work of Wilson in 1946,[16] to describe the low-energy excitations of a thin spherical shell. The first microscopic calculations for the bubble structure appeared around the 1970s.[17] Although the nuclear bubbles have not been detected yet, several experimental programs were proposed, such as the electron scattering on exotic beams and direct reactions in inverse kinematics.[13,18,19] Experimental study on the possibility of bubble nuclei is important and interesting, which can be used to test the validity of various nuclear models.

On the bubble formation, studies suggested that pairing correlations would lead to a significant occupancy of the $s_{1/2}$ state,[13,14] thus weakening the bubble structure. The tensor force, pairing correlation and bubble nuclei are all important aspects in the study of exotic nuclear structure. Therefore, it is interesting to discuss the bubble structure combining the tensor force and pairing correlation, which may deepen our understanding of the nucleon-nucleon interaction deeply. In this paper, we will discuss the tensor force and pairing correlation effects on the proton bubble structure of ^{46}Ar within the SHF and SHFB approaches.

2. Theoretical Framework

In the Skyrme interactions, the tensor force is the most complicated term which not only depends on the momenta of the two interacting nucleons but also on their spins. The triplet-even and triple-odd zero-range tensor terms are expressed as

$$\begin{aligned} v_T = \frac{T}{2} & \left[(\sigma_1 \cdot \mathbf{k}') (\sigma_2 \cdot \mathbf{k}') - \frac{1}{3} (\sigma_1 \cdot \sigma_2) \mathbf{k}'^2 \right] \delta (\mathbf{r}_1 - \mathbf{r}_2) \\ & + \frac{T}{2} \delta (\mathbf{r}_1 - \mathbf{r}_2) \left[(\sigma_1 \cdot \mathbf{k}) (\sigma_2 \cdot \mathbf{k}) - \frac{1}{3} (\sigma_1 \cdot \sigma_2) \mathbf{k}^2 \right] \\ & + U \left[(\sigma_1 \cdot \mathbf{k}') \delta (\mathbf{r}_1 - \mathbf{r}_2) (\sigma_2 \cdot \mathbf{k}) \right] \\ & - \frac{1}{3} U (\sigma_1 \cdot \sigma_2) \left[\mathbf{k}' \cdot \delta (\mathbf{r}_1 - \mathbf{r}_2) \mathbf{k} \right], \end{aligned} \tag{1}$$

where the operator $\mathbf{k}=(\nabla_1-\nabla_2)/2i$ acts on the right and $\mathbf{k'}=-(\nabla_1-\nabla_2)/2i$ acts on the left. The coupling constants T and U are the strengths of the triplet-even and triple-odd tensor interactions, respectively. The tensor force leads to the contributions to the energy density and spin-orbit potential. The details can be found from many research papers.[3,4,6–9] In this work, all calculations were performed using the HFBRAD code [20] with the Skyrme force in the particle-hole channel. The box and mesh sizes are selected as 30 fm and 0.1 fm, respectively.

3. Results and Discussions

We have performed calculations for the proton density distributions by using the SHF approach with the SLy5, SLy5+T and SLy5+T$_w$ interactions. The calculated proton densities with and without the tensor force as a function of the distance r are shown in Fig. 1. From Fig. 1, one can see that the density distributions are almost the same and the bubble structure does not exist for the SLy5 and SLy5+T interactions. This means that the density distribution is not influenced by the tensor force from the SLy5+T interaction. However, as to the SLy5+T$_w$ interaction, it causes a serious depletion in the central density region, which can be observed clearly from Fig. 1. In order to explain the bubble formation, the single proton energies by the SHF approach with the SLy5, SLy5+T and SLy5+T$_w$ interactions are plotted in Fig. 2. As can be seen from Fig. 2, the $2s_{1/2}$ state is below the Fermi energy for the SLy5 and SLy5+T interactions so that this state is fully occupied by protons. Thus no bubble structure can be observed by the two interactions. But for SLy5+T$_w$ interaction, the $1d_{3/2}$ state goes down below the $2s_{1/2}$ state due to the strong tensor force, which leads to the vacancy of the $2s_{1/2}$ state. As a result, the central density drops strongly. Note that the reason that the single-particle orbital shifts driven by the tensor force is not given, which can be found from our previous works.[21–23]

Next let us discuss the pairing correlation on the bubble formation with different pairing interactions by the SHFB approach. In the pairing channel, we use the mixing pairing interaction, which is written as[20]

$$V = \left(t_0' + \frac{t_3'}{6}\rho^{\gamma'}\right)\delta, \tag{2}$$

where $t_3' = -18.75t_0'$, $\gamma' = 1$, and t_0' is the pairing strength parameter, which is adjustable.

The calculated proton density distributions of ^{46}Ar with the SLy5+T$_w$ interaction for different t_0' values are plotted in Fig. 3. From Fig. 3, one can

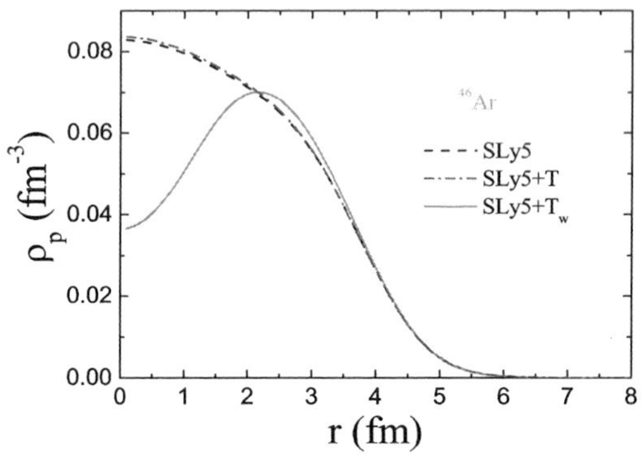

Fig. 1. Proton density distributions of ^{46}Ar by the SHF approach with the SLy5, SLy5+T, and SLy5+T$_w$ interactions.

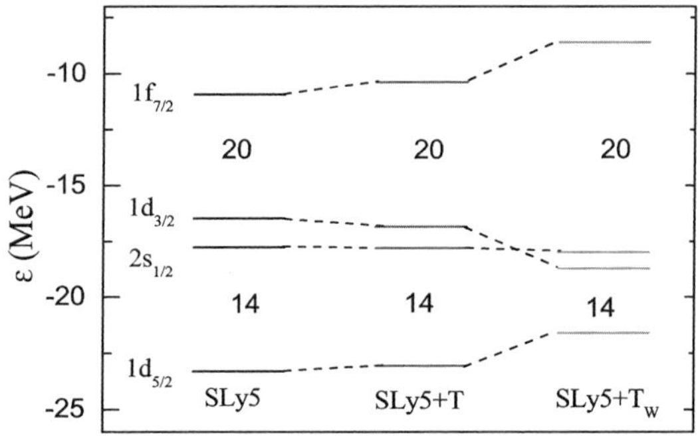

Fig. 2. Proton single-particle spectra of ^{46}Ar by the SHF approach with the SLy5, SLy5+T and SLy5+T$_w$ interactions.

see that the depletion in the nuclear interior becomes weaker and weaker with the increment of the pairing strength. The density depletion in the center of the nucleus disappears when $t_0^{'}$ value is about -350.0. We have known that the $2s_{1/2}$ state moves significantly above the $1d_{3/2}$ state for the SLy5+T$_w$ interaction. Because of the pairing correlation the scattered protons could populate the $2s_{1/2}$ state. The stronger the pairing correlation, the more occupation probability of the $2s_{1/2}$ state. So the depletion in the

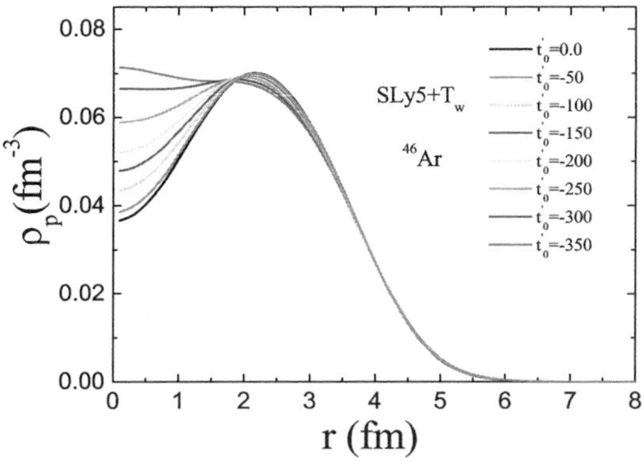

Fig. 3. Proton density distributions of ^{46}Ar for different t_0' values in the case of the SLy5+T$_w$ interaction.

nuclear interior is hindered by different pairing correlations. We know that the t_0' value is usually determined by the empirical pairing gap, which can be extracted from the formula of odd-even mass difference (e.g., the 5-point formula). But in addition to the pairing interaction, the residual interaction can not be canceled by the formula. So it is very difficult to know the precise value of pairing energy gap. If the pairing correlation is not strong in ^{46}Ar, the existence of bubble structure could be possible. According to the discussions mentioned above, it is not difficult to know that the bubble formation in ^{46}Ar originates from the competition between the tensor force and pairing correlation.

4. Conclusions

In this paper, we have discussed the tensor force effect on the proton bubble structure of ^{46}Ar by using the SHF approach with the SLy5, SLy5+T and SLy5+T$_w$ interactions. It is shown that the tensor force in the SLy5+T$_w$ interaction favors its bubble structure due to the $2s_{1/2}$-$1d_{3/2}$ inversion. However, the tensor force in the SLy5+T interaction does not induce any bubble structure. In addition, the pairing correlation effect on the bubble formation has been studied by the SHFB approach with the SLy5+T$_w$ interaction. It is found that the bubble structure in ^{46}Ar stems from the competition between the tensor force and pairing correlation.

5. Acknowledgments

We are thankful to Professors P. Ring, N. Van Giai, J. N. Ginocchio, X. Z. Zhang, Z. Y. Ma, L. G. Cao, S. G. Zhou, J. Q. Li, and F. S. Zhang for helpful discussions. This work was supported by the National Natural Science Foundation of China (Grants 11275271, 11305109), the funds for Creative Research Groups of China (Grant 11321064), the Natural Science Foundation of Hebei Province of China (Grants A2012210043, A2014210005), the Research Foundation for College Outstanding Young Scholars of Hebei Province (Grant BJ2014052), the Specialized Research Foundation of Shijiazhuang TieDao University (Grant Z110197), and the Research Foundation for Excellent Young Scholars of Shijiazhuang TieDao University (Grant Z110306).

References

1. R. B. Wiringa, V. Fiks, and A. Fabrocini, Phys. Rev. C **38,** 1010 (1988).
2. M. Zalewski, P. Olbratowski, M. Rafalski *et al.*, and R. A. Wyss, Phys. Rev. C **80,** 064307 (2009).
3. C. L. Bai, H. Q. Zhang, H. Sagawa *et al.*, and F. R. Xu, Phys. Rev. Lett. **105,** 072501 (2010).
4. L. G. Cao, G. Coló, H. Sagawa *et al.*, Phys. Rev. C **80,** 064304 (2009).
5. T. Otsuka, T. Matsuo, and D. Abe, Phys. Rev. Lett. **97,** 162501 (2006).
6. G. G. Coló, H. Sagawa, S. Fracasso, and P. F. Bortignon, Phys. Lett. **B646,** 227 (2007).
7. D. M. Brink and F. Stancu, Phys. Rev. C **75,** 064311 (2007).
8. M. Moreno-Torres, M. Grasso, H. Liang *et al.*, Phys. Rev. C **81,** 064327 (2010).
9. Y. Z. Wang, J. Z. Gu, J. M. Dong, and X. Z. Zhang, Phys. Rev. C **83,** 054305 (2011).
10. F. Stancu, D. M. Brink and H. Flocard, Phys. Lett. **B68,** 108 (1977).
11. M. Bender, K. Rutz, P.-G. Reinhard *et al.*, Phys. Rev. C **60,** 034304 (1999).
12. B. G. Todd-Rutel, J. Piekarewicz and P. D. Cottle, Phys. Rev. C **69,** 021301(R) (2004).
13. E. Khan, M. Grasso, J. Margueron, N. Van Giai, Nucl. Phys. **A800,** 37 (2008).
14. M. Grasso, L. Gaudefroy, E. Khan *et al.*, Phys. Rev. C **79,** 034318 (2009).
15. M. Grasso *et al.*, Int. J. Mod. Phys. **E18,** 2009 (2009).
16. H. A. Wilson, Phys. Rev. **69,** 538 (1946).
17. X. Campi and D. W. L. Sprung, Phys. Lett. **B46,** 291 (1973).
18. J. Liu, Y. Y. Chu, Z. Z. Ren, Z. Z. Wang, Chin. Phys. C **36,** 48 (2012).
19. Z. Z. Wang, Z. Z. Ren, T. K. Dong, Chin. Phys. C **38,** 024102 (2014).
20. K. Bennaceur and J. Dobaczewski, Comput. Phys. Commun. **168,** 96 (2005).
21. Y. Z. Wang, J. Z. Gu, X. Z. Zhang *et al.*, Phys. Rev. C **84,** 044333 (2011).
22. Y. Z. Wang, J. Z. Gu, X. Z. Zhang *et al.*, Chin. Phys. Lett. **28,** 102101 (2011).
23. Y. Z. Wang, J. Z. Gu, Z. Y. Li *et al.*, Eur. Phys. J. A **49,** 15 (2013).

COVARIANT DENSITY FUNCTIONAL ANALYSIS OF SHAPE EVOLUTION IN $N = 40$ ISOTONES

Zhi Heng Wang, Jian Xiang, and Wen Hui Long*

School of Nuclear Science and Technology, Lanzhou University, Lanzhou 730000, China

**E-mail: longwh@lzu.edu.cn*

Zhi Pan Li†

School of Physical Science and Technology, Southwest University, Chongqing 400715, China

†E-mail: zpliphy@swu.edu.cn

The structure of low-lying excitation states of even-even $N = 40$ isotones is studied using a five-dimensional collective Hamiltonian with the collective parameters determined from the relativistic mean-field plus BCS method with the PC-PK1 functional in the particle-hole channel and a separable paring force in the particle-particle channel. The theoretical calculations in mean-field level present a picture of spherical-oblate-prolate shape transition in the N = 40 isotones and the systematics of the low-lying states along the isotonic chain is reproduced.

Keywords: N=40 isotones, covariant density functional, collective Hamiltonian, shape evolution.

1. Introduction

In the last decades the evolution of shapes and underlying shell structure in the $N = 40$ isotones have attracted lots of attentions. In a simplistic shell-model,[1] $N = 40$ is predicted to be a subshell since neutrons fill up the fp shell and leave the orbit $1g_{9/2}$ remaining empty. This is supported by the measurements of the low-lying excited states and transition rates in ^{68}Ni, namely a rather high 2_1^+ state, the first excited 0^+ lower than 2_1^+ state[2] and very small $B(E2; 2_1^+ \rightarrow 0_1^+)$.[3] However, even more experiments provide evidences for the collectivity of $N = 40$ isotones.[4–9] On the theoretical side, a number of models, for example shell model [3,10] and five-dimensional collective Hamiltonian (5DCH) based on the Harteee-Fock-Bogoliubov (HFB) model with the Gogny force D1S,[11] suggest emergency of collectivity in $N = 40$ isotones.

Since 1980s, the covariant density functional theory (CDFT)[12,13] becomes more and more popular in exploring the ground state and excitation properties of both spherical and deformed nuclei all over the nuclear chart. One of the typical candidates is the relativistic mean-field (RMF) theory[14–17] based on the finite-range meson exchange picture. Recently under the limit of zero range, the CDFT with a point-coupling interaction has attracted more and more attentions[18] because of its advantages in the extensions for nuclear low-lying excited states by implementing projection techniques,[19] the generator coordinate methods[20] and collective Hamiltonian.[24] Specifically, the CDFT-based 5DCH has achieved great success in describing the low-lying states of finite nuclei ranging from $A \sim 40$ to superheavy regions, including the spherical, transitional, and well-deformed ones.[22,23] In this work, the shape evolution and low-lying states will be studied along the isotonic chain of $N = 40$ by the 5DCH based on CDFT. The formalisms of CDFT based on point-coupling density functionals and the framework of 5DCH have been elaborated in a variety of literatures mentioned before and will not be repeated here.

2. Results and Discussion

Figure 1 displays the potential energy surfaces (PESs) of even-even $N = 40$ isotones from ^{62}Ti to ^{80}Zr. Similar results have been obtained by the HFB calculations with the Gogny force D1S.[11] The PESs nicely illustrate the rapid shape evolution along the isotonic chain. Below ^{68}Ni the nuclei are spherical, and ^{70}Zn becomes rather soft along the β direction although the global minimum locates at spherical point. When adding more protons, a triaxial minimum at ($\beta \approx 0.25, \gamma \approx 36°$) emerges in ^{72}Ge, whereas in ^{74}Se an extended oblate minimum is found. Further the coexistence of an extended oblate minimum and a large deformed prolate minimum occurs in ^{76}Kr and these two minima are connected by a rather low triaxial barrier. Extending to $Z = 38$ and 40, the prolate minima become deeper and the triaxial barriers between the prolate and spherical minima rise higher. Thus, a picture of spherical-oblate-prolate shape transition is presented along the isotonic chain.

Two of the distinct features for shape transitions are the excitation energy of the first 2^+ state and the reduced $E2$ transition probability $B(E2; 2_1^+ \to 0_1^+)$, which are shown in Fig. 2 for the $N = 40$ isotones. In Ref.[11] , similar calculations have been done using the Gogny force D1S

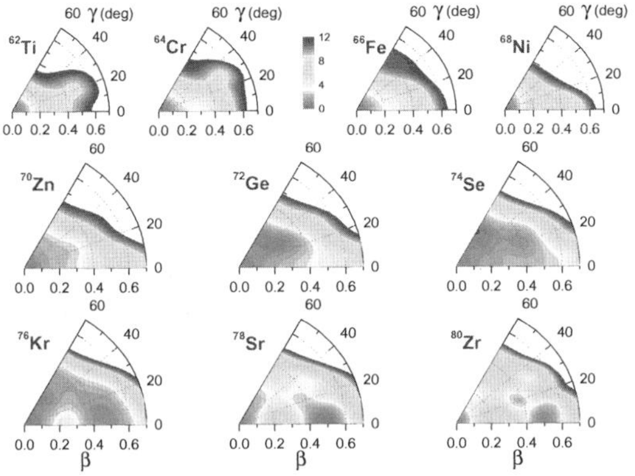

Fig. 1. (Color online) Potential energy surfaces of even-even $N = 40$ isotones in the $\beta-\gamma$ plane extracted from the constrained RMF calculations using the PC-PK1 functional plus BCS pairing. All the energies are normalized to the absolute minimum. The energy difference between neighboring contour lines is 0.5 MeV. See the text for details. Taken from 26.

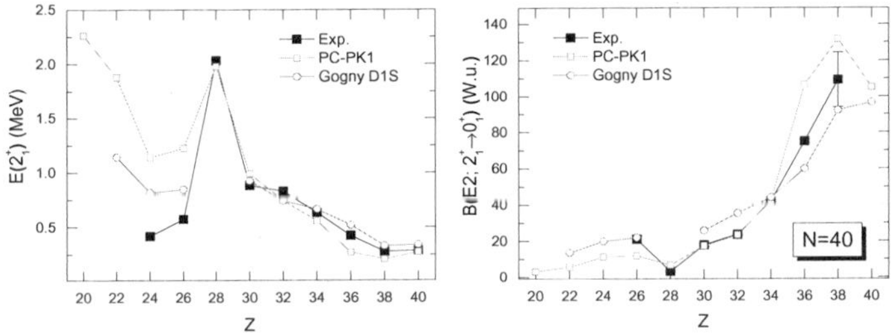

Fig. 2. (Color online) Excitation energies of 2_1^+ states and transition probabilities $B(E2; 2_1^+ \rightarrow 0_1^+)$ as a function of proton number for the $N = 40$ isotones, in comparison with the similar calculations using the Gogny force D1S[11] and the available data.[27] See the text for details. Taken from 26.

and it has been also adopted in a global study of the low-lying states of the nuclei with proton numbers ranging from $Z = 10$ to $Z = 110$ and neutron numbers $N \leqslant 200$.[25] For comparison, the corresponding results are plotted in Fig. 2. It can be seen that the 5DCH calculations based on the PC-PK1 functional give results similar to those obtained by the Gogny force D1S and both reproduce the systematics of the low-lying states for the isotones with $Z > 28$, particularly for the onset of the enhanced collectivity around $Z = 38$. For the neutron rich side, both calculations overestimate the excitation energies of 2_1^+ states, leading to underestimated collectivity.

3. Summary

In summary, we have presented a systematic beyond RMF study of the low-lying states in the even-even $N = 40$ isotones. The excitation energies and electric transition strengths have been obtained by solving a five-dimensional collective Hamiltonian with the collective parameters determined from the RMF calculations using the PC-PK1 force. The results have been compared with experimental data and those obtained by similar calculations but using the Gogny force D1S. The theoretical calculations give a picture of spherical-oblate-prolate shape transition in the $N = 40$ isotones and the systematics of the low-lying states along the isotonic chain is well reproduced.

References

1. M. G. Mayer, J. H. D. Jensen, Elementary theory of nuclear shell structure vol 58 (Wily New York) (1955).
2. M. Bernas *et al*, Phys. Lett. B **113** 279-282 (1982).
3. O. Sorlin *et al*, Phys. Rev. Lett **88** 092501 (2002).
4. W. F. Mueller *et al*, Phys. Rev. Lett **83** 3163 (1999).
5. O. Sorlin *et al*, Nucl. Phys. A **682** 183 (2001).
6. A. Gade *etal*, Phys. Rev. C **81** 051304 (2010).
7. O. B. Tarasov *et al*, Phys. Rev. Lett **102** 142501 (2009).
8. J. Ljungvall *et al*, Phys. Rev. Lett **100** 102502 (2008).
9. E. Clément *et al*, Phys. Rev. C **75** 054313 (2007).
10. S. M. Lenzi *et al*, Phys. Rev. C **82**, 054301 (2010).
11. L. Gaudefroy *et al*, Adv. Nucl. Phys. **16**, 064313 (2009).
12. J. D. Walecka *et al*, Ann. Phys. (NY) **83**, 491(1974).
13. B. D. Serot *et al*, Nucl. Part. Phys. **26**, 1 (1986).
14. P. G. Reinhard *et al*, Rep. Prog. Phys. **52**, 439 (1989).
15. P. Ring *et al*, Prog. Part. Nucl. Phys. **37** 193 (1996).
16. D. vretenar *et al*, Phys. Rep. **409** 101 (2005).
17. J. Meng *et al*, Prog. Part. Nucl. Phys.**57** 470 (2006).
18. T. Nikšić *et al*, Prog. Part. Nucl. Phys.**66**, 519 (2011).
19. J. M. Yao *et al*, Phys. Rev. C **79**, 044312 (2009).
20. J. M. Yao *et al*, Phys. Rev. C **81**, 044311 (2010).
21. T. Nikšić *et al*, Phys. Rev. C**79**, 034303 (2009).
22. P. W. Zhao *et al*, Phys. Rev. C **82**, 054319 (2010).
23. Z. P. Li *et al*, Phys. Rev. C **79**, 054301 (2009).
24. T. Nikšić *et al*, Phys. Rev. C**79**, 034303 (2009).
25. J. P. Delaroche *et al*, Phys. Rev. C**81**, 014303 (2010).
26. Z. H. Wang *et al*, Preprint arXiv: 1411.3821 (2014).
27. Brookhaven database, http://www.nndc.bnl.gov/.

THE STRUCTURE OF STRANGE STARS WITH A NEW QUARK MASS SCALING

C. J. Xia*

State Key Laboratory of Theoretical Physics, Institute of Theoretical Physics, Chinese Academy of Sciences, Beijing 100190, China
**E-mail: cjxia@itp.ac.cn*

On application of a newly obtained quark mass scaling which contains both confinement and perturbative interactions, we study the structure of strange stars. We find that the equation of state becomes stiffer when the perturbative interactions are dominant at higher density, and, correspondingly, the maximum mass of strange stars is as large as two times the solar mass if strange quark matter is absolutely stable or metastable.

Keywords: Equiparticle model; confinement; leading-order perturbative interaction; strange stars.

In 1971, Bodmer suggested a low energy nuclear state called "collapsed nuclei".[1] Subsequently, Witten conjectured the absolutely stablity of strange quark matter (SQM).[2] Since then, the study of SQM has been of great significance in exploring the mechanism of QCD phase transition, and the applications to astrophysics, astronomy, and cosmology, etc.

It has been suggested a long time ago that pulsars are, in fact, strange (quark) stars composed of SQM, rather than neutron stars.[2–5] The neutron star could be converted to a strange star or mixed star due to leptonic weak interactions, or seeded with strangelets by the self-annihilating weakly interacting massive particles.[6] The structure of a strange star depends strongly on the stability of SQM which are still under active investigations.[7] Presently, many aspects on quark matter are still open. Among them the equation of state (EoS) is of special interest.

In principle, the properties of SQM can be studied based on the fundamental theory of strong interactions, i.e., quantum chromodynamics (QCD). However, due to the known difficulties in the nonperturbative regime, phenomenological models are essential for the study of SQM. The most well known phenomenological model is the MIT bag model, where the fundamental assumption is that the vacuum has a constant energy density, i.e., the bag constant B which provides a negative pressure to confine quarks.

Another kind of models achieve quark confinement by the density dependence of quark masses (CDDM).[8] In recent years, this type of models are extensively applied to investigating various aspect of SQM.[9–19]

There are several different versions distinguished by the quark mass scaling and thermodynamic treatment. Normally, the equivalent quark masses take the following form:

$$m_q = m_{q0} + m_{\mathrm{I}}(n, T), \tag{1}$$

where m_{q0} is the current mass of quark flavor q, and the baryon number density (n) and/or temperature (T) dependent term m_{I} takes account of the strong interactions between quarks. Originally the quark mass scaling was proposed to be inversely proportional to the density,[9] which was later extended to finite temperature.[20] Based on the in-medium chiral condensates and linear confinement, an inversely cubic-root scaling was derived,[10,21] and the effects of finite temperature[13] and one-gluon-exchange interaction[22,23] were also studied. Recently, we have obtained a new quark mass scaling with the inclusion of both confinement and leading-order perturbative interactions by expanding the equivalent mass to a Laurant series and taking the leading term in both directions.[19] In this paper, we apply it to study the structure of strange stars.

Considering both the linear confinement and leading order perturbative interactions, the quark mass scaling can be expressed as[19]

$$m_q = m_{q0} + \frac{D}{n^{1/3}} + C n^{1/3}, \tag{2}$$

where C is the perturbative strength parameter and D the confinement strength parameter. The parameter C is linked to the strong coupling constant by $C \approx C_1\sqrt{\frac{2}{3}\alpha}$ and C_1 takes constant values. Here m_{q0} is the current mass of quark flavor q. Since the strong coupling α runs logarithmically, the running rate is thus much slower and the parameter C can be taken as a constant. According to the analytic coupling expression,[24] the maximum value of α is $1/\beta_0$, then we have $C < \left(\frac{3\pi^2}{N_{\mathrm{f}}}\right)^{1/3}\sqrt{\frac{2}{3\beta_0}} \approx 1.1676$. Note that if we take C to be a negative value the scaling in Eq. (2) coincide with the mass formula obtained by including the one-gluon-exchange interaction.[22]

In a model with density-dependent particle masses, the most disputable issue is the thermodynamic inconsistency problem. According to the fundamental thermodynamic differential equations, we proved a fully self-consistent thermodynamic treatment.[8,19] Then the energy density of SQM at zero temperature is

$$E = \Omega_0(\mu_i^*, m_i) + \sum_i n_i \mu_i^*, \tag{3}$$

where μ_i^* is an effective chemical potential while the energy density and particle number density take the same form as those of a free particle system. However, the pressure P has an additional term due to the density dependence of quark masses, i.e.,

$$P = -\Omega_0(\mu_i^*, m_i) + n \frac{\mathrm{d}m_\mathrm{I}}{\mathrm{d}n} \frac{\partial \Omega_0}{\partial m_\mathrm{I}}. \tag{4}$$

Here Ω_0 takes the same form as the thermodynamic potential density of a free system with the particle masses $m_i(n)$ and chemical potentials μ_i^*:

$$\Omega_0(\mu_i^*, m_i) = -\sum_i \frac{g_i}{24\pi^2} \left[\mu_i^* \nu_i (\nu_i^2 - \frac{3}{2} m_i^2) + \frac{3}{2} m_i^4 \ln \frac{\mu_i^* + \nu_i}{m_i} \right]. \tag{5}$$

The degeneracy factor g_i is $g_u = g_d = g_s = 6$ and $g_e = 2$. For the actual chemical potential μ_i, it is connected to the effective one by $\mu_i = \mu_i^* + \frac{1}{3}\frac{\mathrm{d}m_\mathrm{I}}{\mathrm{d}n}\frac{\partial\Omega_0}{\partial m_\mathrm{I}}$, where the effective chemical potential μ_i^* can be linked to the Fermi momentum ν_i by $\mu_i^* \equiv \sqrt{\nu_i^2 + m_i^2}$. Because the electron mass is constant, the corresponding effective chemical potential is the same as the true chemical potential: $\mu_e = \mu_e^*$. The particle number density is $n_i = \frac{g_i \nu_i^3}{6\pi^2}$, and the baryon number density is given by $n = \frac{1}{3}\sum_i n_i$.

For stable SQM, the chemical potentials should satisfy the weak equilibrium condition $\mu_u + \mu_e = \mu_d = \mu_s$. Thus, for given baryon number density n, we first obtain the equivalent masses by applying Eq. (2), then the effective chemical potentials can be obtained by adopting the weak equilibrium condition and the charge neutrality condition $\sum_i q_i n_i = 0$ ($q_u = 2/3$, $q_d = q_s = -1/3$ and $q_e = -1$) with baryon number conservation $n = \frac{1}{3}\sum_i n_i$. The energy density and pressure can then be determined with Eq. (3) and Eq. (4).

Based on previous discussions, the EoS of SQM for various combinations of parameters is obtained. The results are presented in Fig. 1. According to Fig. 1, we have two general observations: (1) the EoS becomes stiffer for parameters within the "SQM stable" area; (2) the stiffness increases with increasing C or decreasing D. For the running coupling case of the curve labeled with '$C_1 = 2.145$', we adopt the fast convergent expression[25] and take the leading term.[19] With the obtained EoS, the structure of a strange

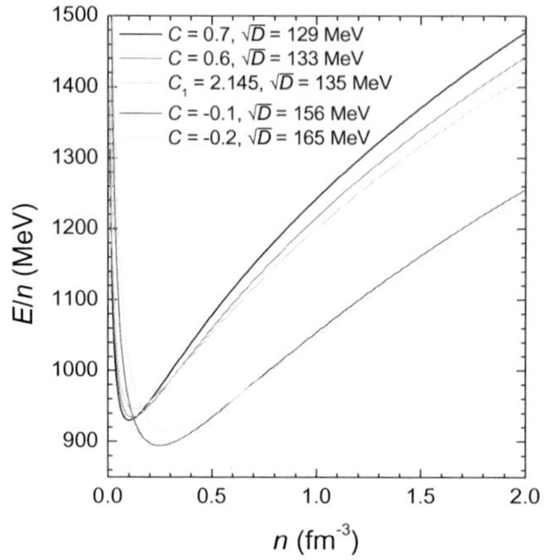

Fig. 1. Equation of state of SQM at $m_{u0} = 5$ MeV, $m_{d0} = 10$ MeV and $m_{s0} = 100$ MeV with different D and C values indicated in the legend. The curves are taken from Refs. 19,26.

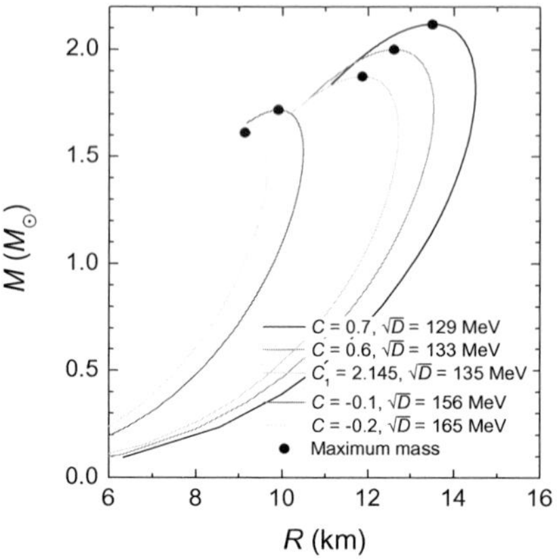

Fig. 2. Mass-radius relation of strange stars for various parameter sets. The curves are taken from Refs. 19,26, for more details please refer to these references.

star can be investigated by solving the Tolman-Oppenheimer-Volkov (TOV) equation

$$\frac{\mathrm{d}P}{\mathrm{d}r} = -(P+E)\frac{G}{r^2}\frac{4\pi r^3 P + M}{1-2GM/r}, \tag{6}$$

$$\frac{\mathrm{d}M}{\mathrm{d}r} = 4\pi r^2 E. \tag{7}$$

The mass-radius relations are then plotted in Fig. 2 where the maximum star masses are marked with full dots. For various $(C, \sqrt{D})$ pairs, the maximum masses are presented in Fig. 3, as well as the range of parameters C and D. It is found that the maximum star mass increases with increasing perturbative strength parameter C, but decreases with increasing confinement strength parameter D. By going towards the lower-right direction in Fig. 3 within the "SQM stable" area, the maximum mass of strange stars can be as large as $2M_\odot$. This is a strong indication that perturbative interactions play an important role in compact objects.

It should be noted that when a comparatively large C value is used to produce a large maximum mass, the density at the surface of the star becomes very small, which corresponds to the density that gives zero pressure in Fig. 1. The possibility that the surface density is small, even smaller than the normal nuclear saturation density, is clearly a signal for the phase transition to nuclear matter. Therefore, further investigations on the QCD

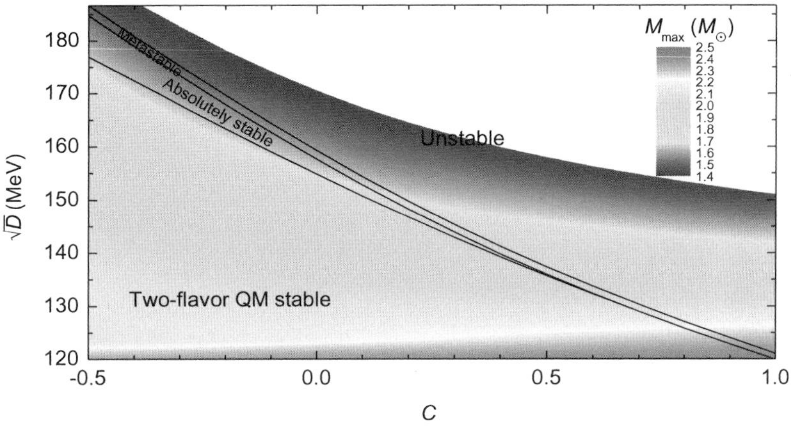

Fig. 3. The maximum strange-star masses $M_{\max}$ for the parameter sets of $(C, \sqrt{D})$ pairs. The 3 solid curves divide the whole area into 4 regions. In the upper-right region, SQM is unstable, while the lower-left is the forbidden region where two-flavor quark matter (QM) is stable. Between those curves are the SQM absolutely stable and metastable regions.

phase diagram[8] are needed in future works with the new quark mass scaling. For this purpose, a refined nuclear model, e.g. the relativistic mean-field model is definitely necessary.

Based on the self-consistent thermodynamic treatment, and taking into account the linear confinement and perturbative interactions in a new quark mass scaling, we have studied the properties of SQM and the structures of strange stars. It is found that the perturbative interactions make the EoS stiffer and, correspondingly, the maximum strange-star mass as large as two times the solar mass, which is consistent with the masses and radii of PSR J1614-2230 and PSR J0348+0432. Therefore, the perturbative interactions play an important role in studying compact stars due to the high density in the inner part. Nevertheless, the density at the surface region is comparatively small, indicating a possible phase transition from quark matter to nuclear matter, which will be studied in the future.

Acknowledgments

It is a great pleasure to thank S. W. Chen, Z. Y. Lu, and J. F. Xu for helpful discussions. The financial supports from National Natural Science Foundation of China (Grant Nos. 11135011, 11121403, 11175252, 11120101005,

11211120152, 11275248, and 11475110), National Key Basic Research Program of China (Grant No. 2013CB834400), and the Knowledge Innovation Project of the Chinese Academy of Sciences (Grant No. KJCX2-EW-N01) are appreciated.

References

1. A. R. Bodmer, *Phys. Rev. D* **4**, 1601 (1971).
2. E. Witten, *Phys. Rev. D* **30**, 272 (1984).
3. N. Itoh, *Prog. Theor. Phys.* **44**, 291 (1970).
4. C. Alcock, E. Farhi and A. Olinto, *Astrophys. J.* **310**, 261 (1986).
5. P. Haensel, J. L. Zdunik and R. Schaeffer., *Astron.Astrophys.* **160**, 121 (1986).
6. M. A. Perez-Garcia, J. Silk and J. R. Stone, *Phys. Rev. Lett.* **105**, 141101 (2010).
7. V. Dexheimer, J. Torres and D. Menezes, *Eur. Phys. J. C* **73**, 1 (2013).
8. G. X. Peng, A. Li and U. Lombardo, *Phys. Rev. C* **77**, 065807 (2008).
9. G. N. Fowler, S. Raha and R. M. Weiner, *Z. Phys. C* **9**, 271 (1981).
10. G. X. Peng, H. C. Chiang, J. J. Yang, L. Li and B. Liu, *Phys. Rev. C* **61**, 015201 (1999).
11. G. Lugones and J. E. Horvath, *Int. J. Mod. Phys. D* **12**, 495 (2003).
12. X. Zheng, X. Liu, M. Kang and S. Yang, *Phys. Rev. C* **70**, 015803 (2004).
13. X. J. Wen, X. H. Zhong, G. X. Peng, P. N. Shen and P. Z. Ning, *Phys. Rev. C* **72**, 015204 (2005).
14. M. Modarres and H. Gholizade, *Int. J. Mod. Phys. E* **17**, 1335 (2008).
15. A. Li, G. X. Peng and J. F. Lu, *Research in Astron. Astrophys.* **11**, 482 (2011).
16. J. R. Torres and D. P. Menezes, *Europhys. Lett.* **101**, 42003 (2013).
17. P. C. Chu and L. W. Chen, *Astrophys. J.* **780**, 135 (2014).
18. C. J. Xia, S. W. Chen and G. X. Peng, *Sci. Chin. Phys. Mech. Astron.* **57**, 1304 (2014).
19. C. J. Xia, G. X. Peng, S. W. Chen, Z. Y. Lu and J. F. Xu, *Phys. Rev. D* **89**, 105027 (2014).
20. Y. Zhang and R. K. Su, *Phys. Rev. C* **65**, 035202 (2002).
21. G. X. Peng, H. C. Chiang, B. S. Zou, P. Z. Ning and S. J. Luo, *Phys. Rev. C* **62**, 025801 (2000).
22. S. W. Chen, L. Gao and G. X. Peng, *Chin. Phys. C.* **36**, 947 (2012).
23. S. W. Chen and G. X. Peng, *Commun. Theor. Phys.* **57**, 1037 (2012).
24. D. V. Shirkov and I. L. Solovtsov, *Phys. Rev. Lett.* **79**, 1209 (1997).
25. G. X. Peng, *Phys. Lett. B* **634**, 413 (2006).
26. C. J. Xia, Thermodynamic consistency and properties of quark matter, PhD thesis, University of Chinese Academy of Sciences 2014.

TRANSVERSE KINEMATICAL ENERGY SPECTRA OF K^+ MESONS IN HEAVY ION COLLISIONS

Yong-Zhong Xing*
Department of physics, Tianshui Normal University, Tianshui 741000, Gansu, China
**E-mail: yzxing@tsnc.edu.cn*

Yu-Ming Zheng
China Institute of Atomic Energy, P. O. Box 275(18), Beijing 102413, China

Pornrad Srisawad
Department of physics, Faculty of Science, Naresuan University, Thailand

The transverse kinematical energy spectra of K^+ mesons from ${}^{58}_{28}\mathrm{Ni} + {}^{58}_{28}\mathrm{Ni}$ collisions at 1.93 AGeV are investigated within the quantum molecular dynamics model based on covariant kaon dynamics. We observe that the KaoS data can only be reasonably described by calculated results with a repulsive K^+N potential. This indicates that one can abstract the information on the in-medium kaon potential in a nuclear medium from the analysis of the transverse kinematical energy spectra of K^+ mesons. It is found that the transverse kinematical energy spectrum of K^+ mesons is a sensitive observable to probe the nuclear equation of state.

Keywords: Heavy ion collision; kaon production; transverse kinematical energy spectra; covariant kaon dynamics.

1. Introduction

One of the hot points in nuclear physics is to study the kaon production in heavy ion collisions. This is important for a better understanding of both, a possible restoration of chiral symmetry in dense hadronic matter and the properties of nuclear matter at high densities.

It is known from chiral models[1] that the kaon mean field is related to chiral symmetry breaking. The in-medium effects give rise to an attractive scalar potential inside the nuclear medium which is in mean field approximation proportional to the kaon-nucleon Sigma term Σ_{KN}. A second part of the mean field originates from the interaction with vector mesons. The vector potential is repulsive for kaon K^+ and, due to G-parity conservation, attractive for antikaon K^-.

Kaon is produced in the early stage of heavy ion collisions at which the nuclear density in the reaction zone is much higher than the saturation density ($\rho_0 = 0.16 fm^{-3}$). After its production K^+ meson escapes near-freely from the reaction zone due to the relatively low K^+N scattering cross section ($\sim 10 milibarn$) and the absence of the absorption channel of a K^+ meson on a nucleon in strong interaction. So, kaon is proposed to be a sensitive probe to study the nuclear equation of state (EOS) in dense hadronic matter.[2]

In this work we apply the quantum molecular dynamics (QMD) model based on the covariant kaon dynamics[3] to simulate the $^{58}_{28}$Ni + $^{58}_{28}$Ni collisions at 1.93 AGeV, to analyze the transverse kinematical energy spectra of K^+ mesons, to discuss the influence of the in-medium kaon potential on the transverse kinematical energy spectra of K^+ mesons, and to make a comparison between calculated results and KaoS data. We observe that the KaoS data can only be reasonably described by calculated results with a repulsive K^+N potential. This indicates that one can abstract the information on the in-medium kaon potential in a nuclear medium from the analysis of the transverse kinematical energy spectra of K^+ mesons. It is found that the transverse kinematical energy spectrum of K^+ mesons is a sensitive observable to probe the nuclear equation of state.

2. The Model

In our model the nuclear system is described by the QMD model.[4]

The natural framework to study the interaction between pseudoscalar mesons and baryons at low energies is the chiral perturbation theory (ChPT). From the chiral Lagrangian the field equations for the $K^\pm$–mesons are derived from the Euler-Lagrange equations[5]

$$\left[\partial_\mu \partial^\mu \pm \frac{3i}{4f_\pi^{*2}} j_\mu \partial^\mu + \left(m_k^2 - \frac{\Sigma_{\rm KN}}{f_\pi^{*2}} \rho_s\right)\right] \phi_{{\rm K}^\pm}(x) = 0. \tag{1}$$

Here the mean field approximation has already been applied. In Eq. (1) ρ_s is the baryon scalar density, j_μ is the baryon four-vector current, f_π^* is the in-medium pion decay constant. Introducing the kaonic vector potential

$$V_\mu = \frac{3}{8f_\pi^{*2}} j_\mu, \tag{2}$$

Eq. (1) can be rewritten in the form[6]

$$\left[(\partial_\mu \pm iV_\mu)^2 + m_K^{*2}\right] \phi_{K^\pm}(x) = 0 \ . \tag{3}$$

Thus, the vector field is introduced by minimal coupling into the Klein-Gordon equation. The effective mass m_K^* of the kaon is then given by[6,7]

$$m_K^* = \sqrt{m_K^2 - \frac{\Sigma_{\mathrm{KN}}}{f_\pi^{*2}}\rho_s + V_\mu V^\mu}, \tag{4}$$

where $m_K = 0.496$ GeV is the bare kaon mass. Due to the bosonic character, the coupling of the scalar field to the mass term is no longer linear as for the baryons but quadratic and contains an additional contribution originating from the vector field. The effective quasi-particle mass defined by Eq. (4) is a Lorentz scalar and is equal for K^+ and K^-.

The $K^\pm$ single-particle energy is expressed as

$$\omega_{K^\pm}(\mathbf{k}, \rho) = \sqrt{\mathbf{k}^{*2} + m_K^{*2}} \pm V_0 \tag{5}$$

where $k^* = k \mp V$ is the kaon effective momentum, $k_\mu = (k_0, \mathbf{k})$, $V_\mu = (V_0, \mathbf{V})$. The kaon vector field is introduced by minimal coupling into the Klein-Gordon with opposite signs for K^+ and K^-. m_K^* is the kaon effective (Dirac) mass. The kaon (antikaon) potential $U_{K^\pm}(\mathbf{k}, \rho)$ is defined as

$$U_{K^+}(\mathbf{k}, \rho) = \omega_{K^\pm}(\mathbf{k}, \rho) - \omega_0(\mathbf{k}), \tag{6}$$

where

$$\omega_0(\mathbf{k}) = \sqrt{\mathbf{k}^2 + m_K^2}. \tag{7}$$

The covariant equations of motion are obtained in the classical (testparticle) limit from the relativistic transport equation for the kaons which can be derived from Eqs. (3). They are analogous to the corresponding relativistic equations for baryons and read[6]

$$\frac{dq^\mu}{d\tau} = \frac{k^{*\mu}}{m_K^*} \quad , \quad \frac{dk^{*\mu}}{d\tau} = \frac{k_\nu^*}{m_K^*} F^{\mu\nu} + \partial^\mu m_K^* \quad . \tag{8}$$

Here $q^\mu = (t, \mathbf{q})$ are the coordinates in Minkowski space and $F^{\mu\nu} = \partial^\mu V^\nu - \partial^\nu V^\mu$ is the field strength tensor for K^+. For K^- where the vector field changes sign. The equation of motion are identical, however, $F^{\mu\nu}$ has to be replaced by $-F^{\mu\nu}$. Following Ref.,[3] in this paper we use the Brown & Rho parametrization (BRP): $\Sigma_{\mathrm{KN}} = 450$ MeV, $f_\pi^{*2} = 0.6 f_\pi^2$ for the vector field and $f_\pi^{*2} = f_\pi^2$ for the scalar part given by $-\Sigma_{\mathrm{KN}}/f_\pi^{*2}\rho_s$. Up to saturation density the Brown and Rho potential is $U_{K^+}(\rho_0) \approx 30\,\mathrm{MeV}$.

For the nuclear forces we use the standard momentum dependent Skyrme interactions corresponding to a soft EOS (the compression modulus K is 200 MeV) and a hard EOS (the compression modulus K is 380 MeV). For the determination of the kaon mean field we adpot the corresponding covariant scalar-vector description of the non-linear $\sigma\omega$ model. Here we use the parametrization of Ref.[8] which corresponds to identical soft/hard nuclear EOSs. The shift of the production thresholds of the kaons by the in-medium potentials is taken into account as described in.[6] The hyperon fields are scaled according to SU(3) symmetry.

3. Results and Discussions

Fig. 1 shows the transverse kinematical energy spectra of K^+ mesons as a function of the transverse kinetic energy $(m_T - m)$ for three equally wide bins in rapidity. The transverse energy is given by $m_T = \sqrt{p_T^2 + m^2}$ with the transverse momentum p_T and the rest mass m of K^+ from the $^{58}_{28}$Ni + $^{58}_{28}$Ni collisions at 1.93 AGeV by using the soft EOS with and without the in-medium kaon potential U_K for impact parameter $b \leq 4.5$ fm. In this figure the spectra are extracted in the following c.m. rapidity intervals (from top to bottom): -0.69 $< y_{c.m.} <$ -0.54, 0.54 $< y_{c.m.} <$ -0.39 and -0.39 $< y_{c.m.} <$ -0.24. The scaling factors 10^2, 10^1 and 10^0 are applied to the spectra from top to bottom. The experimental data are taken from Ref.[9] It is seen clearly from this figure that without the U_K, the transverse kinematical energy spectra of K^+ mesons are too narrow with too high small-momentum part and too low large-momentum part. After taking into account the U_K, which is positive for K^+ mesons, K^+ mesons are pushed away from nucleons. In other words, the K^+ mesons are accelerated during their propagation. Thus, the number of kaons with smaller momentum decreases and the number of kaons with larger momentum increases, leading to good fit to the experimental data.

This feature can also be seen obviously from Fig. 2, in which the hard EOS is used. This indicates that the U_K should be take into account in theoretical simulations of the kaon production in heavy ion collisions in order to reasonably describe the experimental data.

The transverse kinematical energy spectra of K^+ mesons as a function of the transverse kinetic energy $(m_T - m)$ from the $^{58}_{28}$Ni + $^{58}_{28}$Ni collisions at 1.93 AGeV by using the soft and hard EOS with the in-medium kaon potential U_K for impact parameter $b \leq 4.5$ fm are given in Fig. 3. In this figure the spectra are extracted in the following c.m. rapidity intervals (from top to bottom): -0.69 $< y_{c.m.} <$ -0.54, 0.54 $< y_{c.m.} <$ -0.39 and -0.39

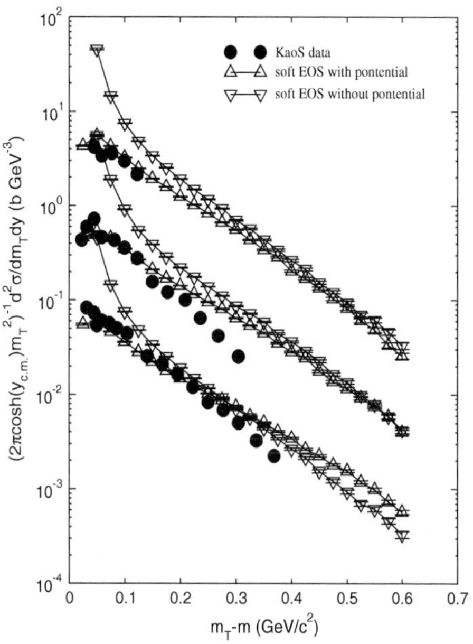

Fig. 1.

$< y_{c.m.} < -0.24$. The scaling factors 10^2, 10^1 and 10^0 are applied to the spectra from top to bottom. Data are taken from Ref.[9] It is known from this figure that the theoretical results given by the soft EOS are in better agreement with data. It is known that nuclear matter with a softer EOS can be more easy compressed. For a given incident energy, in the early stage of reaction process one gets the formed dense matter with higher density for the nuclear matter with a soft EOS. The K^+ mesons produced at this stage feel a stronger in-medium K^+ potential due to that the in-medium K^+ potential increases with increasing of density. The K^+ mesons are accelerated more greatly during their propagation. Thus, in the final state the number of kaons with smaller momentum decreases more and the number of kaons with larger momentum increases more, leading to form a more flat distribution, which better fits to the experimental data. This means that the dense matter created in this reaction has the property with the soft EOS.

In summary, in heavy ion collisions at intermediate energies, i.e. at energies around the threshold region, strangeness is generally produced in the early and high density phase of the reaction. Due to strangeness

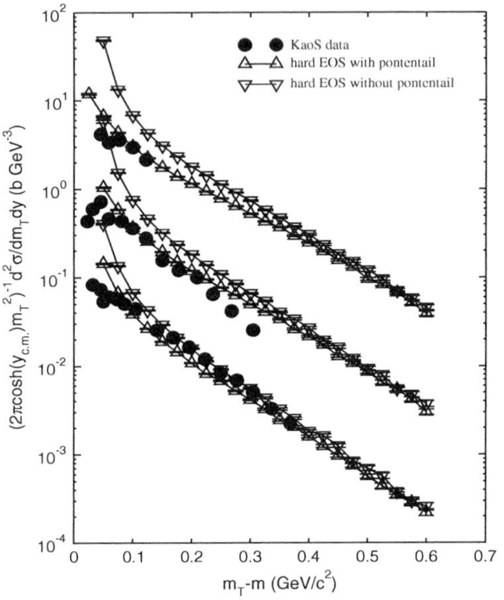

Fig. 2. The same as that in Fig. 1, but by using the hard EOS.

conservation K^+ mesons cannot be reabsorbed by the surrounding nucleons and their chemical freeze-out takes place early. The final state interactions, i.e. the elastic scattering or charge exchange reactions and the influence of the optical kaon-nucleon potential change their dynamical pattern but not the abundances. This makes K^+ mesons to be a suitable probe to study the in-medium properties of hadrons and the EOS in dense nuclear matter created in a heavy ion collision. We have investigated the transverse kinematical energy spectra of K^+ mesons as a function of the transverse kinetic energy $(m_T - m)$ from the $^{58}_{28}$Ni + $^{58}_{28}$Ni collisions at 1.93 AGeV within the QMD model based on the covariant kaon dynamics. We found that this observable can be described by calculated result with the in-medium kaon potential $(U_{K^+}(\rho_0) \approx 30MeV)$ of the Brown & Rho parametrization and the soft EOS (the compression modulus K is 200 MeV). This means that the transverse kinematical energy spectrum of K^+ mesons is a sensitive observable to probe the in-medium kaon potential and the nuclear equation of state in dense nuclear matter.

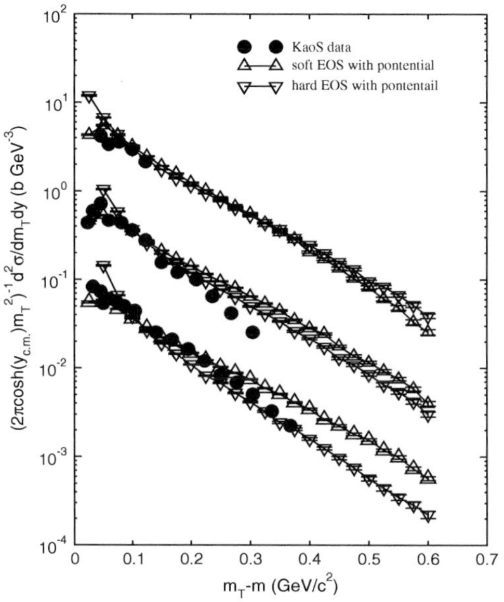

Fig. 3.

Acknowledgments

This work is supported in part by the National Natural Science Foundation of China (NSFC) under Grant No. 11265013 and by the National Research Concil of Thailand (NRCT).

References

1. B. D. Kaplan and A. E. Nelson, Phys. Lett. **B175**, 57 (1986);
   G.E. Brown and M. Rho, Nucl. Phys. **A596**, 503 (1996);
   J. Schaffner, J. Bondorf, I. N. Mishustin, Nucl. Phys. **A625**, 325 (1997);
   T. Waas, N. Kaiser, and W. Weise, Phys. Lett. **B365**, 12 (1996);
   M. Lutz, Phys. Lett. **B426**, 12 (1998).
2. J. Randrup and C. M. Ko, Nucl. Phys. **A343**, 519 (1980).
3. Y. M. Zheng, C. Fuchs, A. Faessler, K. Shekhter, Y. P. Yan, and C. Kobdaj, Phys. Rev. **C69**, 034907 (2004).
4. Y. Z. Xing, P. Srisawad, and Y. M. Zheng, Modern Physics **4**, 50 (2014).
5. G. Q. Li, C. M. Ko, Nucl. Phys. **A594**, 460 (1995);
   C.M. Ko, J. Phys. **G27**, 327 (2001).

6. C. Fuchs, Amand Faessler, E. Zabrodin, Y. M. Zheng, Phys. Rev. Lett. **86**, 1974 (2001);
C. Fuchs, Prog. Part. Nucl. Phys. **56**, 1 (2006).
7. Y. M. Zheng, Z. L. Chu, C. Fuchs, Amand Faessler, W. Xiao, D. P. Hua, Y. P. Yan, Chin. Phys. Lett. **19**, 926 (2002).
8. C.M. Ko, G.Q. Li, J. Phys. **G22**, 1673 (1996);
X. S. Fang, C.M. Ko, G.Q. Li, Y. M. Zheng, Nucl. Phys. **A575**, 766 (1994).
9. P. Senger et al., Nucl. Instr. Meth. Phys. Res. **A327**, 393 (1993).

FURTHER INVESTIGATION OF ISOSPIN ASYMMETRY IN PSEUDOSPIN SYMMETRY

Q. Xu*

College of Physics and Optoelectronics, TaiYuan University of Technology, TaiYuan 030024, P. R. China
**E-mail: xuqiang01@tyut.edu.cn*

Following a paper [R. Lisboa, Phys. Rev. C 67 054305 (2003)], we reinvestigate the role of the Coulomb and ρ potentials in the isospin asymmetry of nuclear pseudospin. By comparing the contributions of these potentials to the splitting of pseudospin doublets, we found the effect of the ρ potential is small and the Coulomb potential gives the main contribution to the observed isospin asymmetry of the psuedospin wavefunction splittings.

Keywords: Pseudospin symmetry; isospin asymmetry; relativistic mean field.

1. Introduction

About half a century ago, Hecht et al.[1] and Arima et al.[2] respectively discovered a quasidegeneracy between two single-particle states with quantum numbers $(n, l, j = l + 1/2)$ and $(n - 1, l + 2, j = l + 3/2)$ in heavy nuclei. In order to describe this phenomena better, this quasidegeneracy was renamed as pseudospin symmetry (PSS) by introducing $j = \tilde{l} \pm \tilde{s}$ with pseudo orbital angular momentum $\tilde{l}$ and pseudospin angular momentum $\tilde{s}$. Many efforts have been made on exploring origin of the PSS.[3,4] Until 1997, Ginocchio made magnificent progress in understanding origin of the PSS by pointing out the PSS is a relativistic symmetry and the PSS is exact when the sum of vector potential V and scalar potential S equals zero.[5] Meng et al. further extended condition of exact PSS as the sum of the scalar S and vector V potentials is equal to a constant[6] and revealed that quality of the PSS approximation in real nuclei is connected with the competition between pseudocentrifugal barrier and pseudospin orbital potential.[7] However, the conditions Σ=0 or $d\Sigma/dr$=0 cannot be realized in nuclei, as was stated in some papers.[5,8] Therefore, it is necessary to study the PSS for relativistic nuclear potential. Many researchers such as Alberto,[9,10] Guo[11] and Xu[12] researched nature of the pseudospin breaking in nuclei by solving Dirac equation with vector and scalar potentials of Woods-Saxon type and harmonic oscillator type, respectively. From the behavior of the splitting with various parameters of these potentials, they showed the PSS has a dynamical character. However, these research is limited within the scope of neutron energy levels. Moreover, Alberto et al. continued to study isospin asymmetry in the PSS and indicated the reason for the different behavior of the PSS of neutrons and protons lies on the ρ meson interaction.[9] *He* also pointed out the ρ potential is more important than the Coulomb potential in playing the role in the isospin asymmetry in the pseudospin splittings.[13] As we know, the PSS not only means single-particle energy levels are quasi degenerate but also means the lower components of pseudospin doublets are equal in magnitude.[14–16] It is necessary to reinvestigate influence of the ρ potential and the Coulomb potential on the splitting of wavefunction between pseudospin doublets in the framework of the relativistic mean field (RMF) theory.

2. Theoretical Framework

We first sketch the theoretical framework of our work. In the conventional RMF theory, the Dirac Hamiltonian describing a nucleon with mass M moving in an attractive scalar potential $S(r)$ and a repulsive vector potential $V(r)$ can be written as:

$$H = \vec{\alpha} \cdot \vec{p} + \beta(M + S) + V, \tag{1}$$

where the S and V potentials have the following forms:

$$\begin{cases} S(\vec{r}) &= g_\sigma \sigma, \\ V(\vec{r}) &= g_\omega \omega_0 + g_\rho \tau_3 \rho_0 + \frac{1}{2} e(1 - \tau_3) A_0, \end{cases} \tag{2}$$

$g_\sigma \sigma$, $g_\omega \omega_0$, and $g_\rho \tau_3 \rho_0$ in the formula above represent interactions of the scalar meson σ, the vector meson ω^μ, and the isovector-vector meson $\vec{\rho}^\mu$ with g_σ, g_ω, and g_ρ as the coupling constants. $\frac{1}{2}e(1-\tau_3)A_0$ represents the Coulomb potential. For simplicity, $g_\sigma \sigma$, $g_\omega \omega_0$, $g_\rho \tau_3 \rho_0$, and Coulomb potential can be substituted with V_σ, V_ω, V_ρ, and V_c, respectively. For spherical nuclei, the Dirac spinor $\psi(\vec{r})$ can be written by the upper component f and lower component g as

$$\psi_{n\kappa}(\vec{r}) = \begin{pmatrix} f_{n\kappa} \\ g_{n\kappa} \end{pmatrix} = \frac{1}{r} \begin{pmatrix} F_{n\kappa}(r) Y^l_{jm}(\theta, \phi) \\ iG_{n\kappa}(r) Y^{\tilde{l}}_{jm}(\theta, \phi) \end{pmatrix}, \tag{3}$$

when angular part was split off, radial wavefunction satisfy following equations

$$\left(\frac{d}{dr}+\frac{\kappa}{r}\right)F_{n\kappa}(r)=\left[M+\varepsilon_{n\kappa}-V(r)+S(r)\right]G_{n\kappa}(r), \tag{4}$$

$$\left(\frac{d}{dr}-\frac{\kappa}{r}\right)G_{n\kappa}(r)=\left[M-\varepsilon_{n\kappa}+V(r)+S(r)\right]F_{n\kappa}(r), \tag{5}$$

where $\Sigma(r) = V(r) + S(r)$ and $\Delta(r) = V(r) - S(r)$. we can obtain single-particle energy level and corresponding wavefunctions $F_{n\kappa}(r)$ and $G_{n\kappa}(r)$ by solving the formula (4) and (5) numerically with a relativistic potential of Woods-Saxon type.

3. The Numerical Calculations and Discussions

With the formalism presented in Sec. 2, it is convenient for us to explore the isospin asymmetry by calculating proton single-particle energy splitting and neutron single-particle energy splitting with the operator H_p. In the RMF theories, the isovector-vector potential V_ρ and Coulomb potential are main contributions to the isospin breaking in nuclei.[?] Accordingly, the $\Sigma(r)$ and $\Delta(r)$ potentials for neutrons and protons can be respectively written as

$$\begin{cases} \Sigma_n = V_\omega + V_\sigma - V_\rho & for\ neutrons \\ \Sigma_q = V_\omega + V_\sigma + V_\rho & for\ protons\ without\ V_\rho \\ \Sigma_p = V_\omega + V_\sigma + V_\rho + V_{Coul} & for\ protons, \end{cases} \tag{6}$$

$$\begin{cases} \Delta_n = V_\omega - V_\sigma - V_\rho & for\ neutrons \\ \Delta_q = V_\omega - V_\sigma + V_\rho & for\ protons\ without\ V_\rho \\ \Delta_p = V_\omega - V_\sigma + V_\rho + V_{Coul} & for\ protons, \end{cases} \tag{7}$$

All these potentials above can be derived from the RMF calculations for a certain nucleus. As in Refs. 10,12, we also adopted a Woods-Saxon type for $\Sigma(r)$ and $\Delta(r)$ potentials, i.e., $\Sigma(r)$=$\Sigma_0 f(a_\Sigma, R_\Sigma, r)$ and $\Delta(r)$=$\Delta_0 f(a_\Delta, R_\Delta, r)$ with

$$f(a,R,r) = \frac{1}{1+\exp(\frac{r-R}{a})}, \tag{8}$$

There are six parameters, i.e., the central depths, Σ_0 and Δ_0, two radii, and two diffuseness parameters by using this potential for Σ_n and Δ_n. In order to make the Woods-Saxon potential closer to the nuclear mean field, the parameters in Woods-Saxon potential are determined by fitting the $\Sigma_n(r)$ and $\Delta_n(r)$ potentials derived from the RMF calculations where S_n isotope chain are chosen as reference, as was calculated in the Ref. 13. In order to analyse the effect of V_{Coul} and V_ρ on pseudospin splittings, we calculated the pseudospin energy splittings with Woods-Saxon potential whose six parameters are redetermined by fitting Σ_q and Δ_q potentials in formula (6) and (7). Because Σ_p and Δ_p have one more item, i.e., V_{Coul} than that of the Σ_q and Δ_q, we should introduce a new formula which is similar to the Coulomb potential to fit V_{Coul}. This new formula $V^*_{Coul}(r)$ can be written as:

$$V^*_{Coul}(r) = \frac{k}{r+b} + c, \tag{9}$$

Table 1. The parameters in Woods-Saxon potential and $V^*_{Coul}(r)$ potential determined by fitting the scalar potential, the vector potential and the Coulomb potential from the RMF calculations with the PK1 for ^{150}Sn.

	U_0 (MeV)	R (fm)	a (fm)	k (MeV)	b (fm)	c (MeV)
Σ_n	-61.0	6.38	0.71	0	0	0
Δ_n	636.5	6.23	0.54	0	0	0
Σ_q	-77.8	6.55	0.70	0	0	0
Δ_q	619.5	6.21	0.53	0	0	0
Σ_p	-77.8	6.55	0.70	509.5	16.3	-11.4
Δ_p	619.5	6.21	0.53	509.5	16.3	-11.4

Three parameters in Eq. (9), i.e., k, b and c were determined by fitting V_{Coul} potential derived from the RMF calculation for Sn isotope chain. For simplicity, only the parameters obtained form ^{150}Sn are listed in Table 1. With these parameters, the energy splitting and the wavefunction splitting of pseudospin doublet ($2d_{5/2}$,$1g_{7/2}$) are obtained and showed in Fig. 1(a) and Fig. 1(b), respectively. In order to conveniently compare with the results of the Ref. 13, we also calculate the pseudospin doublet $(2d_{5/2}, 1g_{7/2})$ in this paper.

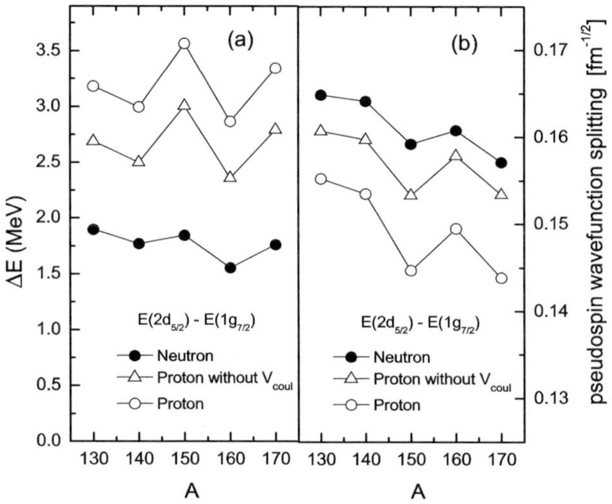

Fig. 1. The effect of V_{Coul} and V_ρ in the neutron and proton pseudospin energy splitting for the $[2d_{5/2}, 1g_{7/2}]$ doublets in the Sn as a function of A.

The results of our calculations for the pseudospin doublet $(2d_{5/2}, 1g_{7/2})$ were displayed in Fig. 1. We will compare these lines in these figures to study the isospin asymmetry of pseudospin symmetry. In Fig. 1(a), we observed the difference between the energy splittings calculated with Σ_n and Σ_q potentials is about 1.5 MeV. However, the difference between the energy splittings calculated with Σ_q and Σ_p potentials is about 1.0 MeV. Therefore, it is clear that the ρ potential gives the main contribution to the observed isospin asymmetry of the psuedospin energy splittings, which is agreeable with the corresponding results presented in the figures of Ref. 13. Moreover, effects of two potentials on the isospin asymmetry of the psuedospin energy splittings are the same to enlarge the isospin asymmetry. Fig. 1(b) presented variation of the pseudospin wavefunction splittings with the function of A for the S_n isotope. It shows that the difference between the wavefuntion splittings calculated with Σ_n and Σ_q potentials is about 0.01 fm$^{-1/2}$. However, the difference between the wavefuntion splittings calculated with Σ_q and Σ_p potentials is about 0.005 fm$^{-1/2}$. So, we concluded that the Coulomb potential gives the main contribution to the observed isospin asymmetry of the psuedospin wavefunction splittings, which is contrary to the conclusion of the psuedospin energy splittings. Furthermore, effects of two potentials on the isospin asymmetry of the psuedospin wavefunction splittings are the same to reduce the isospin asymmetry. This result is contrary to that of the psuedospin energy splittings.

In summary, the role of Coulomb potential and ρ potential in the isospin asymmetry of the PSS are reinvestigated. We found that the effect of the ρ potential is small and the Coulomb potential gives the main contribution to the observed isospin asymmetry of the psuedospin wavefunction splittings, which is contrary to that of the psuedospin energy splittings.

Acknowledgments

The work was supported by the National Natural Science Foundation of China under Grant No. 11205112 and the Youth Foundation of Taiyuan University of Technology under Grant No. 2012L084 and the Qualified Personnel Foundation of Taiyuan University of Technology(No: 800101-02030017).

References

1. K. T. Hecht and A. Adler, *Nucl. Phys. A* **137**, 129 (1969).
2. A. Arima, M. Harvey, and K. Shimizu, *Phys. Lett. B* **30**, 517 (1969).
3. C. Bahri, J. P. Draayer, and S. A. Moszkowski, *Phys. Rev. Lett.* **68**, 2133 (1992).
4. A. L. Blokhin, C. Bahri, and J. P. Draayer, *Phys. Rev. Lett.* **74**, 4149 (1995).
5. J.N. Ginocchio, *Phys. Rev. Lett.* **78** 436 (1997).
6. J. Meng, K. Sugawara-Tanabe, S. Yamaji, P. Ring, and A. Arima, *Phys. Rev. C* **58**, R628 (1998).
7. J. Meng, K. Sugawara-Tanabe, S. Yamaji, and A. Arima, *Phys. Rev. C* **59**, 154 (1999).
8. J.N. Ginocchio, *Phys. Rep.* **315**, 231 (1999).
9. P. Alberto, M. Fiolhais, M. Malheiro, A. Delfino, and M.Chiapparini, *Phys. Rev. Lett.* **86**, 5015 (2001).
10. P. Alberto, M. Fiolhais, M. Malheiro, A. Delfino, and M. Chiapparini, *Phys. Rev. C* **65**, 034307 (2002).
11. J.Y. Guo, X.Z. Fang, F.X. Xu, *Nucl. Phys. A* **757**, 441 (2005).
12. Q. Xu, S.J. Zhu, *Nucl. Phys. A* **768**, 161 (2006).
13. R. Lisboa, M. Malheiro, P. Alberto, *Phys. Rev. C* **67**, 054305 (2003).
14. J.N. Ginocchio, A. Leviatan, *Phys. Rev. Lett.* **87** 072502 (2001).
15. J.N. Ginocchio, *Phys. Rev. C* **66**, 064312 (2002).
16. P.J. Borychi, J.N. Ginocchio, and W. Nazarewicz, *Phys. Rev. C* **68**, 014304 (2003).

DESCRIPTION OF DEFORMED NUCLEI IN A QUARK-MESON COUPLING MODEL WITH TENSOR COUPLING

R.L. Xu* and Z.Z. Ren†

Key Laboratory of Modern Acoustics and Department of Physics, Nanjing University, Nanjing 210093, People's Republic of China
** E-mail: xurenli.phy@gmail.com*
† E-mail: zren@nju.edu.cn

C. Wu‡

Shanghai Institute of Applied Physics, Chinese Academy of Sciences, Shanghai 201800, People's Republic of China
‡ E-mail: wuchenoffd@gmail.com

We extend the quark-meson coupling model to study the ground-state properties of axially deformed even-even nuclei. For this purpose, we introduce ω-N tensor coupling to the model Lagrangian and obtain the parameters from fits to the experimental data for finite nuclei. The present work is mainly focused on the nuclei in the rare-earth region. The ground-state binding energies, quadrupole deformations, shape coexistence, two-neutron separation energies and the root-mean-square charge radii are calculated and compared with the experimental data. It is shown that this extended quark-meson coupling model can give a reasonable description of the axially deformed nuclei.

Keywords: Tensor coupling; shape coexistence; quadrupole deformations.

1. Introduction

The description of both nuclear matter and finite nuclei in the framework of relativistic mean field theory (RMF) has turned out to be considerable successful in the last two decades.[1–6] With a few parameters, the RMF theory is able to provide a quantitative description for the physical properties of spherical and deformed nuclei near and far away from the β-stable line. In RMF model, nucleons are treated as point-like particles, which interact between themselves by the exchange of mesons. The spin-orbit interaction occurs naturally in the relativistic treatment of nucleons. Meanwhile, other classes of effective models, at the hadronic and sub-hadronic levels, have also been widely employed to describe the bulk and single-particle properties of nuclei and nuclear matter. Examples include the relativistic point-coupling (RPC) model,[7–9] the quark-meson coupling (QMC) model,[10–12] the chiral SU(3) quark model,[13] the quark mean field (QMF) model,[14,15] the improved quark mass density-dependent (IQMDD) model,[16–18] etc. Compared to the baryon and meson exchange RMF model, these approaches may give some different insights into the properties of finite nuclei and nuclear matter.

In this paper, we extend the IQMDD model to deformed nuclei. Nuclear deformation has become a familiar topic in nuclear structure physics. There are many deformed or even superdeformed nuclei, ranging from light to heavy systems. The investigation of these nuclei provides us important knowledge of the underlying physics, such as the shape coexistence and shape transition, the evolution of shell structure, etc. It is interesting to carried out deformed calculation based on quark and meson degrees of freedom to investigate finite nuclei and to test its validity and reliability for describing the properties of deformed nuclei.

2. Formulas of the IQMDD Model and Numerical Results

For investigating the finite nuclei, one can construct a relativistic Lagrangian density of IQMDD model at the hadronic level in the following form:

$$\begin{aligned}\mathcal{L} = &\bar{\psi}[i\gamma^\mu\partial_\mu - M_N^*(\sigma) - g_\omega\gamma^\mu\omega_\mu - \frac{g_\rho}{2}\gamma^\mu\tau\cdot\rho_\mu - \frac{e}{2}\gamma^\mu(1+\tau_3)A_\mu \\ &+\frac{f_\omega g_\omega}{2M_N}\sigma^{\mu\nu}\partial_\nu\omega_\mu]\psi + \frac{1}{2}\partial^\mu\sigma\partial_\mu\sigma - U(\sigma) - \frac{1}{4}\Omega^{\mu\nu}\Omega_{\mu\nu} + \frac{1}{2}m_\omega^2\omega^\mu\omega_\mu \\ &-\frac{1}{4}G^{\mu\nu}G_{\mu\nu} + \frac{1}{2}m_\rho^2\rho^\mu\rho_\mu + \eta(g_\rho^2\rho_\mu\rho^\mu)(g_\omega^2\omega_\mu\omega^\mu) - \frac{1}{4}F^{\mu\nu}F_{\mu\nu}. \qquad (1)\end{aligned}$$

The effective nucleon mass is obtained from the bag energy which reads

$$M_N^* = \sum\nolimits_q E_q = \sum\nolimits_q \frac{4}{3}\pi R^3 \frac{\Gamma_q}{(2\pi)^3}\int_0^{K_F^q}\sqrt{m_q^{*2}+k^2}\left(\frac{dN_q}{dk}\right)dk. \qquad (2)$$

The bag radius R is determined by the equilibrium condition for the nucleon bag $\delta M_N^*/\delta R = 0$. Using the Euler-Lagrange equation one obtains the Dirac equation for nucleons, and the equations of motion for the mesons and photons. The set of coupled partial differential equations can be solved self-consistently.

We systematically investigate the ground-state properties of even-even nuclei ranging from $Z = 50$ to $Z = 82$ with this calibrated IQMDD parametrization. This calculated region includes the rare-earth nuclei,

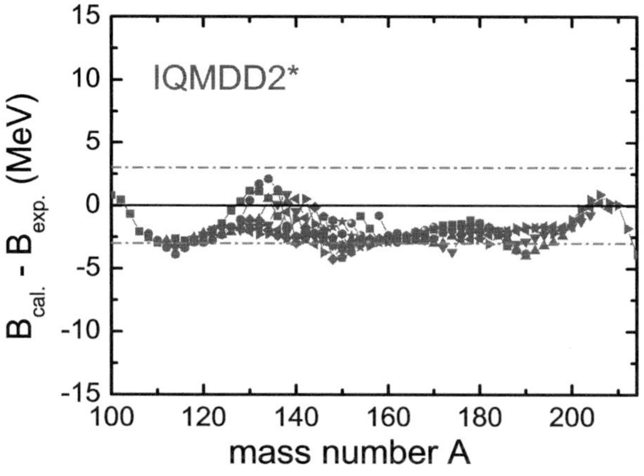

Fig. 1. Binding energy systematics in terms of the errors ΔB (see text) as function of mass number A obtained for IQMDD2*. The horizontal lines indicate the binding energy deviations equaling to zero and ± 3 MeV.

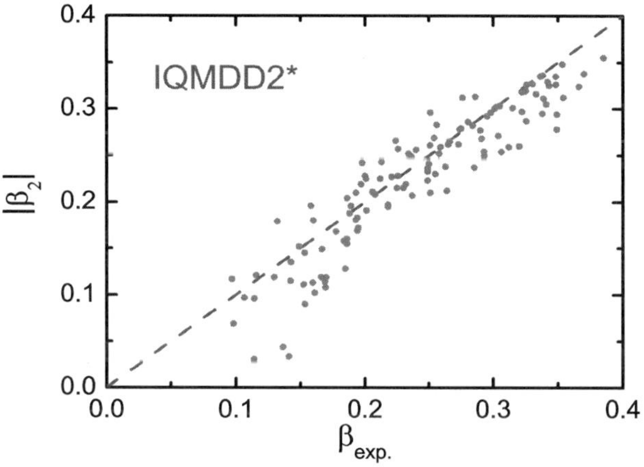

Fig. 2. The calculated absolute values of quadrupole deformation β_2 obtained by IQMDD2*, comparing to the experimental data.

which are known to be typically deformed and have plenty of experimental data available. The results for the binding energy systematics in term of errors $\Delta B(\Delta B = B_{cal.} - B_{exp.})$ are shown in Fig. 1, with $B_{cal.}$ and $B_{exp.}$ being the theoretical and experimental values for the binding energy. The experimental values for the binding energy are taken from Ref. 19.

By using the IQMDD2* model, the root mean square (rms) deviation of the total binding energies from those of the data for all the 278 even-even nuclei is only 2.29 MeV, which indicates that the deformed IQMDD theory with the ω-N tensor coupling can well reproduce the experimental binding energies of the nuclei.

The quadrupole deformation parameter is an important quantity for describing the deformed nuclei. For a better comparison with the data, we show the absolute values of quadrupole deformation parameter β_2 obtained by IQMDD2* in Fig. 2. It can been seen from Fig. 2 that the calculated absolute values are quite close to the experimental data.

In summary, we have systematically studied the ground-state properties of the isotopic chains range from $Z = 50$ to $Z = 82$ by using a set of accurately calibrated parameters in a quark-meson coupling model called IQMDD2*. Based on the sub-hadronic degrees of freedom, the IQMDD2* model can give a good description of the axially deformed nuclei.

References

1. P. Ring, *Prog. Part. Nucl. Phys.* **37**, 193 (1996).
2. G. Lalazissis, J. König and P. Ring, *Phys. Rev. C* **55**, 540 (1997).
3. B. Todd-Rutel and J. Piekarewicz, *Phys. Rev. Lett.* **95**, 122501 (2005).
4. F. Fattoyev, C. Horowitz, J. Piekarewicz and G. Shen, *Phys. Rev. C* **82**, 055803 (2010).
5. J. Meng, K. Sugawara-Tanabe, S. Yamaji, P. Ring and A. Arima, *Phys. Rev. C* **58**, R628 (1998).
6. W. H. Long, J. Meng, N. V. Giai and S. G. Zhou, *Phys. Rev. C* **69**, 034319 (2004).
7. B. A. Nikolaus, T. Hoch and D. G. Madland, *Phys. Rev. C* **46**, 1757 (1992).
8. J. M. Yao, J. Meng, P. Ring and D. Vretenar, *Phys. Rev. C* **81**, 044311 (2010).
9. P. W. Zhao, Z. P. Li, J. M. Yao and J. Meng, *Phys. Rev. C* **82**, 054319 (2010).
10. P. Guichon, *Phys. Lett. B* **200**, 235 (1988).
11. S. Fleck, W. Bentz, K. Shimizu and K. Yazaki, *Nucl. Phys. A* **510**, 731 (1990).
12. D. Menezes, P. Panda and C. Providência, *Phys. Rev. C* **72**, 035802 (2005).
13. P. Papazoglou, S. Schramm, J. Schaffner-Bielich, H. Stöcker and W. Greiner, *Phys. Rev. C* **57**, 2576 (1998).
14. H. Toki, U. Meyer, A. Faessler and R. Brockmann, *Phys. Rev. C* **58**, 3749 (1998).
15. H. Shen and H. Toki, *Phys. Rev. C* **61**, 045205 (2000).
16. G. Fowler, S. Raha and R. Weiner, *Z. Phys. C* **9**, 271 (1981).
17. C. Wu, W. Qian and R. Su, *Phys. Rev. C* **72**, 035205 (2005).
18. C. Wu, W. Qian and R. Su, *Phys. Rev. C* **77**, 015203 (2008).
19. M. Wang *et al.*, *Chin. Phys. C* **36**, 1603 (2012).

CONSTRAINING POSTSADDLE DISSIPATION WITH PARTICLE EMISSION FROM LIGHT DECAYING SYSTEMS

W. Ye*

Department of Physics, Southeast University,
Nanjing 210096, P. R. China
**E-mail: yewei@seu.edu.cn*

In the framework of stochastic model coupled with a statistical model for particle evaporation, we survey the role of spin of compound nuclei in probing postsaddle friction (β) with particle multiplicity. We find that at low spin particle emission displays a greater sensitivity to β. We further study the influence of system size on the sensitivity by a comparison between ^{200}Pb with a high excitation energy and spin and ^{251}Es with a low excitation energy and spin. Calculations show that neutrons (protons and α particles) from ^{200}Pb system are of a comparable (larger) sensitivity to β. These results indicate that to more stringently constrain β through the measurement of prescission particles, it is useful to yield light fissioning systems with large angular momentum and excitation energy.

Keywords: Nuclear dissipation; particle multiplicity; stochastic model; light fissioning systems.

1. Introduction

Evaporation and fission are two principle decay channels when a compound nucleus (CN) de-excites, which can be well described by standard statistical models (SSM). However, numerous experiments[1–3] revealed that with an increase in excitation energy deposited in CNs, deviations between measured prescission particle multiplicity from the estimate of SSM have been observed. This discrepancy is ascribed to dissipation effects that are not accounted for in SSM.[4–10] Due to dissipation, fission is retarded that enhances particle emission and suppresses fission. Stochastic approaches[5–7,10–12] to fission have been demonstrated to be successful to reproduce a large variety of fission data, including particle multiplicity and fission probability.

Till now, many researches have been devoted to constraining presaddle friction, but less efforts are placed on the constraint of postsaddle friction

(β). While heavy systems have a long saddle-to-scission descent, which increases postsaddle emission that favors the determination of β, the measurement of their prescission particle multiplicity has a large uncertainty arising from the strong interference of quasifission channels, particularly when bombarding energy is increased.

By contrast, the population of light systems via heavy-ion collisions is subjected to a little containment from quasifission channels, an advantage for obtaining precise information of particle multiplicity related to fusion-fission processes.

In the present study, we will compare the sensitivity of particle emission to β between light ^{200}Pb having a high excitation energy and spin and heavy ^{251}Es having a low excitation energy and spin.

2. Theoretical Model

We briefly describe the Langevin equation that is coupled to a statistical decay model for particle emission (CDSM)[4,11] The dynamic part of the CDSM is described by the Langevin equation that is expressed by entropy. A one-dimensional overdamped Langevin equation is used to carry out the trajectory calculations:

$$\frac{dq}{dt} = \frac{T}{M\beta}\frac{dS}{dq} + \sqrt{\frac{T}{M\beta}}\Gamma(t). \tag{1}$$

Here q is the dimensionless fission coordinate and is defined as half the distance between the center of mass of the future fission fragments divided by the radius of the compound nucleus, M is the inertia parameter, and β is the dissipation strength. The temperature in Eq. (1) is denoted by T and $\Gamma(t)$ is a fluctuating force with $< \Gamma(t) > = 0$ and $< \Gamma(t)\Gamma(t') > = 2\delta(t-t')$.

The driving force of the Langevin equation is calculated from the entropy $S(q, E^*) = 2\sqrt{a(q)[E^* - V(q)]}$, where E^* is the excitation energy of the system. The entropy is constructed from the Fermi-gas expression, and the q-dependent surface, Coulomb and rotation energy terms are included in the potential $V(q)$.[13,14] In constructing the entropy, the deformation-dependent level density parameter is used: $a(q) = a_1 A + a_2 A^{2/3} B_s(q)$. Here A is the mass number, and $a_1 = 0.073$ and $a_2 = 0.095$ are taken from Ignatyuk *et al.*[15]

In the CDSM calculation, the surface energy $B_s(q)$ and the Coulomb energy $B_c(q)$ as a function of q can be parametrized in the form suggested in Ref. [4]. The deformation dependence of the charged-particle emission barriers is evaluated with formula $V_c(q) = V_\nu \times B_c(q)$ with V_ν being the emission barrier of charged particles for the case of a spherical nucleus.

The formula calculating the emission width of a particle of kind ν ($= n, p, \alpha$) is adopted from Ref. [16]. The particle binding energy B_i ($i = n, p, \alpha$) is calculated as $B_i(q) = M_p(q) - M_d(q) + M_i$, where M_i ($i = n, p, \alpha$) is the mass of the emitted particles. $M_p(q)$ and $M_d(q)$ are respectively the masses of the mother and daughter nuclei. They contain the deformation-dependent surface energy term and Coulomb energy term, and $B_i(q)$ is thus a function of q.[17,18]

Prescission particle multiplicities are calculated by counting the number of corresponding evaporated particle events registered in the dynamic and statistical branch of the CDSM.

3. Results and Discussion

In this work, according to the theoretical estimates and experimental analyses[4,7,19–21] we set the strength of presaddle friction as $3 \times 10^{21}\text{s}^{-1}$.

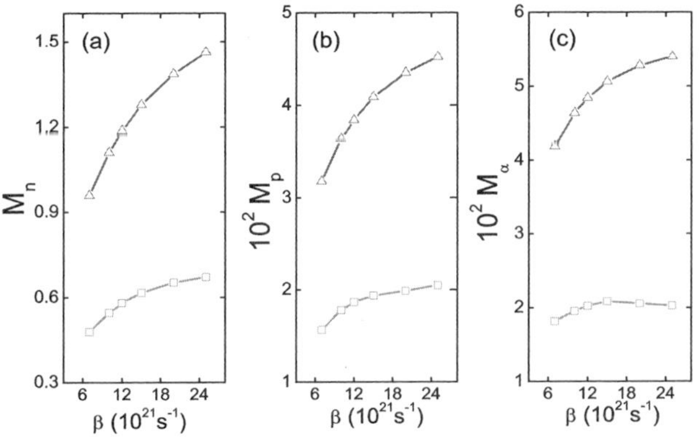

Fig. 1. Evolution of postsaddle neutrons (M_n), protons (M_p) and α-particles (M_α) for ^{200}Pb nuclei with postsaddle friction strength β at an excitation energy of 150 MeV and at critical angular momenta $\ell = 10\hbar$ (squares) and $50\hbar$ (triangles), respectively.

In Fig. 1 we plot M_n, M_p and M_α versus β for ^{200}Pb systems at angular momenta $\ell = 10\hbar$ and $50\hbar$. As can be seen, these particles display a sensitive change to β at a lager ℓ. We explain it in the following way. Fission barrier is a decreasing function of ℓ, meaning a large fission probability and a small presaddle evaporation at large ℓ. As a result, more energy is left for postsaddle emission; A greater postsaddle neutron [Fig. 1(a)] raises its sensitivity to β. A similar picture is observed for the case of charged particles, see Figs. 1(b) and 1(c). One can draw from Fig. 1 that yielding a high-spin light CN can significantly enhance the sensitive dependence of postsaddle particles to friction.

Heavy-ion collisions can deposit a high energy into light CNs, a prominent advantage for probing friction, because the number of particle emitted in fission is a strong function of excitation energy.

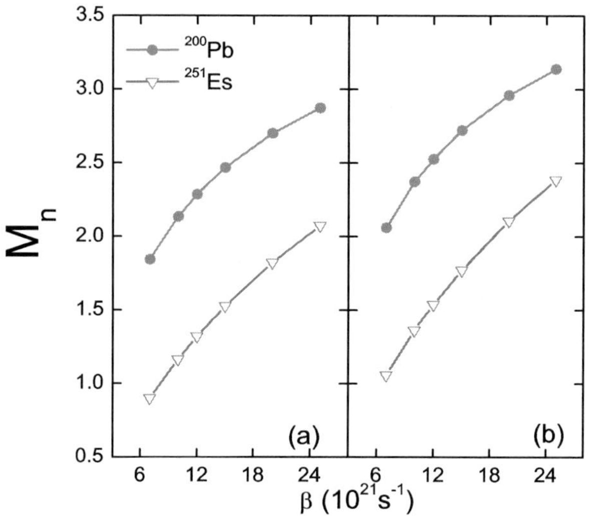

Fig. 2. The sensitive dependence of postsaddle neutrons on β for ^{200}Pb (circles) at E^* = 150 MeV and $\ell = 70\hbar$ and ^{251}Es (triangles) at $E^* = 60$ MeV and $\ell = 40\hbar$ for the case that ignores deformation effect (a) and considers it (b).

Shown in Fig. 2(a) is a comparison for M_n versus β between ^{200}Pb having a large angular momentum and excitation energy, a feature of the formed light CNs in fusion reactions, and ^{251}Es having a low E^* and ℓ, also provided in heavy-ion collisions.[18] One can note that the trend of M_n with β is alike for the two nuclei, indicating that neutrons from light CNs with a high ℓ and E^* are also a good indicator of β.

Accounting for deformation effects in calculation has a minor influence on the sensitivity of neutrons to friction [Fig. 2(b)].

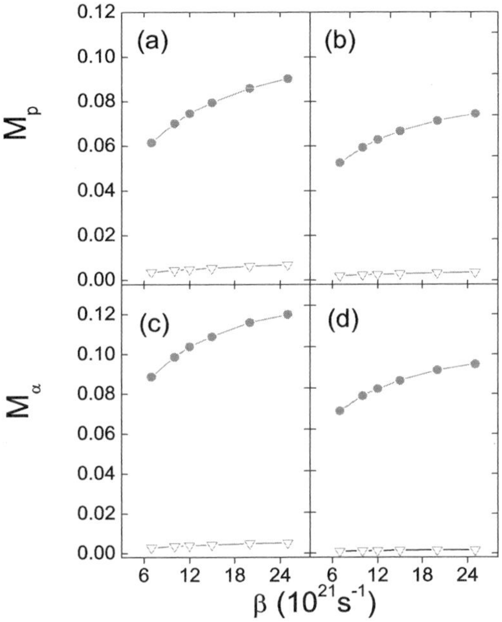

Fig. 3. Postsaddle protons (top panel) and α particles (bottom panel) as a function of β in the absence [(a) and (c)] and in the presence [(b) and (d)] of deformation effects for ^{200}Pb (circles) at $E^* = 150$ MeV and $\ell = 70\hbar$ and ^{251}Es (triangles) at $E^* = 60$ MeV and $\ell = 40\hbar$.

Different from the case of neutrons, the influence of deformation on M_p and M_α is substantial. It increases charged-particle binding energy and lowers their Coulomb emission barrier. As a consequence of these competing factors, M_p and M_α are reduced significantly for heavy ^{251}Es. This greatly affects their sensitivity to β; they are almost insensitive to friction, see Figs. 3(b) and 3(c). This result indicates that charged particles, especially α particles of heavy fissioning nuclei are not a suitable observable of β.

When a CN fissions, it will experience a deformation, and the deformation becomes greater with an increase of the size of decaying systems. It means that in order to better estimate charged particle emission, incorporating deformation effects into theoretical modelling is indispensable, in particular for heavy nuclei. A comparison for the deformation effects in both ^{200}Pb and ^{251}Es systems illustrates that on the experimental side, when charged particles are measured aiming at exploring postsaddle dissipation, yielding a light system with high E^* and ℓ is a very useful approach.

4. Summary and Conclusions

Using the Langevin model that takes into account particle emission along the fission path, we have shown that a high spin can markedly raise the sensitive dependence of particle emission on postsaddle friction. Moreover, Langevin simulation has demonstrated that neutrons from light ^{200}Pb with high excitation energy and spin are of a similar sensitivity to β as that for heavy ^{251}Es with low excitation energy and spin, and revealed a larger sensitivity of changed particles to β for ^{200}Pb system than for ^{251}Es system, in particular when deformation effects are contained in calculations. Our results suggest that, to more severely constrain the strength of postsaddle dissipation with particle multiplicity, especially with charged-particle multiplicity, in experiments, it is optimal to produce a light compound system with a large spin and excitation energy.

Acknowledgements

This work is supported by National Natural Science Foundation of China under Grant No. 11075034.

References

1. D. Hilscher and H. Rossner, *Ann. Phys. Fr.* **17**, 471 (1992).
2. H. Singh *et al.*, *Phys. Rev. C* **78**, 024609 (2008).
3. V. Singh *et al.*, *Phys. Rev. C* **87**, 064601 (2013).
4. P. Fröbrich and I. I. Gontchar, *Phys. Rep.* **292**, 131 (1998).
5. K. Pomorski *et al.*, *Nucl. Phys. A* **605**, 87 (1996).
6. Y. Abe *et al.*, *Phys. Rep.* **275**, 49 (1996).
7. P. N. Nadtochy *et al.*, *Phys. Rev. C* **65**, 064615 (2002).
8. W. Ye, *Phys. Lett. B* **681**, 413 (2009); *Phys. Rev. C* **87**, 051601(R) (2013).
9. D. Mancusi *et al.*, *Phys. Rev. C* **82**, 044610 (2010).
10. J. Sadhukhan and S. Pal, *Phys. Rev. C* **84**, 044610 (2011).
11. P. Fröbrich, *Nucl. Phys. A* **787**, 170c (2007).
12. W. Ye *et al.*, *Phys. Rev. C* **90**, 041604 (R) (2014).
13. W. D. Myers and W. J. Swiatecki, *Nucl. Phys.* **81**, 1 (1966).
14. H. J. Krappe *et al.*, *Phys. Rev. C* **20**, 992 (1979); A. J. Sierk, *Phys. Rev. C* **33**, 2039 (1986).
15. A. V. Ignatyuk *et al.*, *Sov. J. Nucl. Phys.* **21**, 612 (1975).
16. M. Blann, *Phys. Rev. C* **21**, 1770 (1980).
17. J. P. Lestone, *Phys. Rev. Lett.* **70**, 2245 (1993).
18. A. Chatterjee *et al.*, *Phys. Rev. C* **52**, 3167 (1995).
19. G. Chaudhuri and S. Pal, *Eur. Phys. J. A* **14**, 287 (2002).
20. W. Ye *et al.*, *Phys. Rev. C* **77**, 011302 (R) (2008); C**87**, 014610 (2013).
21. C. Schmitt *et al.*, *Phys. Rev. Lett.* **99**, 042701 (2007).

SYSTEMATIC STUDY OF POSITIVE Q-VALUE NEUTRON TRANSFER EFFECT IN FUSION REACTIONS

G. L. Zhang*, X. X. Liu

School of Physics and Nuclear Energy Engineering, Beihang University, Beijing 100191, China
** E-mail: zgl@buaa.edu.cn*
www.buaa.edu.cn

International Research Center for Nuclei and Particles in the Cosmos, Beihang University, Beijing 100191, China

C. J. Lin and H. Q. Zhang

China Institute of Atomic Energy, Beijing 102413, China
www.ciae.ac.cn

The experimental data of fusion cross sections for the different projectile and target nuclei systems are systematically analyzed. In order to explore the positive Q-value neutron transfer ($PQNT$) effect, we systematically analyze their fusion functions $F(x)$ as a function of x, which reduces the barrier height, radius, and curvature of fusion systems. From the comparison of $F(x)$s among them, it is found that after neutron transfer when the deformation of interacting nuclei increases largely, the enhancement of fusion cross sections can be observed at energies below the Coulomb barrier due to the $PQNT$ effect. However, even if the fusion systems show the $PQNT$ values, the fusion cross sections are almost the same due to the small deformation of interaction nuclei after neutron transfer.

Keywords: Fusion function; $PQNT$; deformation.

1. Introduction

Fusion of heavy ions has been an interesting topic for several decades. For heavy-ion sub-barrier fusion, the various questions are still unresolved, such as the breakup effect on the fusion reactions at energies around the Coulomb barrier and the role of the neutron transfer effect in the fusion process. The fusion cross sections for many fusion systems show the enhancement at energies below the Coulomb barrier in comparison with the theoretical calculations. It is indicated that the enhancement is related to the vibrations and deformations of projectile and target nuclei, as well as possibly nucleon transfer degrees of freedom. Large enhancement of fusion reaction yield for many systems with positive Q-value neutron transfer ($PQNT$) has been experimentally observed from the fusion excitation function and the shape of the fusion barrier distribution also shows the different behavior.[1–3] However, in the other experiments for fusion systems with the $PQNT$ values[4,5] the fusion cross sections did not show the enhancement at sub-barrier energies. For example, recent fusion of ^{132}Sn on ^{64}Ni with many more $PQNT$ channels at energies near the Coulomb barrier do not lead to the enhancement when considering the change of nuclear size and barrier height in comparison with the stable Sn+^{64}Ni.[6] In the fusion for the neutron transfer process there are also different explanations.[7,8] So until now the role of the $PQNT$ effect is not yet clear. This coupling is difficult to consider in the coupling channel calculations. It is expected that it will be further studied via the experiments and theories.

2. Methods

Recently, Canto *et al.* proposed the use of fusion functions for fusion processes.[9] It can clearly make conclusions about the deformation effect of collision nuclei and the role of neutron transfer between the interacting nuclei in the fusion process. We will use the reduced fusion function to explore this effect. A dimensionless fusion function $F(x)$ and a dimensionless variable x are defined as

$$F(x) = \frac{2E_{c.m.}}{\hbar\omega R_B^2}\sigma_F, \qquad x = \frac{E_{c.m.} - V_B}{\hbar\omega}, \tag{1}$$

where V_B, R_B, and $\hbar\omega$ correspond to the height, radius, and curvature of the barrier for each fusion system, respectively, and x is the reduced energy. $E_{c.m.}$ and σ_F are the collision energy at the center-of-mass frame and the fusion cross section, respectively. The barrier parameters V_B, R_B, and $\hbar\omega$ are obtained from the CCFULL program.[10] In the program we input the quadrupole excitation of projectiles as well as the quadrupole and octupole of target nuclei. The values of quadrupole and octupole excitation are from the measured transition probabilities $B(E2\uparrow)$[11] and $B(E3\uparrow)$.[12] The deformation parameters of the selected nuclei are shown in Ref.[13] The standard Akyü-Winther (AW) potential[13] is utilized. The imaginary part of the potential is assumed to be zero. The neutron transfer channel is not taken into account.

3. Results and Discussion

We calculate the Q values of the selected fusion systems from the ground-state-to-ground-state transfer, which are shown in Ref. 13. We focus on the fusion systems with the positive-Q-value neutron transfer which is defined as the *PQNT*.

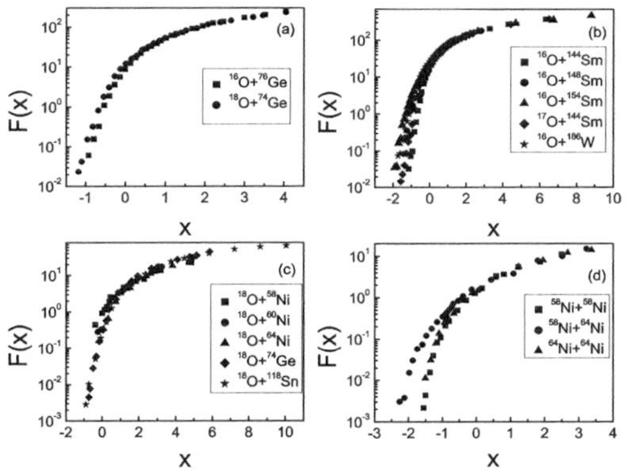

Fig. 1. The fusion function $F(x)$s for the isotopic projectiles on different target nuclei as a function of x.

Figure 1(a) shows that the $F(x)$s of $^{16}O+^{76}Ge$ with the negative Q values of neutron transfer and $^{18}O+^{74}Ge$ with the positive Q_{-2n} value are similar.[14] This is because after the $^{18}O(\beta_2=0.1)+^{74}Ge(\beta_2=0.283)$ $\rightarrow^{16}O(\beta_2=0)+^{76}Ge(\beta_2=0.262)$ neutron transfer reaction, the deformation of the fusion system remains similar. However, after two-neutron transfer in $^{18}O(\beta_2=0.1)+^{74}Ge(\beta_2=0.283)\rightarrow^{16}O(\beta_2=0)+^{76}Ge(\beta_2=0.262)$, the deformation shows a small decrease. As a result, the $F(x)$ of $^{16}O+^{76}Ge$ is slightly suppressed with respect to the reference system, $^{18}O+^{74}Ge$, at sub-barrier energies. It is indicated that the enhancement of the fusion cross section for the fusion systems with the *PQNT* is not observed since after two-neutron transfer the deformation of collision nuclei is slightly changed. In Fig. 1(b), at energies above the Coulomb barrier the $F(x)$s of $^{16}O+^{144,148,154}Sm$, ^{186}W, and $^{17}O+^{144}Sm$[15] show the similarity, and at energies below the Coulomb barrier they have only a small difference due to the deformation effect among $^{144,148,154}Sm$ (β_2=0.09, 0.14, and 0.34, respectively) and ^{186}W (β_2=0.23). Among $^{144,148,154}Sm$ the deformation of ^{154}Sm is largest. The deformation of ^{186}W lies between ^{154}Sm and $^{144,148}Sm$. The trend of their

$F(x)$s shows this result below the Coulomb barrier. The $F(x)$ of ^{17}O+^{144}Sm with positive Q_{-1n} values is similar to that of ^{16}O+^{144}Sm at energies below the Coulomb barrier. The reason is that after the ^{17}O+^{144}Sm→^{16}O+^{145}Sm reaction the deformation of interacting nuclei changes is small. Therefore, even if the fusion system has the *PQNT* value, after neutron transfer, the deformation of interacting nuclei is weak, and the enhancement effect of the fusion cross section is not evident. In Fig. 1(c), the $F(x)$s of ^{18}O+^{58,60,64}Ni,[16,17] ^{74}Ge,[14] and ^{118}Sn[5] have almost no difference at energies above and below the Coulomb barrier. Moreover, among ^{58,60,64}Ni (β_2=0.183, 0.207, and 0.179, respectively), ^{74}Ge (β_2=0.283), and ^{118}Sn (β_2=0.111) the deformation difference is small. For each fusion system after neutron transfer the deformation of collision nuclei does not change much, and the barrier height does not change sharply. In total, their $F(x)$s are almost similar. In Fig. 1(d) ^{58}Ni+^{58}Ni[18] and ^{64}Ni+^{64}Ni[19] are two symmetric fusion systems, and the Q values of the neutron transfers are negative and the difference of their deformation is small, so their $F(x)$s are similar at energies above and below the Coulomb barrier. However, the $F(x)$ of ^{58}Ni+^{64}Ni[19] shows the large deviation at energies below the Coulomb barrier with respect to the ^{58}Ni+^{58}Ni and ^{64}Ni+^{64}Ni systems. Through two-neutron transfer from ^{58}Ni+^{64}Ni to ^{60}Ni+^{62}Ni the mass asymmetry of the interacting fusion system decreases, and the *PQNT* gives an enhancement of the fusion cross section at energies below the Coulomb barrier.

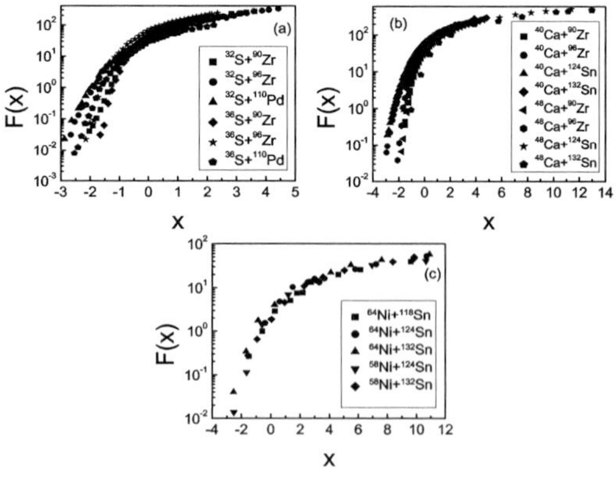

Fig. 2. Similar as Fig. 1(a).

In Fig. 2(a), below the Coulomb barrier the $F(x)$s of $^{32}S+^{96}Zr$ and ^{110}Pd are strongly deviated from that of $^{32}S+^{90}Zr$. The Q_{2n} values of $^{32}S+^{96}Zr$ and ^{110}Pd are positive, that Q value of $^{32}S+^{90}Zr$ is negative. The difference of quadrupole deformation for $^{90,96}Zr$ (β_2=0.09 and 0.08, respectively) is small. Moreover, after neutron transfer for $^{32}S(\beta_2=0.31)+^{96}Zr(\beta_2=0.08)\rightarrow^{34}S(\beta_2=0.25)+^{94}Zr(\beta_2=0.09)$ and $^{32}S(\beta_2=0.31)+^{110}Pd(\beta_2=0.26)\rightarrow^{34}S(\beta_2=0.25)+^{108}Pd(\beta_2=0.24)$, the deformation of the collision nuclei increases, the barrier height lowers, and the fusion cross section increases. Therefore, for fusion systems with the *PQNT* after the neutron transfer when the deformation of collision nuclei increases, the fusion cross section is enhanced below the Coulomb barrier. However, the difference of quadrupole deformation between $^{90,96}Zr$ and ^{110}Pd is large. The deformation of ^{110}Pd is larger than that of ^{96}Zr. As a result, the $F(x)$ of $^{32}S+^{110}Pd$ has less enhancement than that of $^{32}S+^{96}Zr$ at sub-barrier energies. The Q values of neutron transfer for $^{36}S+^{90,96}Zr$ and ^{110}Pd are negative. The $F(x)$ of $^{36}S+^{110}Pd$ is strongly suppressed and enhanced at energies above and below the Coulomb barrier with respect to the systems $^{36}S+^{90,96}Zr$. This is the reason that ^{110}Pd has a much larger deformation than those of $^{90,96}Zr$, and between ^{90}Zr and ^{96}Zr the deformation effect is small. In Fig. 2(b) the phenomenon is similar to that in Fig. 1(a). For the $^{40}Ca(\beta_2=0.12)+^{96}Zr(\beta_2=0.08)\rightarrow^{42}Ca(\beta_2=0.25)+^{94}Zr(\beta_2=0.09)$, $^{40}Ca(\beta_2=0.12)+^{124}Sn(\beta_2=0.095)\rightarrow^{42}Ca(\beta_2=0.25)+^{122}Sn(\beta_2=0.1)$ and $^{40}Ca(\beta_2=0.12)+^{132}Sn(\beta_2=0)\rightarrow^{42}Ca(\beta_2=0.25)+^{130}Sn(\beta_2=0)$ reactions, after neutron transfer the interacting nuclei show the large deformation effect and the mass asymmetry of the fusion systems decreases. So the fusion probability increases and it shows the same phenomenon as that in Fig. 2(a). In Fig. 2(c), the $F(x)$s of $^{48}Ca+^{90,96}Zr$ and $^{124,132}Sn$ are almost similar; below the Coulomb barrier they show only a small difference. All of the Q_{2n} values are negative. Among $^{90,96}Zr$ and $^{124,132}Sn$ from the above values of quadrupole deformation, the difference is small. The deformation effect of ^{132}Sn is smallest, so its fusion cross sections show a suppression below the Coulomb barrier in comparison with the reference systems $^{40}Ca+^{90,96}Zr$ and ^{124}Sn. Finally, we select the data of the $^{64}Ni+^{118,124,132}Sn$ and $^{58}Ni+^{124,132}Sn$ systems[6] to obtain their $F(x)$s and x, which are shown in Fig. 2(c). We can see that the trends of their $F(x)$s are similar at energies above and below the Coulomb barrier even if the $^{58}Ni+^{124,132}Sn$ systems have the larger positive Q_{2n} values than those of $^{64}Ni+^{124,132}Sn$. It means that the *PQNT* does not give a enhancement of fusion cross section at energies below the Coulomb barrier. It is the same for the $^{16,18}O+^{76,74}Ge$

systems. After two-neutron transfer, the deformations of interacting nuclei change slightly. As a result, their Coulomb barrier values are almost the same and their $F(x)$s are similar.

4. Summary

The experimental fusion excitation functions are analyzed by use of the fusion function $F(x)$ with the variable x. The systematic behaviors of the $PQNT$ effect for fusion systems are explored. The fusion enhancements of some fusion systems with positive Q_{2n} values are observed at energies below the Coulomb barrier. It is for this reason that, after two-neutron transfer, the deformation of interacting nuclei increases and the mass asymmetry of reaction systems decreases. However, the fusion enhancement is not observed for the other fusion systems due to the small deformation of interacting nuclei after two-neutron transfer. This effect will be further explored.

References

1. A. M. Stefanini *et al.*, Phys. Rev. C **73**, 034606 (2006).
2. J. J. Kolata *et al.*, Phys. Rev. C **85**, 054603 (2012).
3. H. Q. Zhang *et al.*, Phys. Rev. C **82**, 054609 (2010).
4. A. M. Stefanini *et al.*, Nucl. Phys. **456**, 509 (1986).
5. P. Jacobs *et al.*, Phys. Lett. B **175**, 271 (1986).
6. Z. Kohley *et al.*, Phys. Rev. Lett. **107**, 202701 (2011).
7. M. Beckerman *et al.*, Phys. Rev. Lett. **45**, 1472 (1980).
8. P. H. Stelson, H. J. Kim, M. Beckerman, D. Shapira, and R. L. Robinson, Phys. Rev. C **41**, 1584 (1990).
9. L. F. Canto *et al.*, Nucl. Phys. A **821**, 51 (2009).
10. K. Hagino *et al.*, Comput. Phys. Commun. **123**, 143 (1999).
11. S. Raman *et al.*, At. Data Nucl. Data Tables **78**, 1 (2001).
12. T. Kibédi and R. H. Spear, At. Data Nucl. Data Tables **80**, 35 (2002).
13. G. L. Zhang, X. X. Liu and C. J. Lin, Phys. Rev. C 89, 054602 (2014).
14. H. M. Jia *et al.*, Phys. Rev. C **86**, 044621 (2012).
15. J. R. Leigh *et al.*, Phys. Rev. C **52**, 3151 (1995).
16. C. P. Silva, D. Pereira, L. C. Chamon, E. S. Rossi, G. Ramirez, A. M. Borges, and C. E. Aguiar, Phys. Rev. C **55**, 3155 (1997).
17. A. M. Borges, C. P. da Silva, D. Pereira, L. C. Chamon, E. S. Rossi, and C. E. Aguiar, Phys. Rev. C **46**, 2360 (1992).
18. M. Beckerman, M. Salomaa, A. Sperduto, J. D. Molitoris, and A. Di Rienzo, Phys. Rev. C **25**, 837 (1982).
19. D. Ackermann *et al.*, Nucl. Phys. A **609**, 91 (1996).

CONTRIBUTIONS TO δV_{pn} FOR EVEN-EVEN Ra ISOTOPES BASED ON THE COVARIANT DENSITY FUNCTIONAL THEORY

W. Zhang[1,2,*], Z. P. Li[3] and S. Q. Zhang[4]

[1] *School of Physical Engineering, Zhengzhou University, Zhengzhou 450001, China*
[2] *School of Electrical Engineering and Automation, Henan Polytechnic University, Jiaozuo 454003, PRC*
[3] *School of Physical Science and Technology, Southwest University, Chongqing 400715, PRC*
[4] *State Key Laboratory of Nuclear Physics and Technology, School of Physics, Peking University, Beijing 100871, PRC*
** E-mail: zw76@pku.org.cn*

The δV_{pn} for even-even $^{208-228}$Ra are investigated in the covariant density functional theory and the quadrupole-octupole collective Hamiltonian method. It is shown that the successive introduction of static deformation and collective fluctuation are crucial ingredients for the reproduction of data, especially for $N = 132 - 136$. The properties of the nuclei associated with δV_{pn} at $N = 132$ are analyzed in detail. Microscopically, near the Fermi surfaces in the single-particle level spectrums, levels with similar energies above the gaps $N = 132$ and $Z = 88$ can be found for ^{220}Ra.

Keywords: Proton-neutron interaction; quadrupole and octupole deformation; generator coordinate method.

It is well-known that the proton-neutron interaction plays an important role in nuclear deformations, phase transitions, collective motion, and single-particle levels. Empirically, the mass filter δV_{pn} (also denoted as δV, V_{pn}, and ε_{1p-1n}, δV_{1p-1n} ignoring factor 1/4), which is defined as double energy difference of four neighboring nuclei, has been identified with the proton-neutron interaction. Theoretically, the intriguing properties of δV_{pn} are studied by various models.[1–3]

Particularly for Ra isotopes, the anomaly of δV_{pn} for ^{221,223}Ra with $N = 133, 135$ are confirmed as "Ra puzzle".[4] Such puzzle is speculated with the softness of octupole deformation around $Z = 88$ and $N = 134$, microscopically associated with the interaction between orbital pairs $(\nu 2g_{9/2}, \nu 1j_{15/2})$ and $(\pi 2f_{7/2}, \pi 1i_{13/2})$ with $\Delta l = 3$ and $\Delta j = 3$.[5] Due to the theoretical difficulties in describing odd-A nuclei beyond mean-field,[6] the puzzle has not been fully understood.

On the other hand, several attempts for describing δV_{pn} for even-even Ra isotopes have been carried out.[7,8] It is interesting to investigate the δV_{pn} in the covariant density functional framework. The covariant density functional theory (CDFT) has achieved great successes.[9–11] Based on the CDFT, a variant of generator coordinate method (GCM), which is called quadrupole-octupole collective Hamiltonian (QOCH) method, has been developed.[12]

In this proceeding, the structure of δV_{pn} for even-even Ra isotopes will be analyzed by the CDFT and QOCH methods. The contributions to the δV_{pn} will be decomposed into the deformation energy $E_{\text{def}} = E_{\text{min}}(\beta_2, \beta_3) - E_{\text{sph.}}$, as well as collective correction energy $E_{\text{coll}} = E(0^+) - E_{\text{min}}(\beta_2, \beta_3)$. The point-coupling effective interaction applied is PC-PK1.[13] The pairing strength for neutrons (protons) is also taken from Ref. 13. The major shell for the harmonic oscillator is $N_f = 20$. Based on the mean-field states, after subtracting the zero-point energies from the total energies, the energies of the collective low-lying 0^+ state $E(0^+)$ are calculated by the QOCH code.

The theoretical frameworks can be found in Ref. 12,13 and references therein.

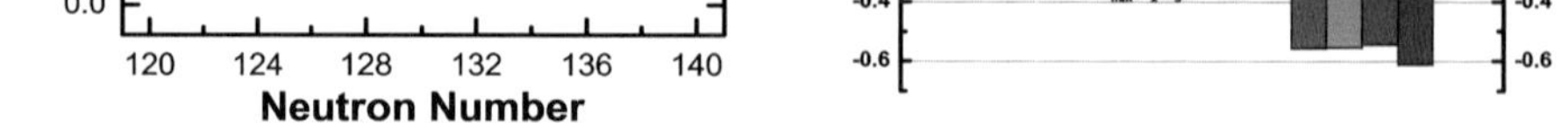

Fig. 1. (Color online) Left: The δV_{pn} for even-even $^{208-228}$Ra obtained by spherical and ground states in the (β_2, β_3) plane CDFT with effective interaction PC-PK1, together with corresponding collective QOCH result and experimental data. Right: The contribution brought to δV_{pn} at $N = 132$ by static deformation and collective fluctuations. For consistency, a factor of $1/4$ is multiplied to each terms.

The δV_{pn} for even-even $^{208-228}$Ra isotopes are plotted in left panel of Fig. 1. In this figure, the experimental δV_{pn} points can be roughly divided into two segments $120 \leqslant N \leqslant 126$ and $128 \leqslant N \leqslant 140$. It decreases slowly and then displays a sudden jump to 288 keV at $N = 128$. In the second segment, a peak 337 keV is formed at $N = 132$, with slow slope from both sides.

In Fig. 1, three theoretical curves reproduce the experimental δV_{pn} in reasonable levels. With spherical constraints, the δV_{pn} in Fig. 1 gives a continuous decreasing slope when $N \geqslant 128$. Furthermore, the introduction of deformation dramatically changes the δV_{pn} points $N = 134, 136, 140$ for about 150 keV, nearly half of its absolute values.

Unfortunately, with the consideration of β_2 and β_3, there are some discrepancies, e.g., the one at $N = 132$. It is shown that the trend as well as the quantitative values from QOCH method well reproduced the data. Taking the δV_{pn} curve with spherical constraints in right panel of Fig. 1 as a baseline, the modifications by the static deformation and dynamical fluctuations are in the opposite positions cancelling each other for $N = 132, 134, 136$. Both the deformation and the collective fluctuation play important roles in giving better description on the fine structure of δV_{pn}. In Ref. 8, in the context of deformed Hartree–Fock–Bogoliubov and GCM theories, good agreement with experimental δV_{pn} for Ra isotopes with $N = 134 - 142$ are obtained after considering the octupole degree of freedom and collective correlation. There is still a discrepancy for the δV_{pn} at $N = 132$.[8]

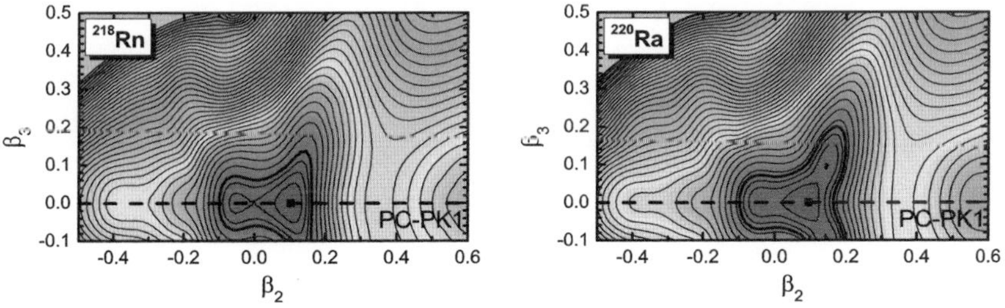

Fig. 2. (Color online) The contour plots of total energies for isotopes ^{218}Rn (left) and ^{220}Ra (right) in (β_2, β_3) plane obtained in CDFT with PC-PK1. The contour lines are plotted every 0.5 MeV up to 2 MeV (thick line) and from there on every 1 MeV. The ground states are denoted by "■".

Taking δV_{pn} at $N = 132$ in the left panel of Fig. 1 as an example, the contributions from E_{def} and E_{coll} are plotted in the right panel. It should be noted that the ground states for the involved nuclei ^{216,218}Rn and ^{218,220}Ra are all β_2-deformed. Considering similar values for ^{216}Rn and ^{218}Ra, the contributions brought to δV_{pn} at $N = 132$ by the static deformation and collective fluctuation are mainly from those of ^{218}Rn and ^{220}Ra.

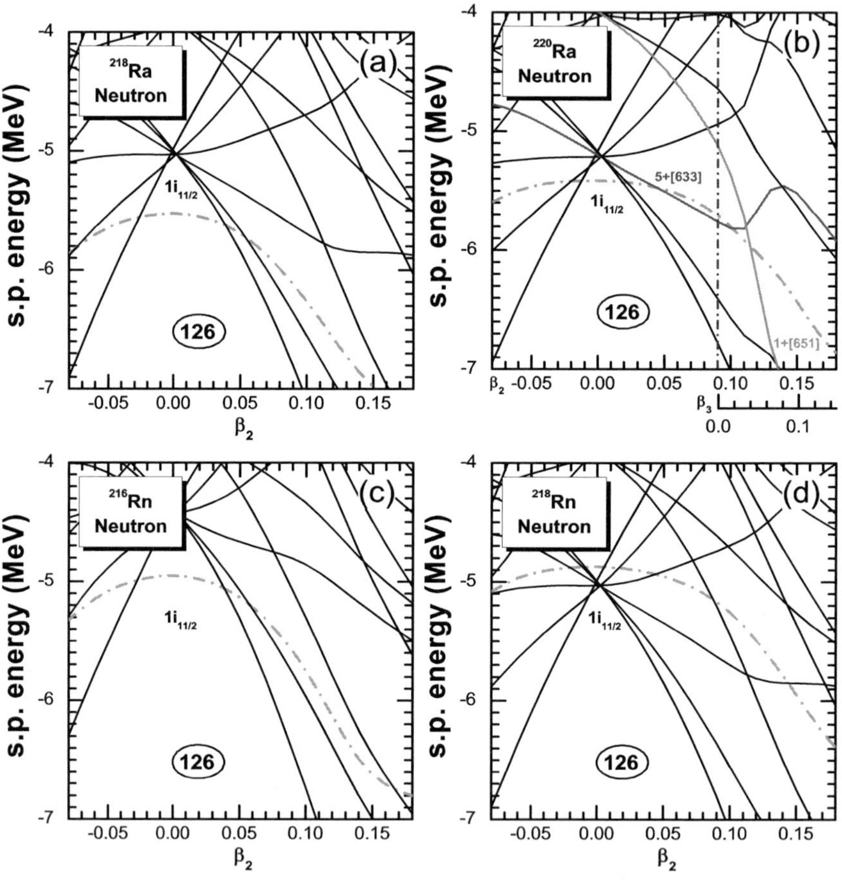

Fig. 3. (Color online) Neutron single-particle levels of ^{218,220}Ra and ^{216,218}Rn in CDFT approach with PC-PK1 as functions of β_2. The dash-dot lines denote the corresponding Fermi surfaces. The states with $\beta_2 \geqslant 0.09$ of ^{220}Ra are also octupole deformed.

To see the difference between ^{218}Rn and ^{220}Ra, the contour plot of the binding energies are plotted in Fig. 2. In the (β_2, β_3) plane, the ground state shape evolves $^{210-214}$Rn and $^{212-216}$Ra with $N = 124 - 128$ are near spher-

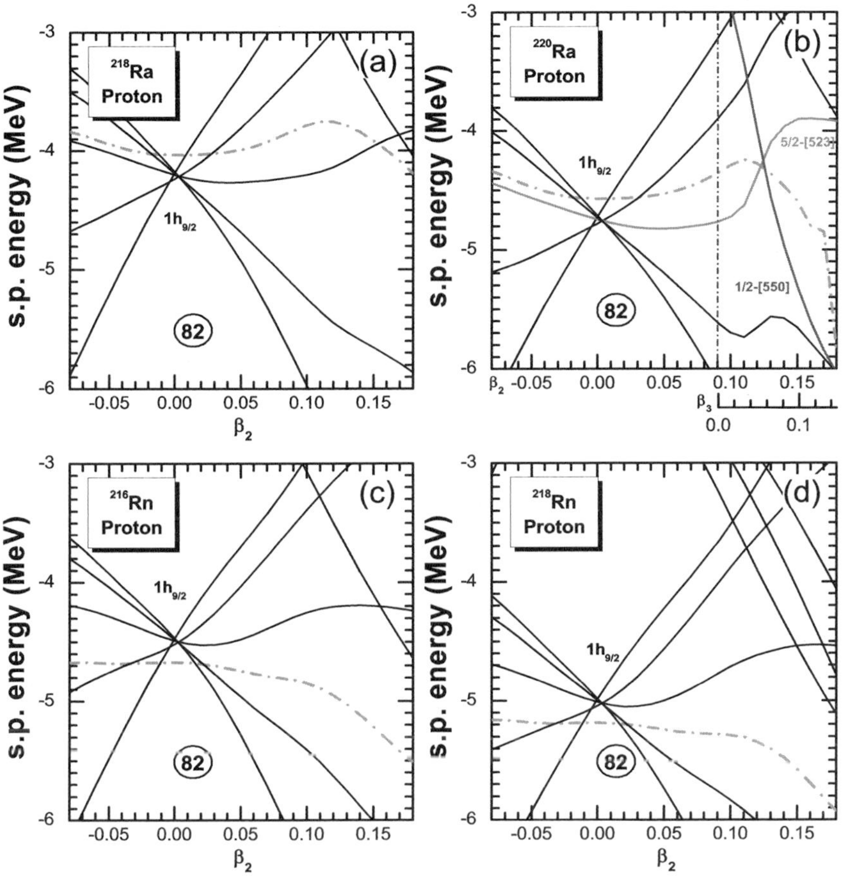

Fig. 4. (Color online) Same as Fig. 3, but for proton.

ical, ^{218,222,226}Rn quadrupole deformed and ^{224,228}Ra octupole deformed while ^{220}Ra displays the transitional property from quadrupole to octupole deformed. For ^{220}Ra in Fig. 2, the quadrupole-octupole minimum is separated from the quadrupole minimum (ground state) by a shallow saddle. There is no quadrupole-octupole minimum for ^{218}Rn.

To understand such behavior microscopically,[14] the neutron and proton single-particle levels for ^{216,218}Rn and ^{218,220}Ra are shown in Figs. 3 and 4. The states for ^{220}Ra with $\beta_2 \geqslant 0.09$ are also β_3-deformed. In Fig. 3, near the Fermi surface for the ground states, an energy gap at $N = 132$ with two crossing levels can be seen. Similar to neutron, an energy gap at $Z = 88$ near the Fermi surface is found in Fig. 4, accompany with two levels just above the Fermi surface. With certain single-particle level structure, the corresponding Fermi surfaces would change little for $N = 132$ and 2 or 4 more neutrons, and for $Z = 88$ and 2 or 4 more protons.

In summary, the δV_{pn} for even-even Ra isotopes are analyzed in the covariant density functional theory and the quadrupole-octupole collective Hamiltonian model. It is shown that the successive introduction of static deformation and collective fluctuation are crucial ingredients for the reproduction of data, especially for $N = 132 - 136$. Microscopically, near the Fermi surfaces in the single-particle level spectrums, levels with similar energies above the gaps $N = 132$ and $Z = 88$ can be found for ^{220}Ra.

This work was supported in part by National Natural Science Foundation of China under Grant Nos. 10775004, 10975007, 10975008, and 11105042, and the Open Fund of Key Laboratory of Precision Navigation and Technology, National Time Service Center, CAS. The calculation was supported by the high-performance grid computing platform of Henan Polytechnic University.

References

1. M. Stoitsov, R. B. Cakirli, R. F. Casten, W. Nazarewicz, and W. Satuła, Phys. Rev. Lett. **98**, 132502 (2007).
2. M. Bender and P.-H. Heenen, Phys. Rev. C **83**, 064319 (2011).
3. D. Bonatsos, S. Karampagia, R. B. Cakirli, R. F. Casten, K. Blaum, and L. Amon Susam, Phys. Rev. C **88**, 054309 (2013).
4. D. Neidherr et al., Phys. Rev. Lett. **102**, 112501 (2009).
5. P. A. Butler and W. Nazarewicz, Rev. Mod. Phys. **68**, 349 (1996)
6. B. Bally, B. Avez, M. Bender, and P.-H. Heenen, Phys. Rev. Lett. **113**, 162501 (2014).
7. L. F. Yu, P. W. Zhao, S. Q. Zhang, and J. Meng, arXiv:1211.0601 [nucl-th]
8. L. M. Robledo, and P. A. Butler, Phys. Rev. C **88**, 051302(R)(2013).
9. P. Ring, Prog. Part. Nucl. Phys. **37**, 193 (1996).
10. Vretenar, A. Afanasjev, G. Lalazissis, and P. Ring, Phys. Rep. **409**, 101 (2005).
11. J. Meng, H. Toki, S. Zhou, S. Zhang, W. Long, and L. Geng, Prog. Part. Nucl. Phys. **57**, 470 (2006).
12. Y. Peng, Z. P. Li, and W. Zhang, in preparation.
13. P. W. Zhao, Z. P. Li, J. M. Yao, and J. Meng, Phys. Rev. C **82**, 054319 (2010).
14. W. Zhang, Z. P. Li, S. Q. Zhang, and J. Meng, Phys. Rev. C **81**, 034302 (2010).

ALGEBRAIC REALIZATION OF THE TRIAXIAL ROTOR DYNAMICS

Y. Zhang*, Q. Y. Li, Y. Zuo and X. X. Wang

Department of Physics, Liaoning Normal University,
Dalian, 116029, China
**E-mail: dlzhangyu_physics@163.com*

A mapping from the triaxial rotor to the SU(3) limit description in the interacting boson model is established. A comparison between the triaxial dynamics generated from the triaxial rotor and those from the IBM image has been made through a numerical example, and the results indicate that the mapping scheme works very well.

Keywords: Triaxial rotor; interacting boson model; SU(3) limit.

1. Introduction

The quantum rotor model has been playing an important role in elucidating rotational excitations in nuclei.[1,2] Collective rotation in nuclei can be also described by the SU(3) limit of the interacting boson model (IBM), in which only the cases in the IBM-1[3] with no distinction between protons and neutrons will be considered. However, similar to Elliott model,[4,5] the SU(3) limit of the IBM only covers the axially symmetric situations if up to two-body interactions are taken into consideration.[3,6] Actually, if high-order interactions are included in the IBM Hamiltonian, a triaxial minimum of the potential energy surface can be found by using the coherent state method.[7] Specifically, a stable, triaxial minimum at $\gamma = 30^\circ$ in the potential energy surface can be induced by the cubic term $(d^\dagger \times d^\dagger \times d^\dagger)^3 \cdot (\tilde{d} \times \tilde{d} \times \tilde{d})^3$.[7,8] However, the triaxial minimum in the potential energy surface is not a sufficient condition for a rigidly triaxial rotation. In view of this, it was further suggested that the rigidly triaxial rotation may be generated by involving higher order interactions in the SU(3) limit.[9] As the triaxial rotation is explicitly defined in the triaxial rotor Hamiltonian, it is necessary to clarify the exact relation between the triaxial rotor Hamiltonian and the corresponding IBM Hamiltonian. The purpose of this work is to construct an exact SU(3) IBM image of the triaxial rotor Hamiltonian and check its validity of the description in the finite-N systems.

2. The IBM Image of the Triaxial Rotor

A general rotor Hamiltonian is given as[2,10]

$$H_{\rm rot} = A_1 L_1^2 + A_2 L_2^2 + A_3 L_3^2 \,, \tag{1}$$

where L_α is the projection of the angular momentum onto the α-th body-fixed principal axis, and A_α is the corresponding inertia parameter. A_α is usually assumed to satisfy $A_2 \leq A_1 \leq A_3$. A Hamiltonian in the IBM framework is constructed from two kinds of boson operators, namely, the s-boson with $J^\pi = 0^+$ and the d-boson with $J^\pi = 2^+$.[3] The total boson number N is taken as the number of valence particle (or hole) pairs in a nucleus. The dynamical symmetry limits in the IBM are characterized by three different chains of the U(6) group:[3]

$$\mathrm{U}(6) \supset \mathrm{U}(5) \supset \mathrm{O}(5) \supset \mathrm{O}(3) \,, \tag{2}$$

$$\mathrm{U}(6) \supset \mathrm{SU}(3) \supset \mathrm{O}(3) \,, \tag{3}$$

$$\mathrm{U}(6) \supset \mathrm{O}(6) \supset \mathrm{O}(5) \supset \mathrm{O}(3) \,. \tag{4}$$

The dynamical symmetries associated with the above three chains correspond to three typical nuclear shapes or collective modes. These are the spherical vibration in the U(5) limit, the axially-symmetric rotation in the SU(3) limit, and the γ-unstable motion in the O(6) limit. Traditionally, the O(6) limit is often considered as the finite-N behavior of the (γ-soft) triaxiality in the IBM if up to two-body interactions are included in the Hamiltonian.[11,12]

Since a rigid rotor Hamiltonian can be exactly mapped into the SU(3) shell model under SU(3) $\supset$ SO(3) integrity basis,[13–15] the similar mapping can also be realized in the SU(3) limit of the IBM. In the IBM, the SU(3) operators are defined as

$$L_u^b = \sqrt{10}(d^\dagger \times \tilde{d})_u^{(1)} \,, \tag{5}$$

$$Q_u^b = 2\sqrt{2}[(d^\dagger \times +s^\dagger \times \tilde{d})_u^{(2)} - \frac{\sqrt{7}}{2}(d^\dagger \times \tilde{d})_u^{(2)}] \,, \tag{6}$$

$$C_2 = \frac{1}{4} Q^b \cdot Q^b + \frac{3}{4} L^b \cdot L^b \,, \tag{7}$$

$$C_3 = -\frac{\sqrt{70}}{72}(Q^b \times Q^b \times Q^b)^{(0)} - \frac{\sqrt{30}}{8}(L^b \times Q^b \times L^b)^{(0)}. \tag{8}$$

It should be noted that the quadrupole operators Q^b_u are multiplied by $2\sqrt{2}$ in comparison to those defined in the original IBM.[3] Then, according to the our recent analysis,[16] the exact IBM image of a rigid triaxial rotor Hamiltonian can be given as

$$H_{\mathrm{IBM}} = H_0 + aL^b \cdot L^b + bX^b_3 + cX^b_4\,, \tag{9}$$

where[15]

$$X^b_3 = \sum_{\alpha\beta} L^b_\alpha Q^b_{\alpha\beta} L^b_\beta = \frac{\sqrt{30}}{6}(L^b \times Q^b \times L^b)^{(0)}\,, \tag{10}$$

$$X^b_4 = \sum_{\alpha\beta\gamma} L_\alpha Q^c_{\alpha\beta} Q^c_{\beta\gamma} L_\gamma \tag{11}$$

$$= \frac{-5\sqrt{3}}{18}[(L^b \times Q^b)^{(1)} \times (L^b \times Q^b)^{(1)}]^{(0)}\,, \tag{12}$$

and

$$H_0 = -tC_2 + \frac{t'}{N^2}(C_2)^2 \tag{13}$$

with $t > 0$ and $t' > 0$ being parameters and N being the total number of bosons. In addition, the parameters a, b, and c depend on the inertia parameters A_α and the SU(3) irrep (λ, μ) with

$$a = \sum_\alpha a_\alpha A_\alpha, \qquad a_\alpha = \lambda_1\lambda_2\lambda_3/D_\alpha\,, \tag{14}$$

$$b = \sum_\alpha b_\alpha A_\alpha, \qquad b_\alpha = \lambda^2_\alpha/D_\alpha\,, \tag{15}$$

$$c = \sum_\alpha c_\alpha A_\alpha, \qquad c_\alpha = \lambda_\alpha/D_\alpha\,, \tag{16}$$

and

$$D_\alpha = 2\lambda^3_\alpha + \lambda_1\lambda_2\lambda_3\,, \tag{17}$$

where

$$\lambda_1 = -(\lambda - \mu)/3\,, \tag{18}$$

$$\lambda_2 = -(\lambda + 2\mu + 3)/3\,, \tag{19}$$

$$\lambda_3 = (2\lambda + \mu + 3)/3\,. \tag{20}$$

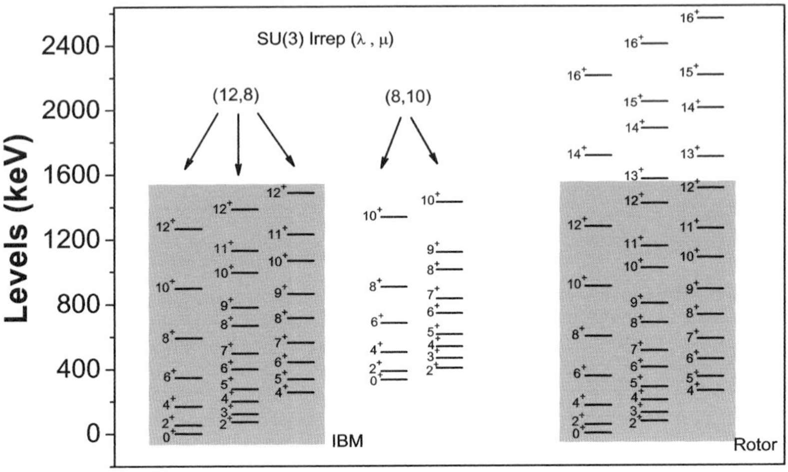

Fig. 1. A typical level pattern of the triaxial rotor described by (1) and that of the corresponding IBM image described by (9) for $N = 14$. In the calculation, the inertia parameters in the triaxial rotor model are taken as $A_1 = 10.0$ keV, $A_2 = 8.0$ keV and $A_1 = 14.0$ keV, and the parameters in the IBM are taken as $t' = 15.0$ keV and $t = 55.7$ keV.

3. Typical Level Pattern

It should be emphasized that the mapping scheme given in (9) is not limit to the triaxial spectrum generated from the quadrupole-deformed rotor, but may also apply for a more general triaxial situation only if the relative difference between the inertial parameters is not too large. In other words, one can choose a suitable SU(3) irrep in the IBM only by adjusting t and t' to generate the corresponding triaxial spectrum. As an explicit illustration, the level pattern calculated from a general triaxial rotor Hamiltonian given in (1) together with the results calculated from its IBM image with the two lowest SU(3) irreps (12,8) and (8,10) for $N = 14$ are shown in Fig. 1. It is clearly shown in Fig. 1 that the mapping scheme works well for the lowest rotational levels with $L^\pi \leq 12^+$ in the rotor model that can be mapped to the IBM as those marked with the dark squares due to the finite N effect. In addition, the results shown in Fig. 1 also indicate that there exists a finite number of rotational levels built on the β and γ vibrational excitations in the IBM. In contrast, only rotational excitation allowed levels, which are infinite in number, appear, while excited bands appearing in the IBM image do not exist in the rotor model. Therefore, the triaxial spectrum generated by the IBM image seems more realistic than that of the triaxial rotor for finite N system.

4. Summary

In summary, the exact IBM mapping scheme of the triaxial rotor has been proposed in the SU(3) limit of the IBM, in which the explicit Hamiltonian form of the algebraic image of a general triaxial rotor is presented. Typical level patterns generated by the triaxial rotor Hamiltonian and the corresponding IBM image has been given to illustrate the validity of the mapping scheme, and the results indicate that the rigid triaxial rotational structure can be well realized in the IBM, which further enriches the IBM interpretation of nuclear collectivity.

5. Acknowledgments

Support from the Natural Science Foundation of China (11375005, 11005056) is acknowledged.

References

1. A. Bohr, Konggl. Dan. Vid. Selsk. Mat.-fys. Medd. **26**, no. 14 (1952).
2. A. S. Davydov and G. F. Filippov, Nucl. Phys. **8**, 237 (1958).
3. F. Iachello and A. Arima, *The Interacting Boson Model* (Cambridge University, Cambridge, England, 1987).
4. J. P. Elliott, Proc. R. Soc, London, Ser. A **245**, 128 (1958).
5. J. P. Elliott and M. Harvey, Proc. R. Soc, London, Ser. A **272**, 557 (1963).
6. Y. Zhang, F. Pan, Y. X. Liu, Y. A. Luo, and J. P. Draayer, Phys. Rev. **C 85**, 064312 (2012).
7. P. Van Isacker and J. Q. Chen, Phys. Rev. **C 24**, 684 (1981).
8. K. Heyde, P. Van Isacker, M. Waroquier, and J. Moreau, Phys. Rev. **C 29**, 1420 (1984).
9. Y. F. Smirnov, N. A. Smirnova, and P. Van Isacker, Phys. Rev. **C 61**, 041302(R) (2000).
10. J. L. Wood, A-M. Oros-Peusquens, R. Zaballa, J. M. Allmond, and W. D. Kulp, Phys. Rev. **C 70**, 024308 (2004).
11. J. Meyer-Ter-Vehn, Phys. Lett. **B 84**, 10 (1979).
12. V. Werner, C. Scholl, and P. von Brentano, Phys. Lett. **B 71**, 054314 (2005).
13. Y. Leschber and J. P. Draayer, Phys. Lett. **B 190**, 1 (1987).
14. O. Castaños, J. P. Draayer, and Y. Leschber, Z. Phys. **A 329**, 33 (1988).
15. H. A. Naqvi, C. Bahri, D. Troltenier, J. P. Draayer, and A. Faessler, Z. Phys. **A 351**, 259 (1995).
16. Y. Zhang, F. Pan, L. R. Dai, and J. P. Draayer, Phys. Rev. **C 90**, 044310 (2014).

AN INVESTIGATION OF FRAGMENTS PRODUCED IN THE ^{58,64}Ni + ^{9}Be FRAGMENTATION REACTIONS

Y. L. Zhang, C. W. Ma*, S. S. Wang, C. Y. Qiao

Department of Physics, Henan Normal University, Xinxiang 453007, China
**E-mail: machunwang@126.com*

The cross sections of the fragments produced in the 140*A* MeV ^{58,64}Ni + ^{9}Be reactions have been studied using the empirical parametrization formula EPAX2 and EPAX3, the modified statistical abrasion-ablation (SAA) model, and the antisymmetrized molecular dynamics (AMD) model. The calculated cross sections of fragments are compared and discussed.

Keywords: EPAX2; EPAX3; modified statistical abrasion-ablation; AMD.

1. Introduction

The cross section of fragment, which can be directly measured in experiments, can be used to construct probes for many important physical parameters in heavy-ion collisions, such as the nuclear symmetry energy, nuclear density, and temperature etc.[1–10] The probe for nuclear symmetry energy basing on cross sections of fragments, for example, theisoscaling,[11–13] isobaric yield ratio,[1,2,5,6,14–16] and isobaric yield ratio difference,[17–19] require high experimental and theoretical quality of data. In theory, many models have been developed to predict the fragments yields, or cross sections of fragment produced in heavy-ion collisions (HICs). In this article, three models are adopted to study the fragments produced in the 140*A* MeV ^{58,64}Ni + ^{9}Be reactions.[31] The frequently used empirical parametrizations EPAX2 and its improved version EPAX3,[20–22] the modified statistical abrasion-ablation (SAA) model,[23–27] and the antisymmetrized molecular dynamics (AMD) model[28–30] are used to calculate cross sections of fragments. In Sec. 2, the EPAX, SAA and AMD models will be briefly introduced. In Sec. 3, the calculated cross sections of fragments will be discussed.

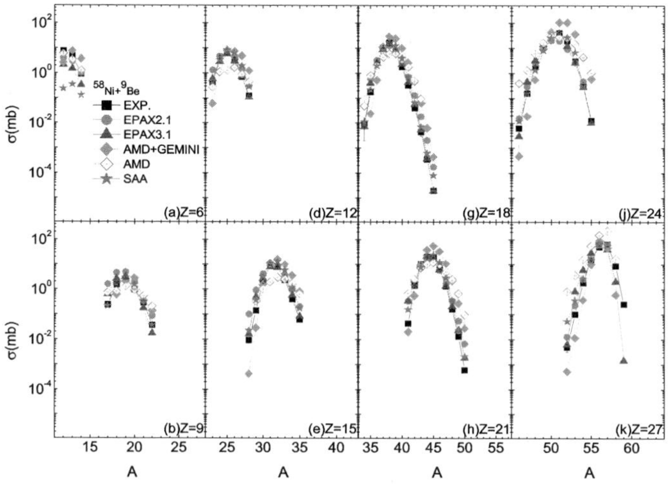

Fig. 1. (Color online) Cross sections of fragments produced in the 140A MeV ^{58}Ni + ^{9}Be reactions. The EPAX2 results are denoted by circles, EPAX3 by triangles, AMD + GEINI by full rhombus, AMD by open rhombus, and SAA by stars.

2. Methods

The EPAX parameterizations was proposed by K. summerer et al.[20] The EPAX2[21] enhances the ability to predict the fragments closing to the projectile. The new EPAX3 improves the ability to predict the most neutron-rich fragments[22] The SAA model is a two-stage model to describe the HICs above the intermediate energies, with the first stage describing the collisions between the overlapping zone of projectile and target nuclei and the second stage calculates the decay of the primary fragments with certain excitation energies. The AMD model was proposed by A. Ono to simulate the HICs of the Fermi energy range. The AMD + GEMINI[32] method is used to obtain the primary and the cold fragments. The fragments are formed using a coalescence radius R_c = 5 fm in the phase space up to the time of 500fm/c, and in the impact parameter range calculated is from 0 to 9 fm.

3. Results and Discussions

The cross sections for fragments in the 140A MeV ^{58}Ni + ^{9}Be reaction are plotted in Fig. 1. Large difference between all the calculated results and experimental date can be found in the small Z (Z =6) fragments. The SAA results are not better agreement with experimental data for the neutron-rich fragments. For the AMD + GEMINI results, an overestimated cross

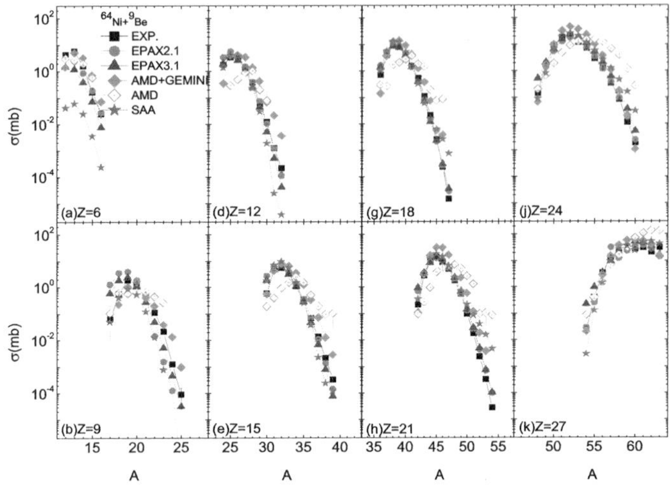

Fig. 2. (Color online) The same as Fig. but for the 140A MeV ^{64}Ni + ^{9}Be reaction.

sections of the neutron-deficient fragments are shown, while the results are underestimated result for the neutron-rich fragments. The EPAX3 has better results than EPAX2 for most of the fragments. Furthermore, the EPAX3 better predict the neutron-rich fragments than EPAX2 due to the improvement of EPAX3.[22] Only in the large-Z fragments, EPAX3 has an overestimation than EPAX2 for the neutron-deficient fragments.

In Fig. 2, the calculated cross sections for fragments in the 140A MeV ^{64}Ni + ^{9}Be reaction are plotted. Compared to the experimental data, large differences are found between the measured results and calcualted results for the small-Z fragments. For the small-Z fragments, the SAA largely underestimates the cross sections of neutron-rich fragments. For the neutron-rich isotopes, the SAA overestimates the cross sections of the large-Z fragments ($Z > 18$). After decaying, the AMD + GEMINI can better reproduce the fragments than AMD. The EPAX3 results very similar to those of EPAX2, with EPAX2 better reproduce the results than EPAX3.

Both in ^{58}Ni+^{9}Be and ^{64}Ni+^{9}Be reactions, the EPAX results agree well with the measured data for most of the fragments. The SAA can also well reproduce the cross sections of large-Z fragments, except some neutron-rich isotopes in the reaction of neutron-rich projectile. The AMD + GEMINI calculated results can reproduce the fragments quite well for most of the fragments. Since AMD is a dynamical model, it is believed much more information can be obtained from its results than EPAX and SAA.

4. Summary

The fragments produced in the 140A MeV ^{58}Ni+^{9}Be and ^{64}Ni+^{9}Be reactions were calculated by the EPAX2, EPAX3, SAA, AMD, AMD + GEMINI models. The SAA model is found to largely underestimate the small-Z fragments. After the decaying calculation, the AMD + GEMINI can reproduce most of the fragments. The EPAX3 shows a better ability to reproduce the neutron-rich fragments in the ^{58}Ni reaction, but is no better than EPAX2 for the fragments in the neutron-rich ^{64}Ni reaction.

References

1. M. Huang *et al.*, *Phys. Rev. C.* **81**, 044620 (2010)
2. C. W. Ma *et al.*, *Phys. Rev. C.* **83**, 064620 (2011)
3. B. A. Li *et al.*, *Phys. Rep.* **464**, 113 (2008)
4. R. W. Minich *et al.*, *Phys. Lett. B.* **118**, 458 (1982)
5. M. Huang *et al.*, *Phys. Rev. C.* **81**, 044618 (2011)
6. P. Marini *et al.*, *Phys. Rev. C.* **85**, 034617 (2012)
7. S. Goyal *et al.*, *Phys. Rev. C.* **83**, 047601 (2011)
8. M. B. Tsang *et al.*, *Phys. Rev. C.* **76**, 041302(R) (2007)
9. C. W. Ma *et al.*, *Phys. Rev. C.* **86**, 054611 (2012)
10. C. W. Ma *et al.*, *Phys. Rev. C.* **88**, 014609 (2013)
11. Y. G. Ma *et al.*, *Phys. Rev. C.* **69**, 064610 (2004); *ibid*, **72**, 064603 (2005)
12. A. S. Botvina *et al.*, *Phys. Rev. C.* **65**, 044610 (2002)
13. M. Colonna *et al.*, *Phys. Rev. Lett.* **110**, 042701 (2013)
14. M. Huang *et al.*, *Phys. Rev. C.* **82**, 054602(R) (2010)
15. W. Lin *et al.*, *Nucl. Sci. Tech.* **24**, 050511 (2013)
16. R. Wada *et al.*, *Nucl. Sci. Tech.* **24**, 050501 (2013)
17. C. W. Ma *et al.*, *Phys. Rev. C.* **87**, 034618 (2013)
18. C. W. Ma *et al.*, *J. Phys. G: Nucl. Part. Phys.* **40**, 125106 (2013)
19. C. W. Ma *et al.*, *Phys. Rev. C.* **89**, 057602 (2014)
20. K. Sümmerer *et al.*, *Phys. Rev. C.* **42**, 2546 (1990)
21. K. Sümmerer *et al.*, *Phys. Rev. C.* **61**, 034607 (2000)
22. K. Sümmerer *et al.*, *Phys. Rev. C.* **86**, 014601 (2012)
23. T. Brohm *et al.*, *Nucl. Phys. A.* **569**, 821 (1994)
24. J. J. Gaimard *et al.*, *Nucl. Phys. A.* **531**, 709 (1991)
25. D. Q. Fang *et al.*, *Phys. Rev. C.* **61**, 044610 (2000)
26. C. W. Ma *et al.*, *Phys. Rev. C.* **79**, 034606 (2009)
27. D. Q. Fang *et al.*, *J. Phys. G: Nucl. Part. Phys.* **34**, 2173 (2007)
28. A. Ono *et al.*, *Phys. Rev. Lett.* **68**, 2898 (1992)
29. A. Ono *et al.*, *Phys. Rev. C.* **53**, 2958 (1996)
30. A. Ono *et al.*, *Phys. Rev. C.* **59**, 853 (1999)
31. M. Mocko *et al.*, *Phys. Rev. C.* **74**, 054612 (2006)
32. R. J. Charity *et al.*, *Nucl. Phys. A.* **483**, 371 (1998)

THE PRODUCTION PROBABILITY OF SUPERHEAVY FRAGMENTS AT DIFFERENT DEFORMATION AND ORIENTATION IN ^{238}U+^{238}U REACTION

Kai Zhao*, Zhuxia Li, Xizhen Wu and Yingxun Zhang

China Institute of Atomic Energy,
Beijing, P. R. China,
**E-mail: zhaokai@ciae.ac.cn*
www.ciae.ac.cn

Ning Wang

Guangxi Normal University,
Guilin, P. R. China
E-mail: wangning@gxnu.edu.cn

The dynamical effects of initial orientation and deformation of projectile and target on the ^{238}U+^{238}U reaction have been studied by using the improved quantum molecular dynamics model. The production probability of superheavy fragments(SHF) with $Z>110$ at incident energies 7-12A MeV are investigated for different orientations and deformations of two colliding nuclei. The inclusion of β_4 deformation in prolate uranium seems to enhance the production probability of SHF in the side-side case as the more stable structure of initial uranium effectively reduces the excitation energy of the composite system. It is suggested that the optimal initial condition for production of SHF in ^{238}U+^{238}U could be the side-side orientation and prolate deformation with a small β_4 deformation in the ground state of uranium.

Keywords: ImQMD model; deformation; orientation; superheavy nuclei.

1. Introduction

In the last forty years, the superheavy elements from $Z = 107$ to 118 have been synthesized by fusion reactions.[1,2] But compared with the predicted "island of stability" around neutron number $N = 184$, only neutron-deficient isotopes for elements with $Z > 100$ have been synthesized so far. The experimental synthesized nuclei by means of complete fusion is limited by the neutron number of available projectiles and targets and by the very low production cross section as well. It is necessary to search for other ways in addition to the complete fusion reaction to approach the region of the island of stability. The strongly damped reactions between very heavy actinide nuclei, such as U+U, could be one possible approach for those purposes.[3–5] In these reactions, large charge and mass transfer probably produce the primary superheavy neutron-rich nuclei. And these produced superheavy nuclei are expected to survive against fission at low excitation energy. However, a nucleus ^{238}U in its ground state exhibits a prolate deformation, which leads to different orientations of projectile and target at the initial time of $^{238}U+ {}^{238}U$ reaction. In recent years, the orientation effect on the potential energy surface was studied by Langevin-type equation approach[4,5] and Di-nuclear model.[6–8] It was pointed out that the orientation effect might play a rather important role in subbarrier reactions of deformed nuclei. In addition, the theoretical description of the strongly damped reactions between very heavy actinide nuclei is one of most difficult problems in nuclear physics due to the extremely large number of degrees of freedom involved. This motivates the use of microscopic approaches. The quantum molecular dynamics type model is a microscopic N-body dynamical model which has been successfully to describe the main features observed in nuclear collisions.[9–12] In our previous studies about the strongly damped reactions,[13,14] the deformation of actinide nuclei was not considered. This may influence the reaction features, such as the mass distribution of fragments, especially the production probability of superheavy nuclei.

In this work, we mainly investigate the effects of orientation and deformation by considering the head-on reaction of $^{238}U + {}^{238}U$. The incident energy dependence of the probability of producing superheavy nuclei under different initial orientations and shapes of projectile and target will be studied. For simplification, we only take nose-nose and side-side cases as initial orientations in the paper.

2. Brief Description of the ImQMD Model

A detailed description of the ImQMD model and its application in low-energy heavy-ion collisions can be found in Refs13–17. Here, we only mention that the potential energy of the system is an integration of the potential energy density functional, and it can be expressed by

$$U = U_{loc} + U_{coul} \tag{1}$$

where the nuclear interaction contribution U_{loc}:

$$U_{loc} = \int V_{loc} d\mathbf{r} \tag{2}$$

and the potential energy density V_{loc} can be derived directly from a zero-range Skyrme interaction:

$$V_{loc} = \frac{\alpha}{2}\frac{\rho^2}{\rho_0} + \frac{\beta}{\gamma+1}\frac{\rho^{\gamma+1}}{\rho_0^{\gamma}} + \frac{g_0}{2\rho_0}(\nabla\rho)^2 + \frac{c_s}{2\rho_0}[\rho^2 - \kappa_s(\nabla\rho)^2]\delta^2 + g_\tau\frac{\rho^{\eta+1}}{\rho_0^{\eta}}, \quad (3)$$

where ρ is the nucleon density. $\delta = (\rho_n - \rho_p)/(\rho_n + \rho_p)$ is the isospin asymmetry, and ρ_n, ρ_p are neutron and proton density, respectively. The same parameters are used as in Refs 13,14.

The Coulomb energy is written as a sum of the direct and the exchange contribution:

$$U_{Coul} = \frac{1}{2}\int\int \rho_p(\mathbf{r})\frac{e^2}{|\mathbf{r}-\mathbf{r}'|}\rho_p(\mathbf{r}')d\mathbf{r}d\mathbf{r}' - e^2\frac{3}{4}\left(\frac{3}{\pi}\right)^{1/3}\int \rho_p^{4/3}d\mathbf{R}. \quad (4)$$

The isospin-dependent in-medium nucleon-nucleon scattering cross sections in the collision term are applied. The phase space occupation number constraint method and the system-size-dependent wave-packet width are adopted as those in the previous work.[13,14]

In this work, we pay a special attention to the initial condition, especially the shape of initial nuclei and the orientation. Here we adopt two different deformations with (1)β_2=0.215 and β_4=0, named init-I and (2) β_2=0.215 and β_4=0.093, named init-II as initial shapes of ^{238}U nucleus. The nucleons are sampled according to the density distribution (the Fermi distribution) which reads

$$\rho(r,\theta) = \frac{\rho_0}{1+\exp[\frac{r-R(\theta)}{a}]}. \quad (5)$$

Here,

$$R(\theta) = R_0[1 + \beta_2 Y_{20}(\theta) + \beta_4 Y_{40}(\theta)] \quad (6)$$

and R_0=1.16$A^{1/3}$ fm, a=0.5 fm. In the work, the orientations of nose-nose (N-N) and side-side (S-S) of two nuclei alone Z axis in the X-Z plane are considered and the initial distance between the centers of mass of the projectile and the target is taken to be 40 fm. Thus, we have 4 different initial conditions, namely, N-N orientation with init-I and init-II and S-S orientation with init-I and init-II.

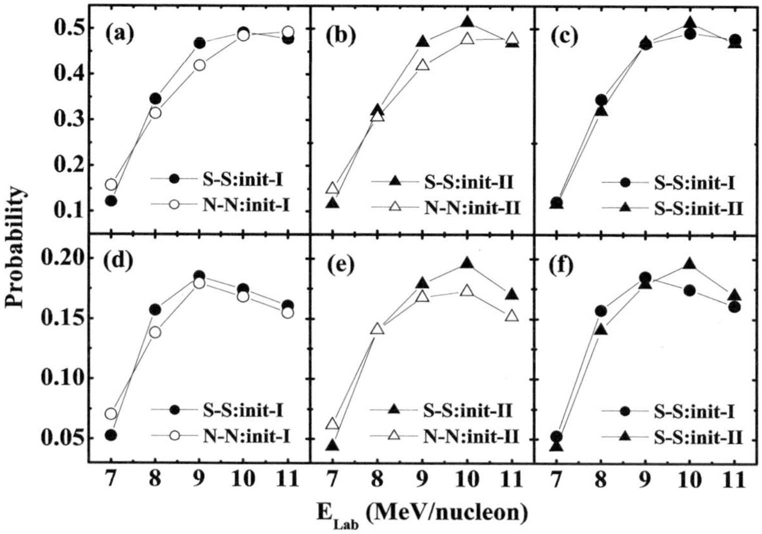

Fig. 1. The energy dependence of the probability of producing SHF for head on ^{238}U + ^{238}U reaction for 4 initial conditions. The subfigures of (a), (b) and (c) are calculated at re-separation time and (d), (e) and (f) at the time of 1000fm/c after re-separation time.

3. Probability of Producing Super-Heavy Fragments

The super-heavy fragments are defined as the fragments with $Z>110$ in this work. The cluster searching process is performed at re-separation time of composite system and at the time of 1000 fm/c after re-separation, respectively. Because only one SHF can be produced in one event at most, the average probability of producing SHF in one event can be calculated by the number of events in which one SHF is produced divided by the number of total events. The probability of producing SHF as a function of incident energy for head on ^{238}U + ^{238}U reaction for 4 initial conditions are presented in Fig.1. The subfigures (a), (b) and (c) are calculated at the re-separation time, and (d), (e) and (f) show the results at the time of 1000 fm/c after re-separation of composite system. One can find from (a), (b), (d) and (e) that the probability of producing SHF for the S-S case at about 7A MeV are lower than that for the N-N case for both initial init-I and init-II, and then the probability increases rapidly for the S-S case and exceeds that for the N-N case at about $8A$ MeV. With an incident energy increasing further the probability of producing SHF increases further till reaching their maximum values. After then the probability of producing SHF decreases with further increase of incident energy. At the time of 1000 fm/c after re-separation, the probability of producing SHF in one event for the S-S case reaches a maximum value at $9A$ MeV for the init-I case (See Fig.1 (d)) and at $10A$ MeV for the init-II case (see Fig.1(e)). One notices that for the init-II condition the maximum value of the probability of producing SHF for the S-S orientation case is obviously higher than that for the N-N case, as shown in Fig.1 (b) and (e). It is also interesting to find from Fig.1 (c) and (f) that for the S-S orientation case the maximum probability of producing SHF is enhanced obviously for the init-II case compared to the init-I case at the time of 1000fm/c after re-separation(see Fig.1(f)) but is only slightly enhanced at re-separation time (see 1 (c)). It implies that the inclusion of β_4 deformation in the initial uranium nucleus might effectively reduce the excitation energy of composite system due to the more stable structure of uranium and thus obviously enhances the probability of producing SHF at the time of 1000fm/c after re-separation. This study suggests that S-S orientation and init-II deformation is probably the most beneficial condition for production of SHF.

4. Summary

In this work the effects of initial deformation and orientation in $^{238}U+^{238}U$ reaction are studied by using ImQMD model. It is found that the influence of the initial orientation on the average probability of SHF production per event are much stronger than that from initial deformation(with and without β_4 deformation). The production probability of SHF at re-separation time firstly increase with energy increasing and at a certain energy they reach their individual maximum values, and then decrease with energy. The production probability of SHF calculated at the time of 1000fm/c after the re-separation for the S-S case reaches a maximum value at $9A$ MeV for init-I and at $10A$ MeV for init-II, respectively, and in the latter case the maximum value of the production probability of SHF is obviously higher than that for the N-N and init-II case(Fig.1(e)). The inclusion of β_4 deformation in prolate uranium seems to enhance the production probability of SHF in the side-side case as the more stable structure of initial uranium might effectively reduce the excitation energy of the composite system. From these studies it is suggested that the fluctuation and dissipation play important roles in the reaction dynamics of two ^{238}U and the optimal initial condition for production of SHF probably is the S-S orientation and init-II deformation.

5. Acknowledgments

This work has been supported by the Chinese National Science Foundation under Grants Nos. (11005155, 11075215, 11275052,10975095,91126010, U1230127) and partly supported by MOST of China (973 Program with Grant No. 2013CB834404).

References

1. S. Hofmann and G. Münzenberg, *Rev. Mod. Phys.* **72**, 733 (2000).
2. Yu. Ts. Oganessian, *et. al.*, *Phys. Rev. C* **70**, 064609 (2004).
3. V. I. Zagrebaev, Yu. Ts. Oganessian, M. G. Itkis, and Walter Greiner, *Phys. Rev. C* **73**, 031602(R) (2006).
4. Valery Zagrebaev and Walter Greiner, *J. Phys. G: Nucl. Part. Phys.* **34**, 1 (2007).
5. Valery Zagrebaev and Walter Greiner, *J. Phys. G: Nucl. Part. Phys.* **34**, 2265 (2007).
6. Nan Wang, Jun-qing Li, and En-guang Zhao, *Phys. Rev. C* **78**, 054607 (2008).
7. Zhao-Qing Feng, Gen-Ming Jin, and Jun-Qing Li, *Phys. Rev. C* **80**, 067601 (2009).
8. Qingfeng Li, Wei Zuo, Wenfei Li, *et. al.*, *Eur. Phys. J. A* **24**, 223 (2005).
9. J. Aichelin, *Phys. Rep.* **202**, 233(1991).
10. H. Feldmeier and J. Schnack, *Rev. Mod. Phys.* **72**, 655(2000).
11. M. Papa,T.Maruyama, and A. Bonasera, *Phys. Rev. C* **64**, 024612(2001)
12. H. Kruse, B. V. Jacak, J.J. Molitoris, G. D. Westfall and H. Stocker, *Phys. Rev. C* **31**, 1770 (1985).
13. Junlong Tian, Xizhen Wu, Kai Zhao, Yingxun Zhang and Zhuxia Li, *Phys. Rev. C* **77**, 064603(2008).
14. Kai Zhao, Xizhen Wu, and Zhuxia Li, *Phys. Rev. C* **80**, 054607 (2009).
15. Ning Wang, Zhuxia Li, and Xizhen Wu, *Phys. Rev. C* **65**, 064608 (2002).
16. Ning Wang, Zhuxia Li, and Xizhen Wu, *et. al.*, *Phys. Rev. C* **69**, 034608 (2004).
17. Ning Wang, Zhuxia Li, and Xizhen Wu, *et. al.*, *Mod. Phys. Lett. A* **20**, 2619 (2005).

POSITIVE-PARITY ROTATIONAL BAND IN ^{158}Tb

J. Zhong[1,2], S. P. Hu[1], H. L. Ma[1], X. G. Wu[1 *], H. B. Sun[2 †], G. S. Li[1], C. Y. He[1], Y. Zheng[1], C. B. Li[1], Q. M. Chen[1], H. W. Li[1,3], Y. H. Wu[1,3], and P. W. Luo[1,2]

[1] *China Institute of Atomic Energy, Beijing 102413, People's Republic of China*

[2] *College of Physics Science and Technology, Shenzhen University, Shenzhen 518060, People's Republic of China*

[3] *College of Physics, Jilin University, Changchun 621010, People's Republic of China*

Excited states of the odd-odd nucleus ^{158}Tb have been studied via the ^{154}Sm(^{7}Li,3n)^{158}Tb reaction at a beam energy of 27 MeV. The new level scheme of ^{158}Tb involving a positive parity rotational band is established up to excitation energy $\sim$ 2.3 MeV and spin $\sim$ 17 $\hbar$. The configuration of this rotational band was assigned as $\pi d_{5/2} \otimes \nu i_{13/2}$ by the previous work, and the signature inversion of this band is also discussed.

1. Introduction

Much of experimental information is available about the high-spin structure in the doubly odd nuclei in rare-earth region. This area has been proven to be fascinating because the variety of structures resulting from the active quasi-particle orbits and the softness of the nuclei with respect to deformation. A lot of high-spin phenomena has been investigated, such as the rotational alignment,[1] and the signature splitting and inversion.[2] In the rare-earth region, a lot of experimental information is available about the high-spin structure and signature inversion in the doubly odd nuclei. But it is still rather meagre mainly for ^{158}Tb[3] due to the fact that it is not easy for relative neutron-rich nuclei to populate at high angular momentum in heavy-ion fusion-evaporation reactions, and it is hard to pick out γ-rays belonging to ^{158}Tb from the contamination of ^{157}Tb.[4] This lack of information is a direct consequence that the existence of the signature in-

*E-mail: wxg@ciae.ac.cn
†E-mail: hbsun@szu.edu.cn

version in ^{158}Tb is unknown. The motivation of the present study was to investigate the properties of the rotational band in ^{158}Tb.

2. Experimental Methods and Results

Excited states in ^{158}Tb were populated using the fusion-evaporation reaction ^{154}Sm(^{7}Li, 3n)^{158}Tb at a beam energy of 27 MeV. The ^{7}Li beam was delivered by the HI-13 tandem accelerator of the China Institute of Atomic Energy (CIAE) in Beijing. A 0.67 mg/cm^2 ^{154}Sm foil, backed on a 1.49 mg/cm^2 gold foil, was used as the target. Nine Compton-suppressed high-purity Ge (HPGe) detectors and two planar HPGe detectors were utilized to detect the deexcited γ-rays from the reaction residues. Three of these detectors were placed at 90°, four at 140° and two at 40° with respect to the beam direction. The detectors were calibrated for γ-ray energies and efficiencies using the ^{133}Ba and ^{152}Eu standard sources. About 1.6×10^7 two folds γ-γ coincidence events were collected in event-by-event mode. The data were sorted off-line into a symmetrical matrix with a dispersion of 0.5 keV/channel. The matrix was analyzed by *GASP*ware programs, developed at National laboratory of Legnaro (Italy). In order to obtain information concerning γ-rays multipolarity, the angular distribution of the oriented states (ADO) ratios were extracted from two asymmetric matrices. These two matrices were created by sorting the data with the detectors at all angles on one axis and the detectors at $\sim$ 140° or $\sim$ 90° on the other axis, respectively. All the matrices were normalized with respect to each other using the overall efficiency of the detectors determined with calibration sources in the target position. By comparing to the ADO ratio of the known transitions in ^{157}Tb[4] (see Fig. 2), the typical ADO ratio for stretched quadrupole transitions was found to be $\sim$ 1.4, for stretched pure dipole transitions $\sim$ 0.8 in the present geometry.

Prior to our work, the low-spin states of ^{158}Tb have been studied by the decay of isomer, ^{159}Tb(d, t)^{158}Tb and ^{157}Gd(^{3}He , d)^{158}Tb reactions. The agreement of these work is the spin and parity of ground state is $I^{\pi}=3^{-}$, while the divergence is the placement of the rotational members of the ground state band. In 1984, D.G.Burke et al. used ^{154}Sm(^{7}Li, 3n)^{158}Tb to revised the low-spin states of ^{158}Tb up to 7^{+} $\hbar$ and 7^{-} $\hbar$ for the positive- and negative-parity levels.[3] It was still a grate pity that no states above the 395μs isomer was reported. Based on D. G. Burke et al.'s results, with better detection conditions, the level scheme of ^{158}Tb is dramatically extended by adding 17 new γ rays and 10 new levels. The rotational band is first

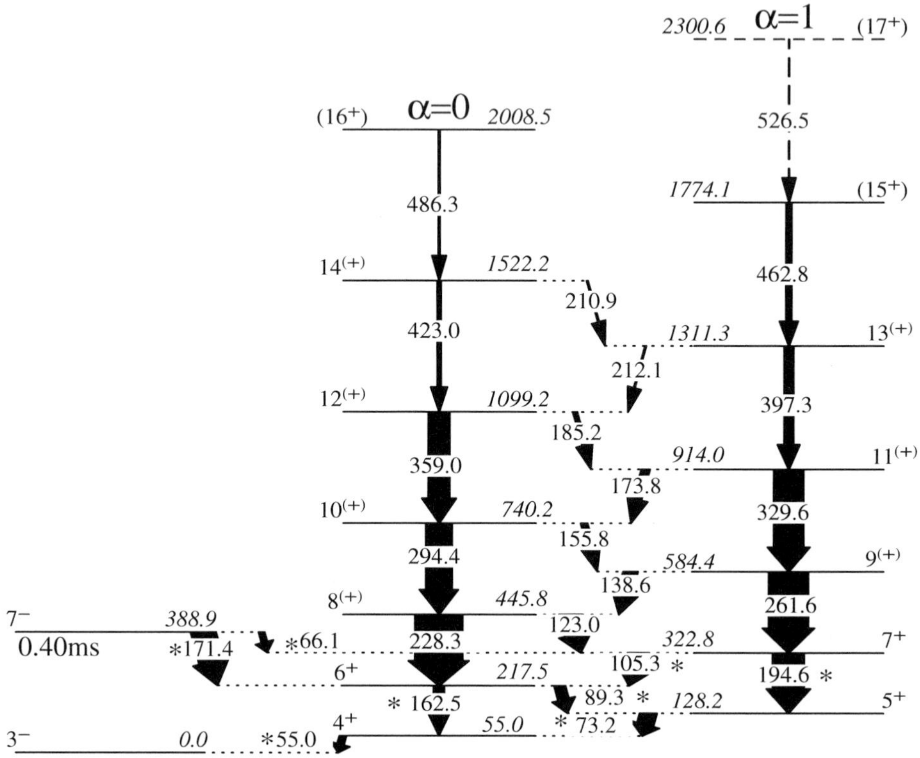

Fig. 1. The level scheme of ^{158}Tb established in the present work. γ-ray transitions with asterisk are obtained from previous work.[3] The decay of the 7^- isomer and the 55.0 keV ($4^+ \rightarrow 3^-$), 66.1 keV ($7^- \rightarrow 7^+$) transitions, although not observed in this study, are included in the level scheme for completeness.

relatively completed up to (I^π)=(17^-). The level scheme of ^{158}Tb proposed from this experiment is shown in Fig. 1.

The present level scheme is built on the previously known 217.5- and 322.8-keV levels. The conversion coefficient of 66.1-keV transition and the relative intensity of 171.1-keV transition related to 66.1-keV transition indicated that these were both E1 transitions. Thus the 322.8 and 217.5-keV levels were assigned I^π=7^+ and I^π=6^+, respectably.[3] As shown in the Fig. 3(a) gated on the 194.6- and (b) on 105.3-keV transitions, the 261.6-, 123.0- and 138.6-keV are both observed. But in the 261.6-keV gating spectrum(Fig. 3(c)), the 123.0- and 138.6-keV transitions can not been seen. It is clear that the 261.6-keV transition is parallel to the 123.0- and 138.6-keV transitions. Fig. 4 shows the spectra recorded in coincidence with the 228.3- and 162.5-keV transitions. In the spectrum setting gate on the

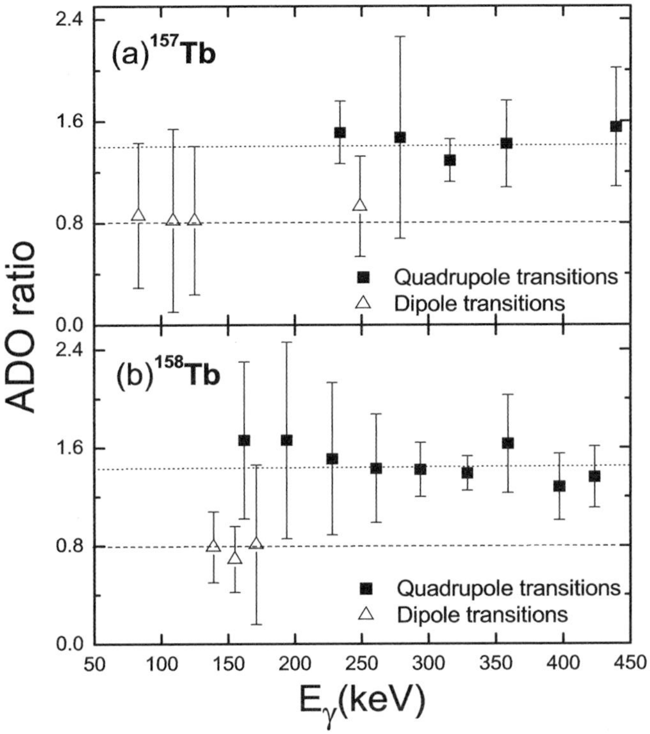

Fig. 2. ADO ratios shown as a function of γ-ray energy for (a) ^{157}Tb and (b) ^{158}Tb. The dotted line and the dashed line indicate the typical values observed for known γ-rays in this experiment for stretched quadrupole ($\Delta I = 2$) and stretched pure dipole transitions ($\Delta I = 1$), respectively.

162.5-keV transition, the 228.3-keV transition is very clear as well as 105.3- and 123-keV transitions. The sum of the 105.3 and 123.0 exactly equals to 228.3, in addition none observations of 105.3- and 123-keV transitions is found in the spectrum setting gate on the 228.3-keV transition. Thus, we fixed the positions of the placements of 445.8- and 584.4-keV levels. The ADO ratios for 261.6- and 228.3-keV transitions are 1.51 and 1.44, which indicate these are ΔI=2 transitions. The spins of 584.4- and 445.8-keV levels are tentatively assigned 9 and 7.

As can be seen in the Fig. 3 and Fig. 4, there are many new-found γ-peaks(labelled with numbers). According to the basis of the coincidence relationships, intensity balances, and energy sums, the rotational band is established. The ADO ratios for 329.6- and 397.3-keV transitions are all

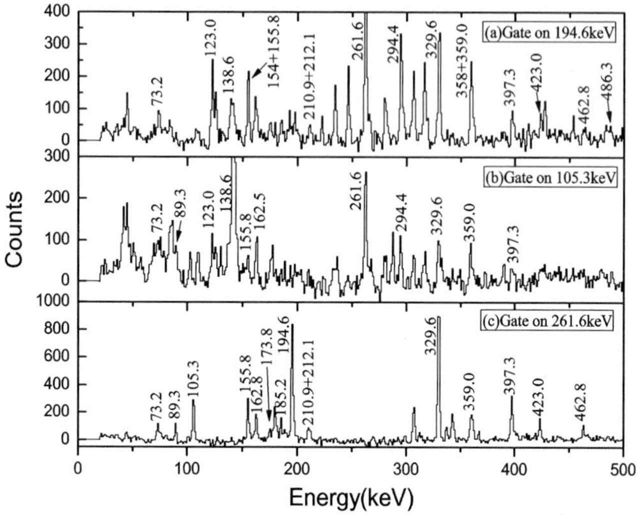

Fig. 3. Background-corrected coincidence spectrum obtained by gating on (a) the 194.6 keV γ-ray transition; (b) the 105.3 keV γ-ray transition; (c) the 261.6 keV γ-rays transition. Peaks associated with ^{158}Tb are labeled with their energies in keV, and 154- and 358-keV γ-rays obtained by gating on (a) the 194.6 keV γ-ray transition are assigned to ^{157}Tb.[4]

consistent with the values expected for stretched $E2$ transition, therefore spin and parity values of $11^{(+)}$ and $13^{(+)}$ are assigned to 914.0- and 1311.3-keV levels. Because the ADO ratios of 162.5-, 228.3-, 294.4-, 359.0-and 423.0-keV transitions are close to 1.4, the placements of the 217.5-keV $6^{(+)}$, 445.8-keV $8^{(+)}$, 740.2-keV $10^{(+)}$, 1099.2-keV $12^{(+)}$, and the 1522.2-keV $14^{(+)}$ states are certain too. Due to the weak intensities of 462.8-, 486.3- and 526.5-keV transitions and some of interband γ-rays, such as 73.2-, 89.3-, 105.3-, 123.0-, 173.8-, 185.2-, 210.9- and 212.1-keV transitions, the ADO ratios for these transitions can not be extracted. The properties of γ-rays observed in this experiment are summarized in Tab. 1. As shown in Fig. 3 and Fig. 4, most transitions of ^{158}Tb observed in the previous study[3] have been confirmed and new γ-rays have been identified, but 125-, 154-, 297-, and 358-keV γ-rays in Fig. 3(a) and Fig. 4(a) belong to another fusion product ^{157}Tb in this experiment.

Several experiments were performed previously to clarify the configurations structure of ^{158}Tb.[3,5] All of those experiments indicated that the configuration assignment of the positive-parity band should be $\pi 3/2^+[411] \otimes \nu 5/2^+[642]$.

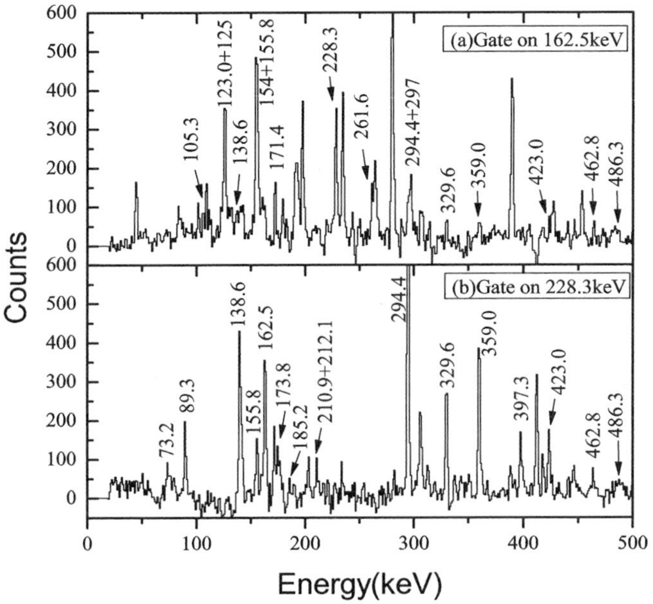

Fig. 4. Background-corrected coincidence spectrum obtained by gating on (a) the 162.5 keV γ-ray transition; (b) 228.3 keV γ-rays transition. Peaks associated with ^{158}Tb are labeled with their energies in keV, and 125-, 154- and 297-keV γ-rays obtained by gating on (a) the 162.5 keV γ-ray transition are assigned to ^{157}Tb.[4]

3. The Signature Inversion of the $\pi d_{5/2} \otimes \nu i_{13/2}$ Band in Tb Isotopes

The signature inversion of the $\pi h_{11/2} \otimes \nu i_{13/2}$ band in doubly odd nuclei was systematically observed in the neighborhood of A $\sim$ 160.[2] The systematic features of the signature inversion of the $\pi h_{11/2} \otimes \nu i_{13/2}$ band in doubly odd nuclei around this region are summarized in ref.[2] For the $\pi h_{11/2} \otimes \nu i_{13/2}$ band, in a chain of isotopes it is clear that with the increasing neutron number, the inversion point shifts to lower spin and the staggering magnitude of the signature dependence below the inversion point decreases. Though the proton orbit in ^{158}Tb is not belongs to $h_{11/2}$ but $d_{5/2}$,[3,5] it is still worth to investigate whether there is signature inversion in the $\pi d_{5/2} \otimes \nu i_{13/2}$ band in ^{158}Tb. The quantity

$$
\begin{aligned}
S(I) =& E(I) - E(I-1) - \\
& [E(I+1) - E(I) + E(I-1) - E(I-2)]/2
\end{aligned}
\tag{1}
$$

Table 1. Level energies, γ-ray energies, relative intensities, ADO ratios, and spin-parity assignments of the initial and final states in ^{158}Tb. Uncertainties in intensities are given in parentheses.

E_{lev}(keV)^a	E_γ(keV)^a	I_γ^b	R_{ADO}	$I_i^\pi \to I_f^\pi$
55.0	$55.0^{c.d.e}$			$4^+ \to 3^-$
128.2	73.2^e	36(4)		$5^+ \to 4^+$
217.5	89.3^e	23(3)		$6^+ \to 5^+$
	162.5	23(4)	1.66(64)	$6^+ \to 4^+$
322.8	105.3^e	42(5)		$7^+ \to 6^+$
	194.6	65(6)	1.66(80)	$7^+ \to 5^+$
388.9	$66.1^{c.d.e}$			$7^- \to 7^+$
	171.4	53(7)	0.81(65)	$7^- \to 6^+$
445.8	123.0^e	47(9)		$8^{(+)} \to 7^+$
	228.3	100(7)	1.51(62)	$8^{(+)} \to 6^+$
584.4	138.6	30(4)	0.74(29)	$9^{(+)} \to 8^{(+)}$
	261.6	82(6)	1.43(44)	$9^{(+)} \to 7^+$
740.2	155.8	14(2)	0.69(27)	$10^{(+)} \to 9^{(+)}$
	294.4	55(2)	1.42(22)	$10^{(+)} \to 8^{(+)}$
914.0	173.8^e	18(4)		$11^{(+)} \to 10^{(+)}$
	329.6	62(2)	1.39(14)	$11^{(+)} \to 9^{(+)}$
1099.2	185.2^e	9.3(1)		$12^{(+)} \to 11^{(+)}$
	359.0	45(3)	1.63(43)	$12^{(+)} \to 10^{(+)}$
1311.3	212.1^e	2.8(1)		$13^{(+)} \to 12^{(+)}$
	397.3	19(1)	1.28(27)	$13^{(+)} \to 11^{(+)}$
1522.2	210.9^e	3.3(2)		$14^{(+)} \to 13^{(+)}$
	423.0	8.0(1)	1.36(25)	$14^{(+)} \to 12^{(+)}$
1774.1	462.8^e	12(1)		$(15^+) \to 13^{(+)}$
2008.5	486.3^e	3.6(1)		$(16^+) \to 14^{(+)}$
2300.6	$526.5^{d.e}$			$(17^+) \to (15^+)$

Note:
^{a}Uncertainties are between 0.3 and 0.7 keV depending upon their intensity.
^{b}Intensities are normalized to the 228.3 keV transition.
^{c}Adopted from previous work.[3]
^{d}Relative intensities could not be determined because of the low intensity of transitions.
^{e}ADO ratios could not be extracted because of the low intensity of transitions.

is shown in Fig. 5 as a function of the spin I. Signature splitting in the $\pi d_{5/2} \otimes \nu i_{13/2}$ band in ^{158}Tb is compared to those in the neighboring isotopes ^{154}Tb[1] and ^{156}Tb.[2,6] As can be seen in Fig. 5, the signature splitting in ^{158}Tb seems to disappear when spin I is less than $11\hbar$, and with increasing neutron number, energy splitting magnitude decrease. If we consider energies in the spin $I < 11$ region as very small signature splitting. It is

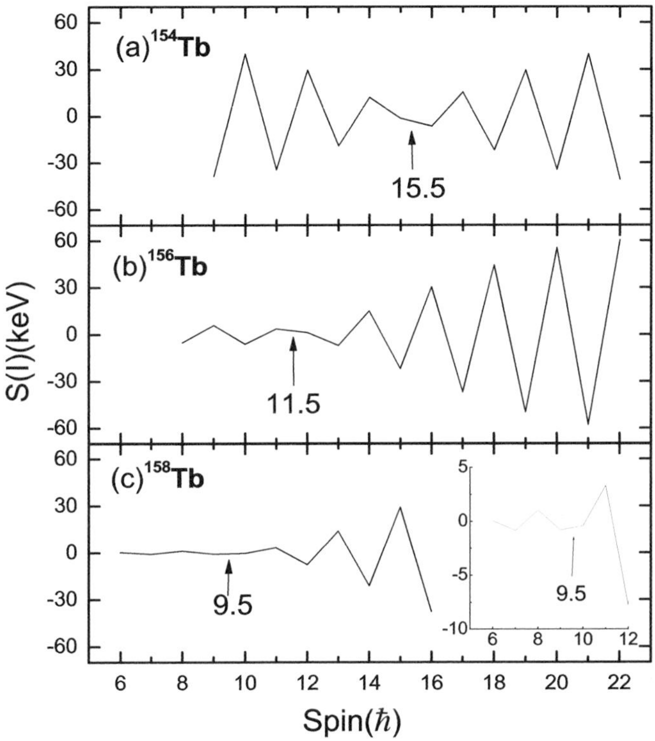

Fig. 5. Experimental energy differences S(I) of the $\pi d_{5/2} \otimes \nu i_{13/2}$ band against spin I for the doubly-odd isotopes with Z = 65. Data sources are: ^{154}Tb,[1] ^{156}Tb[2,6] and ^{158}Tb from this work. The inset in panel (c) shows a vertically expended plot for the spin from $6\hbar$ to $12\hbar$.

interesting to find the staggering phases of positive-parity band in ^{158}Tb changes. The signature inversion point is located approximately at $9.5\hbar$ with an error not larger than $0.5\hbar$ (see partial enlargement in Fig. 5 (c)). So it can be conclude that the inversion point shifts toward lower spin in Tb isotopes. These features are similar with the systematic features of the signature inversion of the $\pi h_{11/2} \otimes \nu i_{13/2}$ bands in doubly odd nuclei around A $\sim$ 160 summarized in Ref.[2]

4. Summary

High-spin states in ^{158}Tb have been studied via the reaction ^{154}Sm(^{7}Li, 3n). A rotational band is established up to excitation energy $\sim$ 2.3 MeV and spin $\sim$ 17 $\hbar$ with the addition of 17 new γ rays and 10 new levels. The configuration of this rotational band was assigned as $\pi d_{5/2} \otimes \nu i_{13/2}$ by the previous work. By comparing to the similar bands with same configuration in neighboring isotopes ^{154}Tb and ^{156}Tb, it is found that with increasing neutron number, the inversion point shifts to lower spin and the staggering magnitude decreases before the inversion point.

5. Acknowledgment

This work is supported by National Natural Science Foundation of China (11375267, 11305269, 11175259, 11475072, 11405274, 11075214, 11205245, 11375266, 10975189, 11405273 and 10775098). The authors would like to thank the crew of the HI-13 tandem accelerator at the China Institute of Atomic Energy for steady operation of the accelerator. We are also grateful to Dr. Q. W. Fan for preparing the target.

References

1. D. J. Hartley *et al.*, Phys. Rev. C **59**, 1171 (1999).
2. Yunzuo Liu *et al.*, Phys. Rev. C **52**, 2514 (1995).
3. D. G. Burke, Y. S. Liang and J. C. Waddington, Phys. Lett. B **146**, 392 (1984).
4. D. J. Hartley *et al.*, Phys. Rev. C **57**, 2944 (1998).
5. H. D. Jones and R. K. Sheline, Phys. Rev. C **2**, 1747 (1970).
6. R. Bengtsson *et al.*, Nucl. Phys. A **389**, 158 (1982).

THE $K\pi$ INTERACTION IN FINITE VOLUME

Dan Zhou*, Er-Liang Cui, Hua-Xing Chen, Li-Sheng Geng and Li-Hua Zhu
School of Physics and Nuclear Energy Engineering and International Research Center for Nuclei and Particles in the Cosmos, Beihang University, Beijing, 100191, China
**E-mail: danzhou@buaa.edu.cn*
hxchen@buaa.edu.cn

We calculate energy levels of the $K\pi$ scattering in the K^* channel in finite volume using chiral unitary theory. We use these energy levels to obtain the $K\pi$ phase shifts and the K^* meson properties. We also investigate their dependence on the pion mass and compare this with Lattice QCD calculations.

Keywords: Chiral unitary theory; Lattice QCD calculation; meson-meson interaction.

1. Introduction

Lattice QCD is developing very fast in these years. One can use this method to evaluate discrete energy levels of the finite box. However, these energy levels can not be directly measured in the experiments. To extract physical informations from these energy levels, one usually uses the Lüscher's approach[1,2] to reconstruct phase shifts of the decay products in the continuum. While, in Ref. 3 the authors proposed one method to estimate them through an effective approach whose parameters are obtained by fitting the experimental data. This method has been applied in Refs. 4–8 to study the meson-baryon and meson-meson interactions in finite volume. In the present work[9] we use this method to study the $K\pi$ interaction in the K^* channel in finite volume.

2. The Chiral Unitary Approach in Infinite and Finite Box

For the $K\pi$ scattering amplitude in P-wave, we use the following Bethe-Salpeter equation in the on-shell factorized form

$$T(s) = \frac{V(s)}{1 - V(s)G(s)}, \quad (1)$$

where the relevant V-matrix for the $K\pi$ scattering is taken from Ref. 10

$$V(s) = -\frac{p^2}{2f^2}(1 + \frac{2G_V^2}{f^2}\frac{s}{M_{K^*}^2 - s}). \tag{2}$$

The dimensional regularization result of G-function for the two-meson propagator is

$$\begin{aligned} G(s) = {} & \frac{1}{(4\pi)^2}\{a(\mu) + \log\frac{m_\pi^2}{\mu^2} + \frac{m_K^2 - m_\pi^2 + s}{2s}\log\frac{m_K^2}{m_\pi^2} \\ & + \frac{Q(\sqrt{s})}{\sqrt{s}}[\log(s - (m_K^2 - m_\pi^2) + 2\sqrt{s}Q(\sqrt{s})) + \log(s + (m_K^2 - m_\pi^2) \\ & + 2\sqrt{s}Q(\sqrt{s})) - \log(-s + (m_K^2 - m_\pi^2) + 2\sqrt{s}Q(\sqrt{s})) \\ & - \log(-s - (m_K^2 - m_\pi^2) + 2\sqrt{s}Q(\sqrt{s}))]\}\,, \end{aligned} \tag{3}$$

and the regularization parameters are chosen to be

$$a(\mu) = -1.0, \qquad \mu = M_{K^*}\,. \tag{4}$$

The two parameters f and G_V are taken from Ref. 10. To fix M_{K^*}, we use the experimental data[11,12] of phase shifts, which are related to the $T(s)$ through:

$$T(E) = \frac{-8\pi E}{p\cot\delta(p) - ip}\,, \tag{5}$$

and the M_{K^*} is fitted to be $M_{K^*} = 919.03$ MeV.

In this paper[9] we shall study the K^* meson in a finite volume. To do this we just use the following G function which is defined in the finite box of side L

$$\tilde{G}(s, L) - G(s) = \lim_{q_{\max}\to\infty}\Big(\frac{1}{L^3}\sum_{q_i}^{q_{\max}} I(q_i) - \int_{q<q_{\max}}\frac{d^3q}{(2\pi)^3}I(q)\Big)\,. \tag{6}$$

where

$$I(q_i) = \frac{1}{2\omega_1(\vec{q})\omega_2(\vec{q})}\frac{\omega_1(\vec{q}) + \omega_2(\vec{q})}{E^2 - (\omega_1(\vec{q}) + \omega_2(\vec{q}))^2}\,, \tag{7}$$

and $\omega_{1,2}(\vec{q}) = \sqrt{m_{1,2}^2 + \vec{q}^2}$. The sum is given by $\vec{q} = \frac{2\pi}{L}\vec{n}$ $(\vec{n} \in \mathcal{Z}^3)$.

3. The Energy Levels in the Chiral Unitary Approach

To calculate the energy levels of the $K\pi$ system, we need to find the poles of the $T(s)$ matrix, which are just solutions of the following equation

$$1 - V(s)\tilde{G}(s, L) = 0\,. \tag{8}$$

From this equation we can clearly see that the energy levels are functions of the cubic box size L. The energy levels are shown in the left panel of Fig. 1, and the lowest one is used to produce the phase shifts shown in the right panel of Fig. 1. Phase shifts can be used to fit the physical quantities:

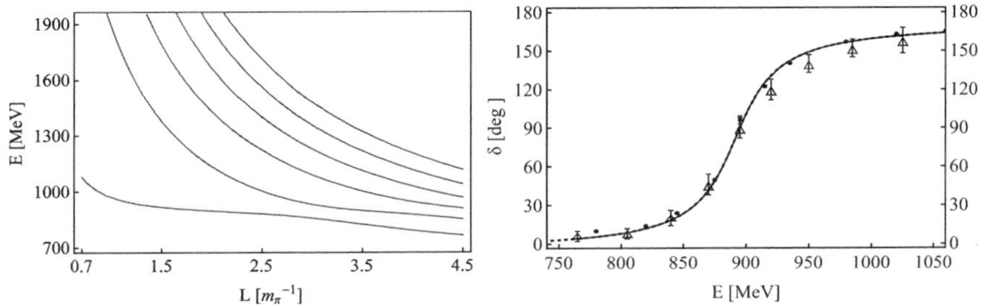

Fig. 1. Energy levels as functions of the cubic box size L, and the phase shift extracted from the lowest energy level,in the infinite space (solid curve) and in the finite space (dashed curve). The experimental data are taken from Refs.[11,12]

m_{K^*}, g_{K^*} and Γ_{K^*}, through the following two equations

$$\cot\delta(s) = \frac{m_{K^*}^2 - s}{\sqrt{s}\Gamma_{K^*}(s)}, \quad \text{and} \quad \Gamma_{K^*}(s) = \frac{p^3}{s}\frac{g_{K^*}^2}{8\pi}. \tag{9}$$

The results from fitting the phase shifts are

$$m_{K^*} = 894.89^{+39.75}_{-37.77}\,\text{MeV}\,, g_{K^*} = 6.48^{+0.13}_{-0.12}\,, \Gamma_{K^*} = 50.68^{+8.24}_{-8.00}\,\text{MeV}\,. \tag{10}$$

To compare our results with the lattice QCD results,[13,14] we should use the same m_π as the references. We note that the parameters of Eq. (2) change with m_π: G_V, f, and m_K. We keep G_V/f unchanged, but the parameter f as a function of m_π is given in Ref.9. The kaon mass m_K also change with the pion mass m_π as the relation in Ref.15.

Moreover, the bare K^* mass, M_{K^*} in Eq. (2), provides the link of the theory to a genuine component of the K^* meson, not related to the $K\pi$ component, and we assume it to be m_π independent.

Now we can compare our results with the Lattice results of Refs.13,14. First, we use $m_\pi = 240$ MeV, $m_K = 548$ MeV, and $f = 3$ fm, and obtain $E_1 = 912.6^{+33.4}_{-33.5}$ MeV, $E_2 = 1166.7^{+5.2}_{-5.1}$ MeV, while the Lattice results[13] are $E_1 = 926.9^{+23.5}_{-10.0}$ MeV, $E_2 = 1171.7^{+40.0}_{-22.5}$ MeV. Then, we use $m_\pi = 266$ MeV, $m_K = 552$ MeV, and $f = 1.98$ fm, and obtain $E_1 = 926.2^{+36.0}_{-36.8}$ MeV, $E_2 = 1511.4^{+9.6}_{-7.5}$ MeV, and $\delta_1 = 158.05^\circ\ ^{+8.52^\circ}_{-8.45^\circ}$, $\delta_2 = 175.52^\circ\ ^{+2.79^\circ}_{-3.62^\circ}$, while the Lattice results are $E_1 = 915.6 \pm 3.0$ MeV,

$E_2 = 1522.3 \pm 7.0$ MeV, and $\delta_1 = 160.61° \pm 0.73°$, $\delta_2 = 177.0° \pm 2.6°$. We can find that the energy levels and the extracted phase shifts are similar.

4. Conclusion

We made a study of the $K\pi$ interaction in P-waves in finite volume using the chiral unitary approach. We obtained energy levels of the $K\pi$ system in the K^* channel in finite volume, and evaluated the relevant phase shifts. Then, we compare our results with the Lattice results,[13,14] and find that the energy levels and the extracted phase shifts are similar.

Acknowledgements

We thank Eulogio Oset for suggesting this problem and valuable help, and Bao-Xi Sun, Chu-Wen Xiao and Xiu-Lei Ren for useful discussion and information. This work is supported by the National Natural Science Foundation of China under Grant No.11205011, 11475015, 11375024 and 11375023, and the Fundamental Research Funds for the Central Universities.

References

1. M. Lüscher, Commun. Math. Phys. **105** (1986) 153.
2. M. Lüscher, Nucl. Phys. B **354** (1991) 531.
3. M. Doring, U. -G. Meissner, E. Oset and A. Rusetsky, Eur. Phys. J. A **47**, 139 (2011).
4. M. Doring, J. Haidenbauer, U. -G. Meissner, A. Rusetsky, Eur. Phys. J. A **47**, 163 (2011).
5. A. Martinez Torres, L. R. Dai, C. Koren, D. Jido and E. Oset, Phys. Rev. D **85**, 014027 (2012).
6. M. Doring, U. G. Meissner, JHEP **1201**, 009 (2012).
7. J. -J. Xie and E. Oset, Eur. Phys. J. A **48**, 146 (2012) [arXiv:1201.0149 [hep-ph]].
8. Hua-Xing Chen and E. Oset, Phys. Rev. D **87**,016014 2013
9. D. Zhou, E. L. Cui, H. X. Chen, L. S. Geng and L. H. Zhu, arXiv:1409.0178 [hep-lat].
10. C. W. Xiao, F. Aceti, and M. Bayar, Eur. Phys. J. A **49**, 22 (2011).
11. P. Estabrooks, R. K. Carnegie, A. D. Martin, W. M. Dunwoodie, T. A. Lasinski and D. W. G. S. Leith, Nucl. Phys. B **133**, 490 (1978).
12. R. Mercer, P. Antich, A. Callahan, C. Y. Chien, B. Cox, R. Carson, D. Denegri and L. Ettlinger *et al.*, Nucl. Phys. B **32**, 381 (1971).
13. Ziwen Fu and Kan Fu, Phys. Rev. D **86**, 094507 (2012).
14. C. B. Lang, Leskovec Luka, Mohler Daniel, Prelovsek Sasa, Phys. Rev. D **86**, 054508 (2012).
15. X. -L. Ren, L. S. Geng, J. Martin Camalich, J. Meng and H. Toki, JHEP **1212**, 073 (2012) [arXiv:1209.3641 [nucl-th]].

THEORETICAL STUDY OF STRUCTURE AND SYNTHESIS MECHANISM OF SUPERHEAVY NUCLEI

Shan-Gui Zhou*

State Key Laboratory of Theoretical Physics, Institute of Theoretical Physics, Chinese Academy of Sciences, Beijing 100190, China
Center of Theoretical Nuclear Physics, National Laboratory of Heavy Ion Accelerator, Lanzhou 730000, China
**E-mail: sgzhou@itp.ac.cn*

The study of superheavy nuclei (SHN) is on the frontier of modern nuclear physics. In recent years, we have carried out theoretical investigations of both the structure properties and the synthesis mechanism of SHN. In this contribution, we briefly review these progresses and focus on the study of potential energy surfaces and fission barriers of actinide nuclei by using the MDC-RMF model and that of the fusion mechanism by using the ImQMD model.

Keywords: Superheavy nuclei; fission barrier; relativistic mean field; fusion; improved quantum molecular dynamics.

1. Introduction

In 2005, *Science* highlighted 125 scientific puzzles and one of them is: Are there stable high-atomic-number elements?[1] This puzzle is related to the study of superheavy nuclei (SHN), *i.e.*, to exlore the mass and charge limits of atomic nuclei, which is at the forefront of modern science.[2] The existence of superheavy nuclei stems from quantum shell effects. In 1960s, an island of stability of SHN centering at $Z = 114$ and $N = 184$ was predicted. So far, elements with $Z \leq 118$ have been synthesized.[3–6] Note that in Ref. 7, one α-decay chain was observed for ^{271}Ds using the gas-filled recoil separator at IMP in Lanzhou and the measured decay properties are consistent with those reported previously.

There are many challenges to synthesize even longer lived or even heavier SHN by using heavy ion fusion reactions. First, the cross section becomes smaller and smaller. Second, with stable nuclei as projectiles and targets, only neutron deficient SHN can be produced. Third, it becomes more difficult to find proper target material. Concerning theoretical study of the structure and the synthesis mechanism of SHN, there are also many open problems: the shell structure and magicity which determines the location of the island of stability, the shape and isomeric states, the potential energy surface (PES) and fission barrier, and the fusion mechanism, *etc.*

In recent years, we have studied several of the above mentioned open problems. In this contribution, we will briefly review some progresses. In Sec. 2, we give an overview of our recent study of structure properties of heavy and superheavy nuclei. In Sec. 3, we review our work on the synthesis mechanism of superheavy nuclei. Finally, a summary is given in Sec. 4.

2. Structure Properties of Heavy and Superheavy Nuclei

2.1. *Rotational spectroscopy of transfermium nuclei and shell structure of SHN*

The location of the island of stability is mainly determined by the single particle shell structure of SHN, *e.g.*, the splitting of $\pi 2f_{5/2,7/2}$ determines whether the next proton shell closure appears at $Z = 114$ or 120. Different models often predict different proton magic number after 82; even with the same model, the predtions may depend on the parameters.[8,9] Experimentally one can not study directly the single particle level structure of SHN with $Z \geq 110$. An indirect way is to investigate lighter nuclei with $Z \approx 100$ and $N \approx 152$. The strongly downsloping orbitals originating from the spherical subshells and active in the vicinity of the predicted shell closures may come close to the Fermi surface in these deformed nuclei. For example, the $\pi 1/2^-[521]$ and $\pi 3/2^-[521]$ orbitals, originating from $\pi 2f_{5/2,7/2}$, are of particular interest. Many high-spin rotational bands in nuclei with $Z \approx 100$ have been established in recent years.[10–17]

Using a cranked shell model (CSM) with the pairing treated by a particle-number conserving (PNC) method,[18] we have systematically studied the rotational bands in nuclei with $Z \approx 100$, including the recent established one in ^{256}Rf.[19–21] The PNC-CSM described well the kinematic moments of inertia ($\mathcal{J}^{(1)}$) for the rotational bands in even-even, odd-A and odd-odd nuclei, and the bandhead energies of the 1-quasiparticle bands in odd-A nuclei. Note that in Ref. 20, we also made predictions for several nuclei. Later a rotational band was established for one of them, ^{246}Fm,[16] and the experimental $\mathcal{J}^{(1)}$ agrees well with our prediction.[22] In Refs. 19,20, by fitting the experimental single-particle spectra in these nuclei, we proposed a new set of Nilsson parameters (κ and μ) with which it was predicted that the center of the superheavy island is at $Z = 120$ and $N = 182$.[22]

2.2. *Fission barriers from the MDC-RMF model*

To give a microscopic and self-consistent description of nuclear PES, we have developed multidimensionally constrained covariant density functional theories[23–25] where all deformations $\beta_{\lambda\mu}$ with even μ are allowed. The functional can be one of the following four forms: the meson exchange or point-coupling nucleon interactions combined with the nonlinear or density-dependent couplings. The pairing is treated with either the BCS approach or the Bogoliubov transformation. We name the MDC-CDFT with the BCS approach for the pairing as the MDC-RMF models[23,24] and those with the Bogoliubov transformation as the MDC-RHB models.[25] The MDC-RMF model has recently been applied to the study of PES's and fission barriers of actinides,[23,24,26–28] the nonaxial-octupole Y_{32} correlations in $N = 150$ isotones,[29] the third minima in light actinides,[30] and shapes of hypernuclei.[31,32]

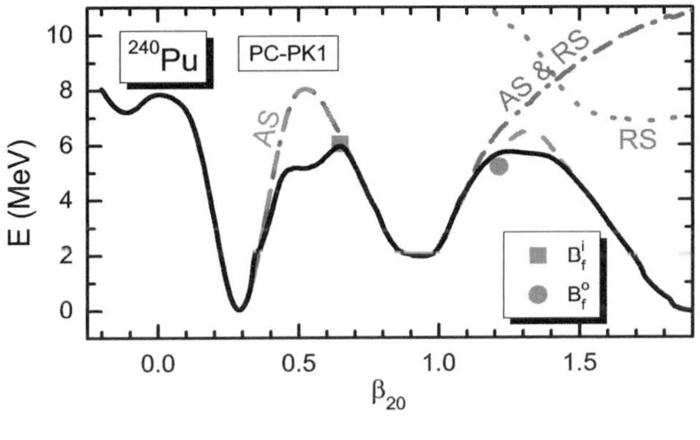

Fig. 1. (Color online) PES's of ^{240}Pu with various self-consistent symmetries imposed: The black solid, red dashed, green dotted, and violet dot-dashed curves represents the lowest fission path with V_4, axial symmetry (AS), reflection symmetry (RS), and AS & RS symmetries imposed. The empirical inner (outer) barrier height $B_{\rm emp}$ is denoted by the grey square (circle). The energy is normalized with respect to the binding energy of the ground state. Taken from Ref. 23.

In Refs. 23,24, one-, two-, and three-dimensional PES's were obtained for ^{240}Pu with the parameter set PC-PK1.[33] In Fig. 1 we show the one-dimensional PES's which are obtained with different self-consistent symmetries imposed in the calculations: the axial (AS) or triaxial (TS) symmetries combined with reflection symmetric (RS) or asymmetric cases. The effect of the triaxial deformation on the inner barrier and that of the octupole deformation on the outer barrier can be clearly seen: The triaxial deformation lowers the inner barrier by more than 2 MeV;[34] the RA shape is favored beyond the fission isomer and lowers very much the outer fission barrier.[35] It was found for the first time that the outer barrier is also considerably lowered by about 1 MeV by the triaxial deformation and a better reproduction of the empirical barrier height is achieved for the outer barrier. This lowering effect can be revealed only when the axial and reflection symmetries are both broken.

3. Synthesis Mechanism of Superheavy Nuclei

3.1. *The capture process in heavy ion collisions*

In the capture process of heavy ion collisions (HIC), the projectile and the target overcome the Coulomb barrier through the barrier tunneling. To get an analytical formula for the barrier penetrability, one often approximates the barrier as an inverted parabola.[36] However, this approximation breaks down at deep sub-barrier energies due to the long-range Coulomb interaction. In Ref. 37, we proposed a new barrier penetration formula for potential barriers containing the Coulomb interaction. As a first attempt, this formula was used to calculate nuclear α decay half-lives and a good agreement with the experiment was obtained. It is desireble to, based on this new formula, derive an expression for calculating the capture cross section ($\sigma_{\rm cap}$). This formula may also be useful when one deals with the distribution of barriers.

For light and medium-heavy reaction systems, all the capture events lead to the formation of a compound nucleus. Therefore the fusion cross section $\sigma_{\rm fus} = \sigma_{\rm cap}$. The breakup of the projectile induces complicated and interesting phenomena in the capture (fusion) process[38] and one of them is the suppression of the complete fusion cross section at energies above the Coulomb barrier. In Ref. 39, we made a systematic study of this suppression effect. More details can also be found in this proceedings.[40]

To learn more about the systematics in the capture process and to propose an empirical way for calculating $\sigma_{\rm cap}$, we have been working on a systematic study of $\sigma_{\rm cap}$ for many years. In Ref. 41, we introduced a barrier distribution of an asymmetric Gassian form to take into account the effects of couplings between the relative and intrinsic motions and of neutron transfer. Empirical formulas were proposed to determine the parameters of the barrier distribution. This empirical model can provide useful information to experimental and theoretical studies on the capture and fusion dynamics, especially for the capture process in synthesis of superheavy nuclei.

3.2. *The fusion process in heavy ion collisions*

The fusion process, in which an excited compound nucleus (CN) is formed after the capture of the projectile by the target, is the most complicated one in the three steps to produce SHN. Experimentally it is very difficult to measure the fusion probability ($P_{\rm CN}$). Theoretically there is a problem that predictions of $P_{\rm CN}$ from various models differ by more than two orders of magnitude.[42] In recent years, we have studied the fusion mechanism with both macroscopic and microscopic models. From macroscopic models, one can calculate the evaporation residue cross sections ($\sigma_{\rm ER}$) for the synthesis of SHN and make predictions for the synthesis of new SHN based on the reproduction of available data. With microscopic theories, one cannot calculate $\sigma_{\rm ER}$ yet, but can get useful information concerning the fusion mechanism and constrain phenomenological models.

Nuclear fusion is a non-equilibrium process where the fluctuation and dissipation play important roles. Under various assumptions, several macroscopic transport models have been introduced to evaluate $P_{\rm CN}$ and one of them is the so called dinuclear system (DNS) model.[43–46] The importance of dynamical deformations of fragments in dissipative HICs has been long known. By taking into account the dynamical deformations of nuclei in each DNS, we have developed a DNS model with a dynamical potential energy surface — the DNS-DyPES model. With this model, hot fusion reactions for synthesizing SHN with $Z = 112$–120 are studied.[47,48] The calculated $\sigma_{\rm ER}$s are in good agreement with available data and predictions were made for synthesizing SHN with $Z = 119$ and 120.

The microscopic mechanism on how two colliding nuclei fuse, *e.g.*, how the relevant kinetic energy dissipates into the intrinsic degrees of freedom (DoF), can been understood with microscopic transport theories. Nowadays, it is becoming feasible to get various information out of microscopic numerical simulations, like time-dependent Hartree-Fock (TDHF) theories,[49–54] the many-body correlation transport (MBCT) theory,[55] and the quantum molecular dynamics (QMD).[56] We have used the TDHF theory to sudy dissipation dynamics in HICs and details can be found in this proceedings[57] and in Refs. 58,59.

The QMD was proposed to study intermediate-energy HIC.[56] An improved QMD (ImQMD) model has been developed in order to study low-energy HICs near the Coulomb barrier.[61,62] Making use of the microscopic information provided by ImQMD simulations, we have investigated how the

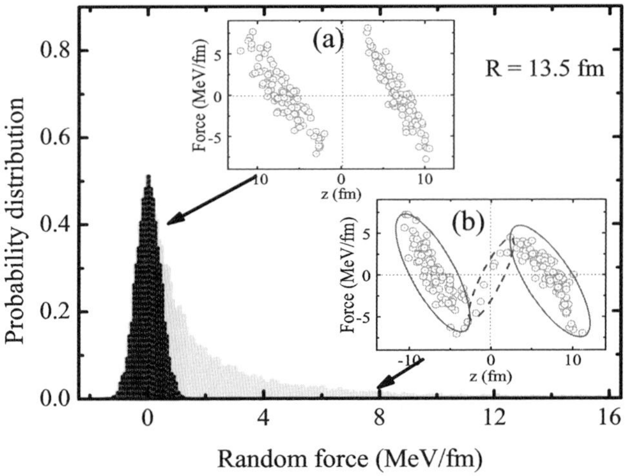

Fig. 2. (Color online) Distribution of the random force $\delta F(R)$ at $R = 13.5$ fm obtained from ImQMD simulations for central collisions of $^{90}Zr+^{90}Zr$ with $E_{c.m.} = 195$ MeV. Two typical events are shown in the insets: The abscissa and the ordinate express relative position z of each nucleon and the force it feels in the z direction. In (b), the distribution is shifted so that the symmetric Gaussian distribution centers at zero of the force. Taken from Ref. 60.

macroscopic fusion dynamics emerges out of the microscopic dynamics.[60,63] We studied central collisions of $^{90}Zr+^{90}Zr$. The system can always be divided into the left- and right-half parts. The relative motion between two centers of mass of these two parts is chosen as the relevant DoF to be described by the Langevin dynamics. It was found the dissipation dynamics of the relative motion between two fusing nuclei is caused by a non-Gaussian distribution of the random force, as is seen in Fig. 2. A typical event in the asymmetric tail is shown in Fig. 2(b) where there appears a small number of nucleons which transfer between the left and right parts. These transferred nucleons play a role to open a window. A clear non-Markovian effect was also observed in the time correlation function of the random force.

Note that in the TDHF theory, two-body collisions are excluded and in the QMD models, the Pauli principle is only approximately taken care and the spin-orbit interaction is often ignored. These shortcomings of the two kinds of transport models were overcomed in the MBCT theory[55] where the n-body correlations are incorporated. However, due to the limitation of the computational capability, this theory has only been used in very light systems.[64] It is highly desirable to extend this theory to the study of heavy systems.

3.3. *The survival probability of an excited compound nucleus*

The survival probability W_{sur} of an excited CN against processes in which the charge number is changed is directly related to the stability of an excited SHN. W_{sur} is an important factor in the study of synthesis mechanism of SHN. Presently most of the calculations focus on the stability of superheavy CN formed in cold and/or hot fusion reactions.[65]

To study the influence of the shell effects on the stability of excited SHN, we have investigated the stability of excited SHN with $100 \leq Z \leq 134$ against fission by using a statistical model.[66] The properties of SHN including the neutron separation energies and shell correction energies were both taken from predictions of the finite range droplet model.[67] We found that the islands of stability of SHN in ground states, e.g., the one around $Z = 115$ and $N = 179$, persists for excited SHN in the sense that the calculated W_{sur} in the 1n-channel of excited SHN on these islands are maximized. This indicates that the survival probability is mainly determined by the nuclear shell effects.

We note that the decay widths in different channels and the survival probabilities of SHN are very sensitive to the nuclear properties such as the separation energy, the ground state and saddle point deformations, the fission barriers, and the level density, etc.. Therefore in the study of the survival process, reliable predictions of these properties from nuclear models is highly needed. We are now using the MDC-RMF model discussed in Sec. 2.2 to study these structure properties for SHN.

4. Summary

In this contribution, we briefly review the progresses we have made in recent years on theoretical investigations of both the structure properties and the synthesis mechanism of SHN. Concerning the structure of SHN, we made a systematic study of low-lying spectra of actinide and transfermium nuclei by using PNC-CSM and investigated PESs and fission barriers of actinide nuclei by using the MDC-RMF model. Concerning the synthesis mechanism of SHN, we studied in detail the three steps (processes) in producing SHN via heavy-ion fusion reactions. For the capture process, a new barrier penetration formula was proposed for potentials containing a long-range Coulomb interaction. The breakup effects on the complete fusion were investigated and a systematic study of capture cross sections was made. For the fusion process, the fusion mechanism was studied with both a phenomenological approch, the DNS-DynPES model, and microscopic transport theories, the TDHF theory and the ImQMD model. For the survival process, we made a systematic study of the survival probability against fission in the 1n-channel of SHN. By using the DNS-DynPES model, predictions were made for synthesizing SHN with $Z = 119$ and 120.

Acknowledgements

I'm grateful to my collaborators Gao-Feng Dai, Paulo Gomes, Lu-Guo, Xiao-Tao He, Lu-Lu Li, Zhu-Xia Li, Bing-Nan Lu, Fumihiko Sakata, Werner Scheid, Bao-Xi Sun, Dario Vretenar, Bing Wang, Nan Wang, Kai Wen, Xi-Zhen Wu, Cheng-Jun Xia, Jin-Yan Zeng, En-Guang Zhao, Jie Zhao, and Wei-Juan Zhao. This work has been supported by the 973 Program of China (Grant No. 2013CB834400), the NSF of China (Grants No. 11121403, No. 11120101005, and No. 11275248), and the Chinese Academy of Sciences (Grant No. KJCX2-EW-N01). The computation of this work was supported by the HPC Cluster of SKLTP/ITP-CAS and the Supercomputing Center, CNIC of CAS.

References

1. D. Kennedy and C. Norman, *Science* **309**, 75 (2005).
2. S.-G. Zhou, *Phys.* **43**, 817 (2014) (in Chinese).
3. S. Hofmann and G. Münzenberg, *Rev. Mod. Phys.* **72**, 733 (2000).
4. K. Morita *et al.*, *J. Phys. Soc. Jpn.* **73**, 2593 (2004).
5. Y. Oganessian, *J. Phys. G: Nucl. Phys.* **34**, R165 (2007).
6. Y. T. Oganessian *et al.*, *Phys. Rev. Lett.* **104**, 142502 (2010).
7. Z.-Y. Zhang *et al.*, *Chin. Phys. Lett.* **29**, 012502 (2012).
8. W. Zhang, J. Meng, S. Zhang, L. Geng and H. Toki, *Nucl. Phys. A* **753**, 106 (2005).
9. A. Sobiczewski and K. Pomorski, *Prog. Part. Nucl. Phys.* **58**, 292 (2007).
10. M. Leino and F. P. Hessberger, *Annu. Rev. Nucl. Part. Sci.* **54**, 175 (2004).
11. D. Ackermann, *Acta Phys. Pol. B* **42**, 577 (2011).
12. R.-D. Herzberg, *J. Phys. G: Nucl. Phys.* **30**, R123 (2004).
13. R.-D. Herzberg and P. Greenlees, *Prog. Part. Nucl. Phys.* **61**, 674 (2008).
14. S. K. Tandel *et al.*, *Phys. Rev. C* **82**, 041301R (2010).
15. P. T. Greenlees, *Acta Phys. Pol. B* **42**, 587 (2011).
16. J. Piot *et al.*, *Phys. Rev. C* **85**, 041301(R) (2012).
17. P. T. Greenlees *et al.*, *Phys. Rev. Lett.* **109**, 012501 (2012).
18. J. Y. Zeng and T. S. Cheng, *Nucl. Phys. A* **405**, 1 (1983).
19. Z.-H. Zhang, J.-Y. Zeng, E.-G. Zhao and S.-G. Zhou, *Phys. Rev. C* **83**, 011304(R) (2011).
20. Z.-H. Zhang, X.-T. He, J.-Y. Zeng, E.-G. Zhao and S.-G. Zhou, *Phys. Rev. C* **85**, 014324 (2012).

21. Z.-H. Zhang, J. Meng, E.-G. Zhao and S.-G. Zhou, *Phys. Rev. C* **87**, 054308 (2013).
22. Z. Zhang, K. Wen, X. He, J. Zeng, E. Zhao and S. Zhou, *Nucl. Phys. Rev.* **30**, 268 (2013) (in Chinese).
23. B.-N. Lu, E.-G. Zhao and S.-G. Zhou, *Phys. Rev. C* **85**, 011301(R) (2012).
24. B.-N. Lu, J. Zhao, E.-G. Zhao and S.-G. Zhou, *Phys. Rev. C* **89**, 014323 (2014).
25. B.-N. Lu et al. to be published.
26. B.-N. Lu, J. Zhao, E.-G. Zhao and S.-G. Zhou, *EPJ Web Conf.* **38**, 05003 (2012).
27. B.-N. Lu, J. Zhao, E.-G. Zhao and S.-G. Zhou, *J. Phys: Conf. Ser.* **492**, 012014 (2014).
28. B.-N. Lu, J. Zhao, E.-G. Zhao and S.-G. Zhou, *Phys. Scr.* **89**, 054028 (2014).
29. J. Zhao, B.-N. Lu, E.-G. Zhao and S.-G. Zhou, *Phys. Rev. C* **86**, 057304 (2012).
30. J. Zhao, B.-N. Lu, D. Vretenar, E.-G. Zhao and S.-G. Zhou, *Phys. Rev. C* **91**, 014321 (2015).
31. B.-N. Lu, E.-G. Zhao and S.-G. Zhou, *Phys. Rev. C* **84**, 014328 (2011).
32. B.-N. Lu, E. Hiyama, H. Sagawa and S.-G. Zhou, *Phys. Rev. C* **89**, 044307 (2014).
33. P. W. Zhao, Z. P. Li, J. M. Yao and J. Meng, *Phys. Rev. C* **82**, 054319 (2010).
34. H. Abusara, A. V. Afanasjev and P. Ring, *Phys. Rev. C* **82**, 044303 (2010).
35. K. Rutz, J. A. Maruhn, P. G. Reinhard and W. Greiner, *Nucl. Phys. A* **590**, 680 (1995).
36. C. Y. Wong, *Phys. Rev. Lett.* **31**, 766 (1973).
37. L.-L. Li, S.-G. Zhou, E.-G. Zhao and W. Scheid, *Int. J. Mod. Phys. E* **19**, 359 (2010).
38. L. F. Canto, P. R. S. Gomes, R. Donangelo and M. S. Hussein, *Phys. Rep.* **424**, 1 (2006).
39. B. Wang, W.-J. Zhao, P. R. S. Gomes, E.-G. Zhao and S.-G. Zhou, *Phys. Rev. C* **90**, 034612 (2014).
40. B. Wang, W.-J. Zhao, P. R. S. Gomes, E.-G. Zhao and S.-G. Zhou, in *Nuclear Structure in China 2014: Proceedings of the 15th National Conference on Nuclear Structure in China (NSC2014)*, 2015.
41. B. Wang et al. to be published.
42. R. S. Naik *et al.*, *Phys. Rev. C* **76**, 054604 (2007).
43. G. G. Adamian, N. V. Antonenko, W. Scheid and V. V. Volkov, *Nucl. Phys. A* **633**, 409 (1998).
44. J.-Q. Li, Z.-Q. Feng, Z.-G. Gan, X.-H. Zhou, H.-F. Zhang and W. Scheid, *Nucl. Phys. A* **834**, 353c (2010).
45. Z.-G. Gan, X.-H. Zhou, M.-H. Huang, Z.-Q. Feng and J.-Q. Li, *Sci. China-Phys. Mech. Astron.* **54 (Supp. 1)**, s61 (2011).
46. A. K. Nasirov, G. Mandaglio, G. Giardina, A. Sobiczewski and A. I. Muminov, *Phys. Rev. C* **84**, 044612(Octtober 2011).
47. N. Wang, E.-G. Zhao, W. Scheid and S.-G. Zhou, *Phys. Rev. C* **85**, 041601(R) (2012).

48. N. Wang, E.-G. Zhao and S.-G. Zhou, in *Nuclear Structure in China 2012: Proceedings of the 14th National Conference on Nuclear Structure in China (NSC2012)*, eds. J. Meng, C.-W. Shen, E.-G. Zhao and S.-G. Zhou (World Scientific, 2012).
49. P. Bonche, S. Koonin and J. W. Negele, *Phys. Rev. C* **13**, 1226 (1976).
50. L. Guo, J. A. Maruhn and P.-G. Reinhard, *Phys. Rev. C* **76**, 014601 (2007).
51. L. Guo, J. A. Maruhn, P.-G. Reinhard and Y. Hashimoto, *Phys. Rev. C* **77**, 041301(R) (2008).
52. K. Washiyama and D. Lacroix, *Phys. Rev. C* **78**, 024610 (2008).
53. K. Washiyama, D. Lacroix and S. Ayik, *Phys. Rev. C* **79**, 024609 (2009).
54. C. Simenel, *Eur. Phys. J. A* **48**, 152 (2012).
55. S.-J. Wang and W. Cassing, *Ann. Phys.* **159**, 328 (1985).
56. J. Aichelin, *Phys. Rep.* **202**, 233 (1991).
57. L. Guo et al., in *Nuclear Structure in China 2015: Proceedings of the 15th National Conference on Nuclear Structure in China (NSC2014)*, 2015.
58. G.-F. Dai, L. Guo, E.-G. Zhao and S.-G. Zhou, *Sci. China-Phys. Mech. Astron.* **57**, 1618 (2014).
59. G.-F. Dai, L. Guo, E.-G. Zhao and S.-G. Zhou, *Phys. Rev. C* **90**, p. 044609 (2014).
60. K. Wen, F. Sakata, Z.-X. Li, X.-Z. Wu, Y.-X. Zhang and S.-G. Zhou, *Phys. Rev. Lett.* **111**, 012501 (2013).
61. N. Wang, Z. Li and X. Wu, *Phys. Rev. C* **65**, 064608 (2002).
62. N. Wang, Z. Li, X. Wu, J. Tian, Y. Zhang and M. Liu, *Phys. Rev. C* **69**, 034608 (2004).
63. K. Wen, F. Sakata, Z.-X. Li, X.-Z. Wu, Y.-X. Zhang and S.-G. Zhou, *Phys. Rev. C* **90**, 054613 (2014).
64. J.-Y. Liu, S.-J. Wang, M. Di Toro, H. Liu, X.-G. Lee and W. Zuo, *Nucl. Phys. A* **604**, 341 (1996).
65. A. Zubov, G. Adamian and N. Antonenko, *Phys. Part. Nucl.* **40**, 847 (2009).
66. C.-J. Xia, B.-X. Sun, E.-G. Zhao and S.-G. Zhou, *Sci. China-Phys. Mech. Astron.* **54 (Suppl. 1)**, s109 (2011).
67. P. Möller, J. R. Nix, W. D. Myers and W. J. Swiatecki, *At. Data Nucl. Data Tables* **59**, 185 (1995).

THE MOLECULAR PICTURE OF AXIAL-VECTOR MESONS

Yu Zhou*, Xiu-Lei Ren, Hua-Xing Chen, and Li-Sheng Geng

School of Physics and Nuclear Energy Engineering and International Research Center for Nuclei and Particles in the Cosmos, Beihang University, Beijing, 100191, China
**E-mail: zhouyu.1516@163.com*
www.buaa.edu

The axial-vector mesons $a_1(1260)$, $b_1(1235)$, $f_1(1285)$, $h_1(1170)$, $h_1(1380)$, and $K_1(1270)$ are dynamically generated in the unitized chiral perturbation theory. Such a picture has been tested extensively in the past few years. In this talk, we report on a recent study of the $\mathcal{O}(p^2)$ chiral potentials and their impact on the dynamically generated axial-vector states. In anticipation of future lattice QCD simulations, we calculate the scattering lengths and the pole positions as functions of the pion mass, with the light-quark mass dependence of the pseudoscalar and the vector mesons determined by the $n_f = 2+1$ lattice QCD simulations of the PACS-CS Collaboration.

Keywords: Chiral Lagrangians; lattice QCD calculations; meson-meson interactions.

1. Introduction

It was shown in Refs.1,2 that the $a_1(1260)$, $b_1(1235)$, $f_1(1285)$, $h_1(1170)$, $h_1(1380)$, and $K_1(1270)$ can be built from the interactions between the pseudoscalar octet of the pion and the vector nonet of the ρ within the so-called unitized chiral perturbation theory (UChPT) approach. It should be noted that the studies of Refs.ref[1,2] are both based on the leading order (LO) chiral potentials. It is not clear whether the scenario of these states being dynamically generated will change when higher order kernels are included. Given the fact that lately there have been attempts to study these states on the lattice,[3,4] such a study is urgently needed. The main purpose of the present work is to include the contributions of the next-to-leading order (NLO) chiral potentials, and to study their impact on the LO results and on the light-quark mass dependence of the scattering lengths and the pole positions.

2. Theoretical Framework

The unitarized amplitude obtained from the Bethe-Salpeter approach[2] is

$$T = [1 + V\hat{G}]^{-1}\vec{\epsilon}\cdot\vec{\epsilon}\,', \tag{1}$$

where $\vec{\epsilon}$ ($\vec{\epsilon}\,'$) is the spatial polarization vector of the initial (final) vector meson and $\hat{G} = G(1 + \frac{1}{3}\frac{q_l^2}{M_l^2})$ is a diagonal matrix with the lth element, G_l being the loop function containing a vector and a pseudoscalar meson. V is the kernel, provided by chiral Lagrangians. The LO amplitudes of pseudoscalar meson-vector meson interactions are calculated with the following chiral Lagrangians

$$\mathcal{L}_{VVPP} = -\frac{1}{4f^2}\mathrm{Tr}([V^\mu, \partial^\nu V_\mu][P, \partial_\nu P]), \tag{2}$$

where f is the pion decay constant in the chiral limit and P and V are the $SU(3)$ matrices containing the octet of pseudoscalar and the nonet of vector mesons. From the Lagrangian of Eq. (2) one obtains the S-wave amplitude:

$$V_{ij}(s) = -\frac{\epsilon\cdot\epsilon'}{8f^2}C_{ij}[3s - (M^2 + m^2 + M'^2 + m'^2) - \frac{1}{s}(M^2 - m^2)(M'^2 - m'^2)], \tag{3}$$

where ϵ (ϵ') stands for the polarization four-vector of the incoming (outgoing) vector meson. The masses M (M'), m (m') correspond to the initial (final) vector mesons and initial (final) pseudoscalar mesons, respectively. The C_{ij} coefficients can be found in Ref. 2 .

The chiral Lagrangians up to chiral order $\mathcal{O}(p^2)$ reads[5]

$$\begin{aligned}\mathcal{L}_{VV}^{(2)} =& \frac{1}{2}\delta_1\mathrm{Tr}(V_\mu V^\mu u_\nu u^\nu) + \frac{1}{2}\delta_2\mathrm{Tr}(V_\mu u_\nu V^\mu u^\nu)\\ &+ \frac{1}{2}\delta_3\mathrm{Tr}(V_\mu V_\nu u^\mu u^\nu) + \frac{1}{2}\delta_4\mathrm{Tr}(V_\mu V_\nu u^\nu u^\mu)\\ &+ \frac{1}{2}\delta_5\mathrm{Tr}(V_\mu u^\mu V_\nu u^\nu + V_\mu u_\nu V^\nu u^\mu)\\ &+ \frac{1}{2}\kappa_V\mathrm{Tr}(V_\mu V^\mu \chi_+),\end{aligned} \tag{4}$$

where $u_\mu = i\{u^\dagger\partial_\mu u - u\partial_\mu u^\dagger\}$ and $\chi_\pm = u^\dagger\chi u^\dagger \pm u\chi^\dagger u$ with $\chi = \mathrm{diag}(m_\pi^2, m_\pi^2, 2m_K^2 - m_\pi^2)$.

At $\mathcal{O}(p^2)$, the contributions proportional to the low-energy constants (LECs) $\delta_{3,4,5}$ are suppressed compared to those proportional to δ_1 and δ_2.

As a result, the NLO chiral potentials, after projection into S wave, are

$$V^{NLO} = \left[\frac{\delta_1 D_{ij}^{(1)} + \delta_2 D_{ij}^{(2)}}{f^2} E_P E_{P'} + \frac{\kappa_V}{2f^2} D_{ij}^{(3)}\right] \epsilon \cdot \epsilon', \tag{5}$$

where E_P and $E_{P'}$ are the energy of the initial and final pseudoscalar mesons, and $D_{ij}^{1,2,3}$ are the corresponding Clebsch-Gordan coefficients, which can be found in Ref. 6.

3. Results and Discussion

This section gives the main results. One can get more details in Ref. 6.

Figure 1 compares the experimental axial-vector masses with the corresponding UChPT results. It is seen that the inclusion of the NLO chiral potentials can slightly improve the description, but the overall agreement with the experimental masses is similar to the LO results.

Figure 2 shows the scattering lengths of the dominant channels in each isospin, strangeness, and G-parity sector as a function of the pion mass, with the strangeness quark mass fixed at its physical value. The difference between the results obtained with the LO kernel and the NLO kernel indicates inherent theoretical uncertainty. The singularities indicate the energies at which the pole positions cross the corresponding thresholds (see Fig. 3).

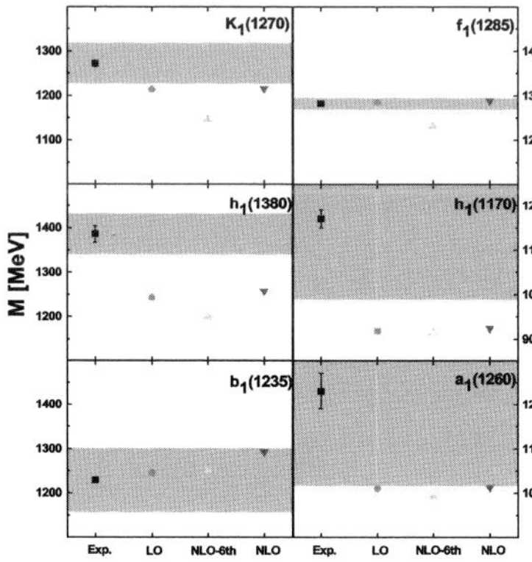

Fig. 1. The experimental axial-vector masses[7] in comparison with the corresponding UChPT results.

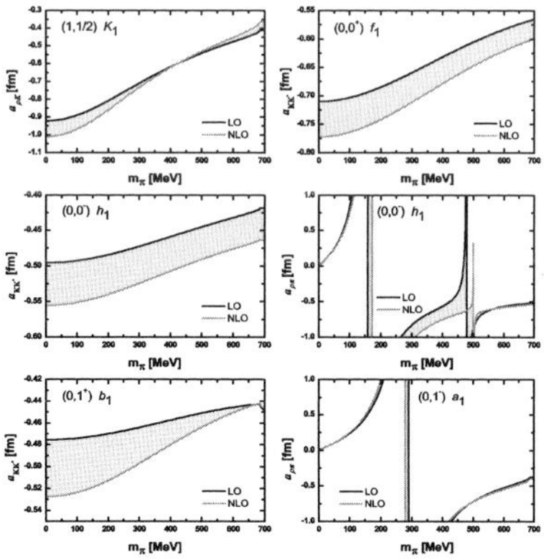

Fig. 2. Scattering lengths in the dominant channels of each isospin, strangeness, and G-parity (S, I^G) sector.

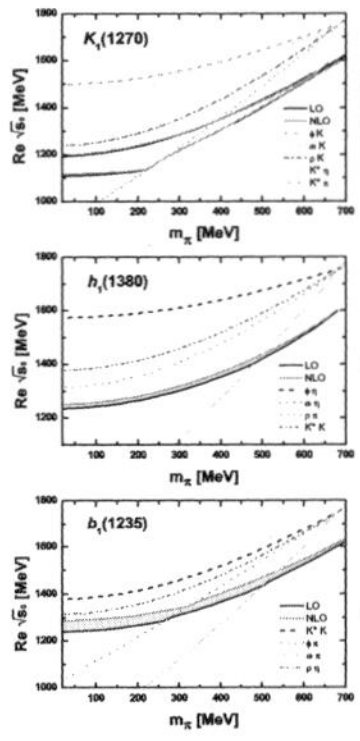

Fig. 3. Real part of the pole positions, $\mathrm{Re}\sqrt{s_0}$, as a function of m_π. The dashed lines indicate the thresholds of the corresponding coupled channels, respectively.

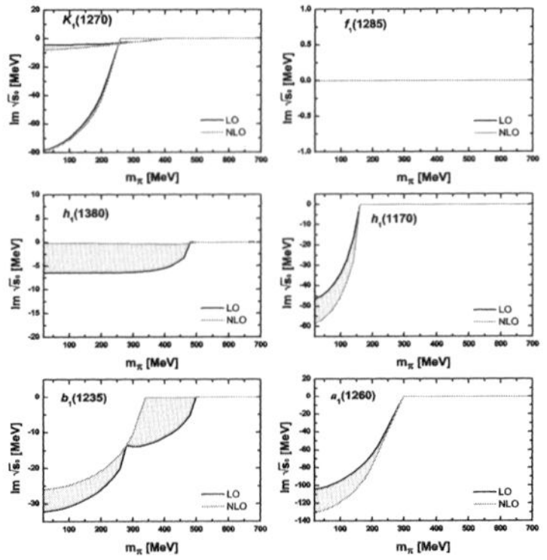

Fig. 4. Imaginary part of the pole positions, $\mathrm{Re}\sqrt{s_0}$, as a function of m_π.

Figure 3 and Figure 4 show the real and imaginary parts of the pole positions corresponding to the two $K_1(1270)$ states, $a_1(1260)$, $b_1(1235)$, $f_1(1285)$, $h_1(1170)$, $h_1(1380)$ as functions of the pion mass. It is seen that as the light quark masses increase, all the resonances become bound states, but they still exist. This tells that even with larger-than-physical light quark masses, LQCD simulations should be able to extract these resonances.

4. Acknowledgments

This work is supported in part by the National Natural Science Foundation of China under Grant Nos. 11375024 and 11205011, and the New Century Excellent Talents in University Program of Ministry of Edu- cation of China under Grant No. NCET-10-0029.

References

1. M. Lutz and E. Kolomeitsev, *Nucl.Phys.* **A730**, 392 (2004).
2. L. Roca, E. Oset and J. Singh, *Phys.Rev.* **D72**, 014002 (2005).
3. S. Prelovsek, C. Lang, D. Mohler and M. Vidmar, *PoS* **LATTICE2011**, 137 (2011).
4. C. Lang, L. Leskovec, D. Mohler and S. Prelovsek, *JHEP* **1404**, 162 (2014).
5. J. Prades, *Z.Phys.* **C63**, 491 (1994).
6. Y. Zhou, X.-L. Ren, H.-X. Chen and L.-S. Geng, *Phys.Rev.* **D90**, 014020 (2014).
7. J. Beringer *et al.*, *Phys.Rev.* **D86**, 010001 (2012).

List of Participants

Chun-Lin Bai (白春林)
Sichuan University
bclphy@scu.edu.cn

Hong-Bo Bai (白洪波)
Chifeng College
hbbai@vip.sina.com

Zhan Bai (摆展)
Peking University

Xiao-Jun Bao (包小军)
Lanzhou University
baoxj11@lzu.edu.cn

Man Bao (鲍曼)
Shanghai Jiaotong University
bymmy@sjtu.edu.cn

Li-Gang Cao (曹李刚)
North China Electric Power University
caolg@ncepu.edu.cn

Chun-Xing Chen (陈春杏)
Guangxi Normal University
1026768600@qq.com

Yong-Jing Chen (陈永静)
China Institute of Atomic Energy
cyj@ciae.ac.cn

Fang-Qi Chen (陈芳祁)
China Institute of Atomic Energy
fangqi0591@qq.com

Qi-Bo Chen (陈启博)
Peking University
qbchen@pku.edu.cn

Jing Chen (陈晶)
Peking University

Qi-Ming Chen (陈启明)
China Institute of Atomic Energy
chenqm@ciae.ac.cn

Shou-Wan Chen (陈寿万)
Anhui University
chenshouwan@ahu.edu.cn

Ruo-Fu Chen (陈若富)
Institute of Modern Physics of Chinese Academy of Sciences
chenruofu@impcas.ac.cn

Xiao-Chi Chen (陈潇驰)
Peking University
xiaochichen@pku.edu.cn

Yong-Shou Chen (陈永寿)
China Institute of Atomic Energy
yschen@ciae.ac.cn

Yang-Ping Chen (谌阳平)
China Institute of Atomic Energy
shenyp09@gmail.com

Jing-Xia Cheng (程锦霞)
Beifang University of Nationalities
lfcjxlyl@163.com

Shan-Jing Cheng (程善敬)
South China University of Technology
owcheng@163.com

Yi-Yuan Cheng (程奕源)
Shanghai Jiaotong University
yiyuancheng86@gmail.com

Peng-Cheng Chu (初鹏程)
Shanghai Jiaotong University
kyois@sjtu.edu.cn

Er-Liang Cui (崔二亮)
Beihang University
seeyounalar@126.com

Zhi-Tao Dai (代智涛)
Shanghai Institute of Applied Physics, CAS
Daizhitao.@sinap.ac.cn

Lian-Rong Dai (戴连荣)
Liaoning Normal University
dailr@lnnu.edu.cn

Ming-Hui Ding (丁明慧)
Peking University

Bao-Guo Dong (董保国)
China Institute of Atomic Energy
dongbg@ciae.ac.cn

Hong-Fei Dong (董鸿飞)
Chifeng College
hongfeidong@126.com

Jian-Min Dong (董建敏)
Institute of Modern Physics, CAS
dongjm07@impcas.ac.cn

Xian-Chao Du (杜先超)
China Institute of Atomic Energy
duxcsss@163.com

Chun-Gui Duan (段春贵)
Hebei Normal University
duancg@mail.hebtu.edu.cn

Zhao-Qing Feng (冯兆庆)
Institute of Modern Physics, CAS
fengzhq@impcas.ac.cn

Guang-Wei Fan (樊广伟)
Anhui University of Science & Technology
gwfan@aust.edu.cn

Chang-Bo Fu (符长波)
Shanghai Jiaotong University
cbfu@sjtu.edu.cn

Guang-Jan Fu (傅冠健)
Shanghai Jiaotong University
XZero_F@163.com

Lin Gan (甘林)
China Institute of Atomic Energy
ganlin2007@hotmail.com

Zao-Chun Gao (高早春)
China Institute of Atomic Energy
zcgao@ciae.ac.cn

Zhi-Gang Ge (葛智刚)
China Institute of Atomic Energy
gezg@ciae.ac.cn

Xin Guan (关鑫)
Liaoning Normal University
ggguanxinnn@163.com

Bing Guo (郭冰)
China Institute of Atomic Energy
guobing@ciae.ac.cn

Chen-Chen Guo (郭琛琛)
Beijing Normal University
guochenchen063@163.com

Chen-Lei Guo (郭晨雷)
Beihang University
davidguo1988@126.com

Jian-You Guo (郭建友)
Anhui University
jianyou@ahu.edu.cn

Lu Guo (郭璐)
University of CAS
luguo@ucas.ac.cn

Shu-Qing Guo (郭树青)
Lanzhou University
guosq09@126.com

Wen-Mei Guo (郭文梅)
Institute of Modern Physics, CAS
guowenmei@impcas.ac.cn

Yuan-Yuan Guo (郭媛媛)
Science China Press
guoyuanyuan@scichina.org

Yun Guo (郭云)
Guangxi Normal University
ahingy@gmail.com

Rui Han (韩蕊)
Peking University
cang.min@163.com

Liu-Chun He (何鎏春)
Beihang University
helc@phys.buaa.edu.cn

Chuang-Ye He (贺创业)
China Institute of Atomic Energy
hechuangye@126.com

Rong-Jiang Hu (胡荣江)
Institute of Modern Physics, CAS
hurj@impcas.ac.cn

Shi-Peng Hu (胡世鹏)
China Institute of Atomic Energy
huspchina@sina.com

Ding-Guang Huang (黄定光)
Institute of Modern Physics, CAS
hdg@impcas.ac.cn

Qing-Yuan Huang (黄庆元)
Sun Yat-sen University
huangcyh20@gmail.com

Xiao-Long Huang (黄小龙)
China Institute of Atomic Energy
huang@ciae.ac.cn

Yan Huang (黄彦)
Tsinghua University
h-y12@mails.tsinghua.edu.cn

Zhen-Zhen Jing (荆珍珍)
South China University of Technology
jingzhenzhen2526@163.com

Hui-Ming Jia (贾会明)
China Institute of Atomic Energy
jiahm@ciae.ac.cn

Li-Juan Jiang (江丽娟)
Lanzhou University
jianglj10@lzu.edu.cn

Hui Jiang (姜慧)
Shanghai Maritime University
huijiang@shmtu.edu.cn

Wei-Zhou Jiang (蒋维洲)
Southeast University
wzjiang@seu.edu.cn

Shi-Lun Jin (金仕纶)
Institute of Modern Physics, CAS
jinshilun@impcas.ac.cn

Gen-Ming Jin (靳根明)
Institute of Modern Physics, CAS
jingm@impcas.ac.cn

Long Jing (景龙)
China Institute of Atomic Energy
jingl24@163.com

Xiao-Yun Le (乐小云)
Beihang University
xyle@buaa.edu.cn

Yang Lei (雷杨)
China Academy of Engineering Physics
leiyang19850228@gmail.com

Ang Li (李昂)
Xiamen University
liang@xmu.edu.cn

Chen-Guang Li (李晨光)
Peking University
phyguang@pku.edu.cn

Cheng Li (李成)
Beijing Normal University
licheng@mail.bnu.edu.cn

Cheng-Bo Li (李成波)
Beijing Radiation Center
lichengbo2008@163.com

Cong-Bo Li (李聪博)
China Institute of Atomic Energy
lcbbch@sina.com

De-Sheng Li (李德胜)
Institute of High Energy Physics, CAS
lidesheng@ihep.ac.cn

Dong-Peng Li (李冬鹏)
Anhui University
lidp007@sina.com

Han-Tao Li (李瀚涛)
Nanjing University
hantao.li.phy@gmail.com

Er-Tao Li (李二涛)
Shenzhen University
let@szu.edu.cn

Guang-Sheng Li (李广生)
China Institute of Atomic Energy
ligs@ciae.ac.cn

Hui Li (李慧)
University of Science and Technology of China
Lihui12@mail.ustc.edu.cn

Jian Li (李剑)
Jiling University
jianli@jlu.edu.cn

Niu Li (李牛)
Anhui University
270403511@qq.com

Song-Lin Li (李松林)
Institute of Modern Physics, CAS
lisl@impcas.ac.cn

Xiu-Fang Li (李秀芳)
Shanghai Institute of Applied Physics, CAS
lixiufang@sinap.ac.cn

Xian Li (李贤)
Leshan Normal University
lixianlxwcq@163.com

Xiao-Hua Li (李小华)
University of South China
lixiaohuaphysics@126.com

Yun-Ju Li (李云居)
China Institute of Atomic Energy
li_yunju@163.com

Zhi-Hong Li (李志宏)
China Institute of Atomic Energy
zhli@ciae.ac.cn

Hao-Zhao Liang (梁豪兆)
RIKEN
haozhao.liang@riken.jp

Wei-Hong Liang (梁伟红)
Guangxi Normal University
liangwh@gxnu.edu.cn

Guang-Rui Liao (廖广睿)
Guangxi Normal University
liaogr@mailbox.gxnu.edu.cn

Cheng-Jian Ling (林承键)
China Institute of Atomic Energy
cjlin@ciae.ac.cn

He-Lei Liu (刘荷蕾)
South China University of Technology
2276904318@qq.com

Chen Liu (刘晨)
Shandong Unviersity-Weihai
chen880710@126.com

Jian Liu (刘健)
China University Of Petroleum
liujiannju@126.com

Lei Liu (刘雷)
Shandong Unviersity-Weihai
liulei20062007@126.com

Min Liu (刘敏)
Guangxi Normal University
lium_816@hotmail.com

Quan Liu (刘泉)
Anhui University
quanliu@ahu.edu.cn

Wen-Peng Liu (刘文鹏)
Southwest University
wenpenglau@163.com

Yan-Xin Liu (刘艳鑫)
Huzhou University
lyx1910@foxmail.com

Yu-Xin Liu (刘玉鑫)
Peking University
yxliu@pku.edu.cn

Yuan-Yuan Liu (刘圆圆)
Nuclear and Radiation Safety Center
178592225@qq.com

Zhong Liu (刘忠)
Institute of Modern Physics, CAS
liuzhong@impcas.ac.cn

Ji-Feng Liu (柳继锋)
Guangxi Normal University
jfliu@gxnu.edu.cn

Wen-Hui Long (龙文辉)
Lanzhou University
longwh@lzu.edu.cn

Jing-Bin Lu (陆景彬)
Jiling University
ljb@jlu.edu.cn

Jun-Xu Lu (陆俊旭)
Beihang University
ljxwohool@sina.com

Xiao Lu (陆晓)
Guangxi Normal University
luxiao@gxnu.edu.cn

Yi Lu (路毅)
Shanghai Jiaotong University
luyi@sjtu.edu.cn

Cui-Juan Lv (吕翠娟)
Henan Normal University
cuijuanlv0403@163.com

Hong-Feng Lv (吕洪凤)
China Agricultural University
hongfeng@cau.edu.cn

Li-Jun Lu (吕立君)
Chifeng College
lulijun0476@sina.com

Xiang-Yun Luo (罗祥云)
Lanzhou University
luoxy@lzu.edu.cn

Yan-An Luo (罗延安)
Nankai University
luoya@nankai.edu.cn

Wei-Xing Ma (马维兴)
Institute of High Energy Physics, CAS
mawx@mail.ihep.ac.cn

Hai-Liang Ma (马海亮)
China Institute of Atomic Energy
mhl624@ciae.ac.cn

Ke-Yan Ma (马克岩)
Jiling University
mky@jlu.edu.cn

Na-Na Ma (马娜娜)
Lanzhou University
mann13@lzu.edu.cn

Nan-Ru Ma (马南茹)
China Institute of Atomic Energy
mnr209@163.com

Peng Ma (马朋)
Institute of Modern Physics, CAS
mapeng@impcas.ac.cn

Chun-Wang Ma (马春旺)
Henan Normal University
machunwang@126.com

Ying-Jun Ma (马英君)
Jiling University
myj@jlu.edu.cn

Yu-Gang Ma (马余刚)
Shanghai Institute of Applied Physics, CAS
ygma@sinap.ac.cn

Jie Meng (孟杰)
Peking University
mengj@pku.edu.cn

Qiu-Hong Mo (莫秋虹)
Guangxi Normal University
517183095@qq.com

Ke-Xiong Mou (牟克雄)
Institute of Theoretical Physics, CAS

Dong-Dong Ni (倪冬冬)
Nanjing University
dongdongnick@gmail.com

Chen-Yang Niu (牛晨阳)
Peking University
niu@pku.edu.cn

Zhong-Ming Niu (牛中明)
Anhui University
zmniu@ahu.edu.cn

Li Ou (欧立)
Guangxi Normal University
liou@gxnu.edu.cn

Guang-Xiong Peng (彭光雄)
National University of Defense Technology /Institute of High Energy Physics,CAS
gxpeng@ucas.ac.cn

Jing Peng (彭婧)
Beijing Normal University
jpeng@bnu.edu.cn

Jun Qi (祁珺)
Inner Mogolia University for The Nationalities
qijun63@126.com

Xiao-Hong Qi (祁晓红)
Lanzhou University
qixh@lzu.edu.cn

Yi-Bin Qian (钱以斌)
Nanjing University of Science and Technology
qyibin@njust.edu.cn

Chun-Yuan Qiao (乔春源)
Henan Normal University
qiaochunyuan919@126.com

Lv-Hua Qiao (乔律华)
China Institute of Atomic Energy
qiaolvhua@163.com

Wei-Wei Qu (屈卫卫)
Beihang University
qww56@126.com

Zhong-Zhou Ren (任中洲)
Nanjing University
zren@nju.edu.cn

Xin-Le Shang (尚新乐)
Institute of Modern Physics, CAS
shangxinle@impcas.ac.cn

Cai-Wan Shen (沈彩万)
Huzhou University
cwshen@hutc.zj.cn

Hong-Tao Shen (沈洪涛)
Guangxi Normal University
shenht@ihep.ac.cn

Peng-Nian Shen (沈彭年)
Institute of High Energy Physics, CAS
shenpn@ihep.ac.cn

Guo-Zhu Shi (石国柱)
Institute of Modern Physics, CAS
sgz12sgz@impcas.ac.cn

Min Shi (仕敏)
Anhui University
adshimin@126.com

Mao Song (宋昴)
Anhui University
Songmao@mail.ustc.edu.cn

Li-Hua Song (宋丽华)
North China University of Science and Technology
songlh@heuu.edu.cn

Yu Song (宋誉)
University of South China
songyu408@sina.com

Jun Su (苏军)
Sun Yat-sen University
sujun04@gmail.com

Jun Su (苏俊)
China Institute of Atomic Energy
junsu@ciae.ac.cn

Bao-Hua Sun (孙保华)
Beihang University
bhsun@buaa.edu.cn

Bao-Yuan Sun (孙保元)
Lanzhou University
sunby@lzu.edu.cn

Ji Sun (孙吉)
Jiling University
agglehpge@163.com

Jun-Jie Sun (孙君杰)
Peking University
sunjunjie198701@163.com

Kai-Jia Sun (孙开佳)
Shanghai Jiaotong University
sunkaijia@sjtu.edu.cn

Wei-Li Sun (孙伟力)
Institute of Applied Physics and Computational Mathematics
Sun_weili@iapcm.ac.cn

Li-Jie Sun (孙立杰)
China Institute of Atomic Energy
sunlijie@ciae.ac.cn

Xiao-Dong Sun (孙小东)
University of South China
xiaodong-sun@foxmail.com

Xiao-Jun Sun (孙小军)
Guangxi Normal University
sxj0212@gxnu.edu.cn

Yu-Liang Sun (孙宇梁)
Huzhou University
sunyl@hutc.zj.cn

Shu-Wen Tang (唐述文)
Institute of Modern Physics, CAS
tangsw@impcas.ac.cn

Jun-Long Tian (田俊龙)
Anyang Normal University
tianjl@163.com

Ya Tu (图雅)
Shenyang Normal University
tuya_sy@126.com

Tmurbagen (特木尔巴根)
Inner Mogolia University For The Nationalities
tmurbagen@163.com

Isao Tanihata
Beihang University
tanihata@rcnp.osaka-u.ac.jp

Feng Wang (王枫)
Beihang University
wangfengbee@126.com

Han-Kui Wang (王韩奎)
Zhoukou Normal University
whk2007@163.com

Jin-Chuan Wang (王金川)
Institute of Modern Physics, CAS
jcwang@impcas.ac.cn

Jun-Ying Wang (王均英)
Institute of Modern Physics, CAS
wangjunying@impcas.ac.cn

Long-Jun Wang (王龙军)
Shanghai Jiaotong University
tommylewis@126.com

Nan Wang (王楠)
Shenzhen University
wangnan@szu.edu.cn

Ning Wang (王宁)
Guangxi Normal University
wangning@gxnu.edu.cn

Bing Wang (王兵)
Zhengzhou University/Institute of Theoretical Physics, CAS
121wb@163.com

Sha Wang (王莎)
Lanzhou University
wangsh2008@lzu.edu.cn

Shi-Tao Wang (王世陶)
Institute of Modern Physics, CAS
wstcn@qq.com

Shou-Yu Wang (王守宇)
Shandong Unviersity-Weihai
sywang@sdu.edu.cn

Shuo Wang (王硕)
Shandong Unviersity-Weihai
wangshuo_wh@sdu.edu.cn

Xiao-Bao Wang (王小保)
Huzhou University
wangxiaobaopku@gmail.com

Xin Wang (王欣)
Shanghai Jiaotong University
xinwang@sjtu.edu.cn

Yan-Zhao Wang (王艳召)
Shijiazhuang Tiedao University
yanzhaowang09@126.com

Yong-Jia Wang (王永佳)
Huzhou University
wangyongjia@hutc.zj.cn

You-Bao Wang (王友宝)
China Institute of Atomic Energy
ybwang@ciae.ac.cn

Zhang-Yin Wang (王章银)
Anhui University
zywang@ahu.edu.cn

Zhi-Heng Wang (王之恒)
Lanzhou University
wangzhh2013@lzu.edu.cn

Dai-Hui Wei (魏代会)
Guangxi Normal University
weidh@mail.ihep.ac.cn

De-Hua Wen (文德华)
South China University of Technology
wendehua@scut.edu.cn

Pei-Wei Wen (温培威)
Beijing Normal University
wenpeiwei@mail.bnu.edu.cn

Qiang-Hua Wu (吴强华)
China Institute of Atomic Energy/
Guangxi Normal University
qh297239367@126.com

Cheng-Jun Xia (夏铖君)
Institute of Theoretical Physics, CAS
cjxia@itp.ac.cn

Qun-Gang Wen (文群刚)
Anhui University
qungang@ahu.edu.cn

Xiao-Guang Wu (吴晓光)
China Institute of Atomic Energy
wxg@ciae.ac.cn

Xin-Yi Wu (吴新义)
Shanghai Jiaotong University
408687379@qq.com

Jian Xiang (向剑)
Lanzhou University
zhgxiangjian@163.com

Zhi-Gang Xiao (肖志刚)
Tsinghua University
xiaozg@tsinghua.edu.cn

Wen-Jie Xie (谢文杰)
Institute of Modern Physics, CAS/
Yuncheng University
wenjiexie@yeah.net

Chun Xiong (熊春)
IOP China
xiongchun27@hotmail.com

Chuan Xu (徐川)
Peking University
chuan@pku.edu.cn

Qiang Xu (徐强)
Taiyuan University of Technology
xuqiang01@tyut.edu.cn

Ren-Li Xu (徐仁力)
Nanjing University
songxue111@163.com

Xin-Xing Xu (徐新星)
China Institute of Atomic Energy
xuxinxing@ciae.ac.cn

Yong-Li Xu (徐永丽)
Datong University
xuyongli776@126.com

Yan Xu (许妍)
Changchun Observatory, National Astronomical Observatories, CAS
xuy@cho.ac.cn

Sheng-Quan Yan (颜胜权)
China Institute of Atomic Energy
yanye7703@sina.com

Ding Yang (杨丁)
Communication University of China
yd9955@sina.com

Dong Yang (杨东)
Jiling University
dyang@jlu.edu.cn

Hua-Bin Yang (杨华彬)
Institute of Modern Physics, CAS
yanghb@impcas.ac.cn

Jie-Cheng Yang (杨杰成)
Beihang University
revolution_y@126.com

Lei Yang (杨磊)
China Institute of Atomic Energy
yang_lei@ciae.ac.cn

Shen Yang (杨申)
Lanzhou University
syang11@lzu.edu.cn

Yong-Xu Yang (杨永栩)
Guangxi Normal University
yyxu@gxnu.edu.cn

Yuan Yao (姚远)
Kaiyuan Institute for Theoretical Physics
yaoyuangz@163.com

Wei Ye (叶巍)
Southeast University
yewei@seu.edu.cn

Shao-Ying Yu (于少英)
Huzhou University
ysy@hutc.zj.cn

Cen-Xi Yuan (袁岑溪)
Sun Yat-sen University
yuancx@mail.sysu.edu.cn

Guo-Mo Zeng (曾国模)
Jiling University
gmzeng@yahoo.com

Sheng Zeng (曾晟)
China Institute of Atomic Energy
llzs@ciae.ac.cn

Da-Li Zhang (张大立)
Huzhou University
zdl@hutc.zj.cn

Dong-Hong Zhang (张冬红)
Beijing Normal University
zhangdonghong@mail.bnu.edu.cn

Fan Zhang (张凡)
Beijing Normal University
zhangfan@mail.bnu.edu.cn

Feng-Shou Zhang (张丰收)
Beijing Normal University
fszhang@bnu.edu.cn

Lin Yu (郁琳)
University of CAS/Institute of Modern Physics, CAS
Yulin@impcas.ac.cn

Gao-Long Zhang (张高龙)
Beihang University
zgl@buaa.edu.cn

Huan-Qiao Zhang (张焕乔)
China Institute of Atomic Energy
huan@ciae.ac.cn

Miao-Jing Zhang (张妙静)
Guangxi Normal University
zhangmiaojing@sina.com

Shuang-Quan Zhang (张双全)
Peking University
sqzhang@pku.edu.cn

Wei Zhang (张炜)
Zhengzhou University
zw76@pku.org.cn

Xin Zhang (张欣)
Nanjing University
zxontt@163.com

Xin-Rui Zhang (张欣蕊)
Guangxi Normal University
773959942@qq.com

Yan-Li Zhang (张艳丽)
Henan Normal University
yanliz12@126.com

Cui-Lan Zhao (赵翠兰)
Inner Mogolia University for The Nationalities
nmdzcl@163.com

Zhen Zhang (张振)
Shanghai Jiaotong University
malkuth@sjtu.edu.cn

Xue-Heng Zhang (章学恒)
Institute of Modern Physics, CAS
zhxh@impcas.ac.cn

Yu Zhang (张宇)
Liaoning Normal University
dlzhangyu_physics@163.com

Kai Zhao (赵凯)
China Institute of Atomic Energy
zhaokai@ciae.ac.cn

Qiang Zhao (赵强)
Lanzhou University
zhaoq08@lzu.edu.cn

Hao Zheng (郑皓)
Shanghai Jiaotong University
zh-i@sjtu.edu.cn

Lei Zheng (郑垒)
Beihang University
bookzhl@163.com

Yong Zheng (郑勇)
Institute of Modern Physics, CAS
zhengyong@impcas.ac.cn

Yu-Ming Zheng (郑玉明)
China Institute of Atomic Energy
zhengymcn@163.com

Yun Zheng (郑云)
China Institute of Atomic Energy
zhengyun@ciae.ac.cn

Jian Zhong (钟健)
China Institute of Atomic Energy
happychinazj@139.com

Dan Zhou (周丹)
Beihang University
danzhou@buaa.edu.cn

En-Fu Zhou (周恩付)
Southwest University
efzhou@126.com

Hou-Bing Zhou (周厚兵)
Guangxi Normal University
zhouhb07@163.com

Wen-Kui Zhou (周文奎)
China Institute of Atomic Energy
zhou_wenkui@163.com

Wen-Ping Zhou (周文平)
Shenyang Institute of Engineering
zhouwp@sie.edu.cn

Xian-Rong Zhou (周先荣)
Xiamen University
xrzhou@xmu.edu.cn

Ying Zhou (周颖)
Shanghai Jiaotong University
YingChow@sjtu.edu.cn

Yu Zhou (周禹)
Beihang University
zhouyu.1516@163.com

Long Zhu (祝龙)
Beijing Normal University
zhulong@mail.bnu.edu.cn

Yao-Yao Zong (宗瑶瑶)
Lanzhou University
zongyy14@lzu.edu.cn

Shan-Gui Zhou (周善贵)
Institute of Theoretical Physics, CAS
sgzhou@itp.ac.cn

Bing-Song Zou (邹冰松)
Institute of Theoretical Physics, CAS
zoubs@itp.ac.cn